Beyond the Kalman Filter

Particle Filters for Tracking Applications

For a listing of recent titles in the *Artech House Radar Library,*
turn to the back of this book.

Beyond the Kalman Filter

Particle Filters for Tracking Applications

Branko Ristic
Sanjeev Arulampalam
Neil Gordon

Artech House
Boston • London
www.artechhouse.com

Library of Congress Cataloging-in-Publication Data

A catalog record for this book is available from the Library of Congress.

British Library Cataloguing in Publication Data

A catalog record of this book is available from the British Library.

Cover design by Yekaterina Ratner

© 2004 DSTO
All rights reserved.

All rights reserved. Printed and bound in the United States of America. No part of this book may be reproduced or utilized in any form or by any means, electronic or mechanical, including photocopying, recording, or by any information storage and retrieval system, without permission in writing from the publisher. All terms mentioned in this book that are known to be trademarks or service marks have been appropriately capitalized. Artech House cannot attest to the accuracy of this information. Use of a term in this book should not be regarded as affecting the validity of any trademark or service mark.

International Standard Book Number: 1-58053-631-x, ISBN 978-1-58053-631-8

10 9 8 7

Contents

Preface	xi
Acknowledgments	xiii

I Theoretical Concepts — 1

Chapter 1	Introduction	3
	1.1 Nonlinear Filtering	3
	1.2 The Problem and Its Conceptual Solution	4
	1.3 Optimal Algorithms	7
	1.3.1 The Kalman Filter	7
	1.3.2 Grid-Based Methods	9
	1.3.3 Beneš and Daum Filters	10
	1.4 Multiple Switching Dynamic Models	11
	1.5 Basics of Target Tracking	14
	1.6 Summary	16
	References	16
Chapter 2	Suboptimal Nonlinear Filters	19
	2.1 Analytic Approximations	19
	2.2 Numerical Methods	22
	2.3 Gaussian Sum Filters	24
	2.3.1 Static MM Estimator	25
	2.3.2 Dynamic MM Filter	26
	2.4 Unscented Kalman Filter	28
	2.4.1 Filtering Equations	29
	2.4.2 The Unscented Transform	30

	2.5	Summary	32
		References	32

Chapter 3 A Tutorial on Particle Filters — 35
- 3.1 Monte Carlo Integration — 35
- 3.2 Sequential Importance Sampling — 37
- 3.3 Resampling — 41
- 3.4 Selection of Importance Density — 45
  - 3.4.1 The Optimal Choice — 45
  - 3.4.2 Suboptimal Choices — 47
- 3.5 Versions of Particle Filters — 48
  - 3.5.1 SIR Filter — 48
  - 3.5.2 Auxiliary SIR Filter — 49
  - 3.5.3 Particle Filters with an Improved Sample Diversity — 52
  - 3.5.4 Local Linearization Particle Filters — 55
  - 3.5.5 Multiple-Model Particle Filter — 57
- 3.6 Computational Aspects — 58
- 3.7 Summary — 61
- 3.8 Appendix: Combination of Quadratic Terms — 61
- References — 62

Chapter 4 Cramér-Rao Bounds for Nonlinear Filtering — 67
- 4.1 Background — 68
- 4.2 Recursive Computation of the Filtering Information Matrix — 71
- 4.3 Special Cases — 73
  - 4.3.1 Additive Gaussian Noise — 73
  - 4.3.2 Linear/Gaussian Case — 75
  - 4.3.3 Zero Process Noise — 76
- 4.4 Multiple-Switching Dynamic Models — 76
  - 4.4.1 Enumeration Method — 77
  - 4.4.2 Deterministic Trajectory — 79
- 4.5 Summary and Further Reading — 80
- References — 80

II Tracking Applications — 83

Chapter 5 Tracking a Ballistic Object on Reentry — 85
- 5.1 Introduction — 85
- 5.2 Target Dynamics and Measurements — 86
- 5.3 Cramér-Rao Bound — 88

	5.4	Tracking Filters	93
	5.5	Numerical Results	94
	5.6	Concluding Remarks	98
		References	101

Chapter 6 Bearings-Only Tracking 103
    6.1 Introduction 103
    6.2 Problem Formulation 104
        6.2.1 Nonmaneuvering Case 104
        6.2.2 Maneuvering Case 106
        6.2.3 Multiple Sensor Case 108
        6.2.4 Tracking with Constraints 108
    6.3 Cramér-Rao Lower Bounds 109
        6.3.1 Nonmaneuvering Case 109
        6.3.2 Maneuvering Case 110
        6.3.3 Multiple Sensor Case 112
    6.4 Tracking Algorithms 113
        6.4.1 Nonmaneuvering Case 113
        6.4.2 Maneuvering Target Case 121
        6.4.3 Multiple Sensor Case 127
        6.4.4 Tracking with Hard Constraints 127
    6.5 Simulation Results 129
        6.5.1 Nonmaneuvering Case 130
        6.5.2 Maneuvering Case 138
        6.5.3 Multiple Sensor Case 145
        6.5.4 Tracking with Hard Constraints 147
    6.6 Summary 148
    6.7 Appendix: Linearized Transition Matrix for MP-EKF 148
        References 150

Chapter 7 Range-Only Tracking 153
    7.1 Introduction 153
    7.2 Problem Description 154
    7.3 Cramér-Rao Bounds 157
        7.3.1 Derivations 157
        7.3.2 Analysis 158
    7.4 Tracking Algorithms 164
    7.5 Algorithm Performance and Comparison 168
    7.6 Application to Ingara ISAR Data 173
    7.7 Summary 176
        References 178

Chapter 8 Bistatic Radar Tracking — 179
- 8.1 Introduction — 179
- 8.2 Problem Formulation — 180
- 8.3 Cramér-Rao Bounds — 183
  - 8.3.1 Derivations — 183
  - 8.3.2 Analysis — 185
- 8.4 Tracking Algorithms — 189
  - 8.4.1 Stage 1 of Tracker — 191
  - 8.4.2 Stage 2 of Tracker — 196
- 8.5 Algorithm Performance — 196
- 8.6 Summary — 199
- References — 201

Chapter 9 Tracking Targets Through the Blind Doppler — 203
- 9.1 Introduction — 203
- 9.2 Problem Formulation — 204
- 9.3 EKF-Based Track Maintenance — 206
- 9.4 Particle Filter-Based Solution — 208
- 9.5 Simulation Results — 210
- 9.6 Summary — 213
- References — 214

Chapter 10 Terrain-Aided Tracking — 215
- 10.1 Introduction — 215
- 10.2 Problem Description and Formulation — 216
  - 10.2.1 Problem Description — 216
  - 10.2.2 Dynamics and Measurement Models for VS-IMM — 219
  - 10.2.3 Dynamic Models for VS-MMPF — 221
- 10.3 Variable Structure IMM — 227
  - 10.3.1 Model Set Update — 229
- 10.4 Variable Structure Multiple-Model Particle Filter — 229
  - 10.4.1 Prediction Step — 230
  - 10.4.2 Update Step — 230
- 10.5 Simulation Results — 231
- 10.6 Conclusions — 236
- References — 237

Chapter 11 Detection and Tracking of Stealthy Targets — 239
- 11.1 Introduction — 239
- 11.2 Target and Sensor Models — 240
  - 11.2.1 Target Model — 240

	11.2.2 Sensor Model	241
	11.3 Conceptual Solution in the Bayesian Framework	242
	11.4 A Particle Filter for Track-Before-Detect	244
	11.5 A Numerical Example	247
	11.6 Performance Analysis	251
	11.6.1 Tracking Error Performance	251
	11.6.2 Detection Performance	254
	11.7 Summary and Extensions	257
	References	258
Chapter 12	Group and Extended Object Tracking	261
	12.1 Introduction	261
	12.2 Tracking Model	263
	12.3 Formal Bayesian Solution	265
	12.4 Affine Model	268
	12.5 Particle Filters	269
	12.5.1 SIR Particle Filter	270
	12.5.2 Rao-Blackwellized Particle Filter	271
	12.6 Simulation Example	273
	12.7 Concluding Remarks	277
	References	284
Epilogue		287
Appendix	Coordinate Transformations for Tracking	289
	A.1 Geodetic to ECEF and Vice Versa	290
	A.2 ECEF to Tangential Plane and Vice Versa	290
	References	292
List of Acronyms		293
About the Authors		295
Index		297

Preface

Target tracking is an important element of surveillance, guidance, or obstacle avoidance systems, whose role is to determine the number, position, and movement of targets. The fundamental building block of a tracking system is a *filter* for recursive target state estimation. This book is devoted to defense applications of nonlinear and non-Gaussian filtering, in the context of target tracking.

Kalman filter is the best known filter, a simple and elegant algorithm formulated more than 40 years ago [1], as an optimal recursive Bayesian estimator for a somewhat restricted class of linear Gaussian problems. Recently there has been a surge of interest in nonlinear and non-Gaussian filtering. A number of techniques for this type of filtering are reviewed in the book, but our main focus are the tools of sequential Monte Carlo estimation, collectively referred to as *particle filters*. Since the seminal paper [2], particle filters have become one of the most popular methods for stochastic dynamic estimation problems and so far one book [3] and a few special journal issues have appeared devoted to the subject. This popularity can be explained by a wave of optimism among the practitioners that for any nonlinear/non-Gaussian dynamic estimation problem one can design an accurate, reliable, and fast recursive Bayesian filter. The computational cost of particle filters has often been considered as their main disadvantage, but with ever faster computers, this argument is becoming less relevant. Clearly we authors believe that beyond the Kalman filter is the particle filter. However, this is not to say that the Kalman filter will disappear altogether. Whenever the conventional methods work fine, there is no need for further complications. In this book we address a few practical defense applications, where the conventional methods are inappropriate (unreliable or inaccurate) and where we indeed had to look beyond the Kalman filtering framework.

The style of writing is suitable for engineers and scientists interested in practical applications of estimation, tracking, and information fusion. The authors have tried in their presentation to balance the mathematical rigor with the conceptual insight into the key ideas, so that the reader can "see both the trees and the forest."

The book is organized as follows. Chapters 1 to 4 form Part I, which reviews the current status of the theory for nonlinear/non-Gaussian filtering and its relevance to target tracking. The problem of nonlinear and non-Gaussian filtering and its conceptual recursive Bayesian solution are formulated in Chapter 1. This chapter also presents optimal algorithms that can be formulated for some specific and limited classes of problems. Chapters 2 and 3 review the common techniques for nonlinear filtering, categorized in four groups: analytic methods, numerical methods, Gaussian sum approaches, and sampling methods. The emphasis is on particle filters (reviewed in Chapter 3), which will be used extensively throughout the book. Chapter 4 describes the development of Cramér-Rao lower bounds for nonlinear filtering.

Chapters 5 to 12 form Part II, which is devoted to target tracking applications. Chapter 5 describes the problem of tracking a ballistic object. The problem has a long history and is particularly difficult because of the highly nonlinear target dynamics. Bearings-only tracking is discussed in Chapter 6. This is another nonlinear filtering problem with a long history, particularly relevant for surveillance with passive sensors (passive sonar, ESM, IRST, passive radar). Chapter 7 is devoted to range-only tracking, a relatively new application motivated by a practical problem with an inverse synthetic aperture radar (ISAR). Bistatic radar tracking using bearings and Doppler measurements is described in Chapter 8. This problem is becoming increasingly important for tracking stealthy targets with a low probability of detection. Chapter 9 describes a particle filter-based technique for tracking targets through the blind Doppler of a surveillance radar. An enemy aircraft would occasionally hide in the blind Doppler in an attempt to cause a loss of track. This application describes how prior information about sensor limitations can be used to improve the track maintenance. Terrain-aided tracking of ground vehicles using a GMTI radar is another application where prior information plays a crucial role. Chapter 10 is devoted to GMTI tracking using terrain information, road maps, and visibility conditions with an objective to improve tracking accuracy. Recent developments of stealthy military aircraft and cruise missiles have emphasized the need for detection and tracking of low signal-to-noise ratio targets. Chapter 11 describes a particle filter-based technique for this application. Finally, Chapter 12 is devoted to group and extended object tracking. The problem is of interest when a sensor (due to its high resolution) resolves the features of an object or, by analogy, if a group of point targets is moving in a formation. The main difficulty in this application is a highly uncertain measurement-to-object association.

References

[1] R. E. Kalman, "A new approach to linear filtering and prediction problems," *Trans. ASME, Journal of Basic Engineering*, vol. 82, pp. 35–45, March 1960.

[2] N. J. Gordon, D. J. Salmond, and A. F. M. Smith, "Novel approach to nonlinear/non-Gaussian Bayesian state estimation," *IEE Proc.-F*, vol. 140, no. 2, pp. 107–113, 1993.

[3] A. Doucet, J. F. G. de Freitas, and N. J. Gordon, eds., *Sequential Monte Carlo Methods in Practice*. New York: Springer, 2001.

Acknowledgments

We would like to thank our employer, Defence Science and Technology Organisation, and in particular Bruce Ward, David Heilbronn, John Percival, and Todd Mansell for supporting the work on this book.

The book is a result of collaboration with many of our colleagues across the world. This collaboration contributed significantly to our improved understanding of particle filters and their tracking applications. In particular, we are indebted to Alfonso Farina (Alenia Marconi Systems), David Salmond (QinetiQ), Adrian Smith (Queen Mary, University of London), Arnaud Doucet (Cambridge University), Alan Marrs, Simon Maskell, Marcel Hernandez (all from QinetiQ), Christian Musso (ONERA), Bill Fitzgerald (Cambridge University), Christophe Andrieu (Bristol University), Neil Shephard (Oxford University), Niclas Bergman (SaabTech Systems), Frederic Gustafsson (Linköping University), Mark Morelande, Subhash Challa, B-N Vo (all from Melbourne University), Yvo Boers (Thales Nederland), Mike Pitt (Warwick University), Dario Benvenuti, Luca Timmoneri (Alenia Marconi Systems), Keith Kastella (Veridian), Mahendra Mallick (Orincon), Lawrence Stone (Metron), and many others.

Chapter 12 includes the contributions from David Salmond (QinetiQ), Martin Robinson (DSTO), and Mark Morelande (Melbourne University); the authors thank them for their help in writing this chapter.

Finally we would like to express our gratitude to Fred Daum (Raytheon) for giving us permission to use a version of his original phrase *beyond Kalman filters* in the title of our book; Mark Rutten (DSTO), for his help with LaTeX and his comments on the manuscript; Martin Robinson, Amanda Bessell, and Steve Zollo (DSTO) for their help in coding various algorithms described in the book; and all our colleagues from the TSF group of the ISR division in DSTO for many helpful discussions. B. Ristic would also like to thank IRIDIA, Université Libre de Bruxelles, for being his host during the final stages of writing the book.

Part I

Theoretical Concepts

Chapter 1

Introduction

1.1 NONLINEAR FILTERING

Nonlinear filtering has been the focus of interest in the statistical and engineering community for more than 30 years [1]. The problem is to estimate sequentially the state of a dynamic system using a sequence of noisy measurements made on the system. We adopt the state-space approach to modeling dynamic systems and we focus on the discrete-time formulation of the problem (unless otherwise stated). Thus, difference equations will be used to model the evolution of the system over time, and measurements are assumed to be available at discrete times. For dynamic state estimation, the discrete-time approach is both widespread and convenient.

The state-space approach to time-series modeling focuses attention on the state vector of a system. The state vector contains all relevant information required to describe the system under investigation. For example, in tracking problems this information could be related to the kinematic characteristics of the target. Alternatively, in an econometrics problem it could be related to monetary flow, interest rates, inflation, and so forth. The measurement vector represents (noisy) observations that are related to the state vector. The measurement vector is generally (but not necessarily) of lower dimension than the state vector. The state-space approach is convenient for handling multivariate data and nonlinear/non-Gaussian processes and it provides a significant advantage over traditional time-series techniques for these problems.

In order to analyze and make inferences about a dynamic system, at least two models are required: first, a model describing the evolution of the state with time (the system or dynamic model), and second, a model relating the noisy measurements to the state (the measurement model). We shall assume that these models are available in a probabilistic form. The probabilistic state-space formulation and the requirement for the updating of information on receipt of new measurements are ideally suited for the Bayesian approach. This provides a rigorous general framework for dynamic state estimation problems.

In the Bayesian approach to dynamic state estimation one attempts to construct the *posterior* probability density function (pdf or density) of the state, based on all available information, including the sequence of received measurements. If either the system or measurement model is nonlinear, the posterior pdf will be non-Gaussian. Since this pdf embodies all available statistical information, it may be regarded to be the complete solution to the estimation problem. In principle, an optimal (with respect to any criterion) estimate of the state may be obtained from the posterior pdf. A measure of the accuracy of the estimate may also be obtained. For many problems an estimate is required every time a measurement is received. In this case a recursive filter is a convenient solution. A recursive filtering approach means that received data can be processed sequentially rather than as a batch, so that it is not necessary to store the complete data set nor to reprocess existing data if a new measurement becomes available. Such a filter consists of essentially two stages: prediction and update. The prediction stage uses the system model to predict the state pdf forward from one measurement time to the next. Since the state is usually subject to unknown disturbances (modeled as random noise), prediction generally translates, deforms, and broadens the state pdf. The update operation uses the latest measurement to modify (typically to tighten) the prediction pdf. This is achieved using Bayes theorem, which is the mechanism for updating knowledge about the target state in the light of extra information from new data.

1.2 THE PROBLEM AND ITS CONCEPTUAL SOLUTION

To define the problem of nonlinear filtering, let us introduce the target state vector $\mathbf{x}_k \in \mathbb{R}^{n_x}$, where n_x is the dimension of the state vector; $\mathbb{R}$ is a set of real numbers; $k \in \mathbb{N}$ is the time index; and $\mathbb{N}$ is the set of natural numbers. Here index k is assigned to a continuous-time instant t_k, and the "sampling interval" $T_{k-1} \triangleq t_k - t_{k-1}$ may be time-dependent (i.e., a function of k). The target state evolves according to the following discrete-time stochastic model:

$$\mathbf{x}_k = \mathbf{f}_{k-1}(\mathbf{x}_{k-1}, \mathbf{v}_{k-1}), \tag{1.1}$$

where $\mathbf{f}_{k-1}$ is a known, possibly nonlinear function of the state $\mathbf{x}_{k-1}$ and $\mathbf{v}_{k-1}$ is referred to as a process noise sequence. Process noise caters for any mismodeling effects or unforeseen disturbances in the target motion model. The objective of nonlinear filtering is to recursively estimate $\mathbf{x}_k$ from measurements $\mathbf{z}_k \in \mathbb{R}^{n_z}$. The measurements are related to the target state via the measurement equation:

$$\mathbf{z}_k = \mathbf{h}_k(\mathbf{x}_k, \mathbf{w}_k), \tag{1.2}$$

where $\mathbf{h}_k$ is a known, possibly nonlinear function and $\mathbf{w}_k$ is a measurement noise sequence. The noise sequences $\mathbf{v}_{k-1}$ and $\mathbf{w}_k$ will be assumed to be white, with known probability density functions and mutually independent. The initial target state is assumed to have a known pdf $p(\mathbf{x}_0)$ and also to be independent of noise sequences.

We seek filtered estimates of $\mathbf{x}_k$ based on the sequence of all available measurements $\mathbf{Z}_k \triangleq \{\mathbf{z}_i, i = 1, \ldots, k\}$ up to time k. From a Bayesian perspective, the problem is to recursively quantify some degree of belief in the state $\mathbf{x}_k$ at time k, taking different values, given the data $\mathbf{Z}_k$ up to time k. Thus, it is required to construct the posterior pdf $p(\mathbf{x}_k|\mathbf{Z}_k)$. The *initial* density of the state vector is $p(\mathbf{x}_0) \triangleq p(\mathbf{x}_0|\mathbf{z}_0)$, where $\mathbf{z}_0$ is the set of no measurements. Then, in principle, the pdf $p(\mathbf{x}_k|\mathbf{Z}_k)$ may be obtained recursively in the aforementioned two stages: prediction and update.

Suppose that the required pdf $p(\mathbf{x}_{k-1}|\mathbf{Z}_{k-1})$ at time $k-1$ is available. The prediction stage involves using the system model (1.1) to obtain the prediction density[1] of the state at time k via the Chapman-Kolmogorov equation:

$$p(\mathbf{x}_k|\mathbf{Z}_{k-1}) = \int p(\mathbf{x}_k|\mathbf{x}_{k-1}) p(\mathbf{x}_{k-1}|\mathbf{Z}_{k-1}) d\mathbf{x}_{k-1}. \quad (1.3)$$

Note that in (1.3), use has been made of the fact that $p(\mathbf{x}_k|\mathbf{x}_{k-1}, \mathbf{Z}_{k-1}) = p(\mathbf{x}_k|\mathbf{x}_{k-1})$ as (1.1) describes a Markov process of order one. The probabilistic model of the state evolution (often referred to as *transitional density*), $p(\mathbf{x}_k|\mathbf{x}_{k-1})$, is defined by the system equation (1.1) and the known statistics of $\mathbf{v}_{k-1}$.

At time step k when a measurement $\mathbf{z}_k$ becomes available, the update stage is carried out. This involves an update of the prediction (or prior) pdf via the Bayes' rule:

$$\begin{aligned} p(\mathbf{x}_k|\mathbf{Z}_k) &= p(\mathbf{x}_k|\mathbf{z}_k, \mathbf{Z}_{k-1}) \\ &= \frac{p(\mathbf{z}_k|\mathbf{x}_k, \mathbf{Z}_{k-1}) p(\mathbf{x}_k|\mathbf{Z}_{k-1})}{p(\mathbf{z}_k|\mathbf{Z}_{k-1})} \\ &= \frac{p(\mathbf{z}_k|\mathbf{x}_k) p(\mathbf{x}_k|\mathbf{Z}_{k-1})}{p(\mathbf{z}_k|\mathbf{Z}_{k-1})} \end{aligned} \quad (1.4)$$

where the normalizing constant

$$p(\mathbf{z}_k|\mathbf{Z}_{k-1}) = \int p(\mathbf{z}_k|\mathbf{x}_k) p(\mathbf{x}_k|\mathbf{Z}_{k-1}) d\mathbf{x}_k \quad (1.5)$$

depends on the likelihood function $p(\mathbf{z}_k|\mathbf{x}_k)$, defined by the measurement model (1.2) and the known statistics of $\mathbf{w}_k$. In the update stage (1.4), the measurement $\mathbf{z}_k$

---

[1] The prediction pdf is often referred to as the (dynamic) prior pdf.

is used to modify the prior density to obtain the required posterior density of the current state.

The recurrence relations (1.3) and (1.4) form the basis for the optimal Bayesian solution.[2] Knowledge of the posterior density $p(\mathbf{x}_k|\mathbf{Z}_k)$ enables one to compute an optimal state estimate with respect to any criterion. For example, the minimum mean-square error (MMSE) estimate is the conditional mean of $\mathbf{x}_k$ [2]:

$$\hat{\mathbf{x}}_{k|k}^{\text{MMSE}} \triangleq \mathbb{E}\{\mathbf{x}_k|\mathbf{Z}_k\} = \int \mathbf{x}_k \cdot p(\mathbf{x}_k|\mathbf{Z}_k) d\mathbf{x}_k, \qquad (1.6)$$

while the maximum a posteriori (MAP) estimate is the maximum of $p(\mathbf{x}_k|\mathbf{Z}_k)$:

$$\hat{\mathbf{x}}_{k|k}^{\text{MAP}} \triangleq \arg\max_{\mathbf{x}_k} p(\mathbf{x}_k|\mathbf{Z}_k). \qquad (1.7)$$

Similarly, a measure of accuracy of a state estimate (e.g., covariance) may also be obtained from $p(\mathbf{x}_k|\mathbf{Z}_k)$.

The recursive propagation of the posterior density, given by (1.3) and (1.4), is only a conceptual solution in the sense that in general it cannot be determined analytically. The implementation of the conceptual solution requires the storage of the entire (non-Gaussian) pdf which is, in general terms, equivalent to an infinite dimensional vector. Only in a restrictive set of cases, described in the next section, the posterior density can be exactly and completely characterized by a sufficient statistic of fixed and finite dimension. Since in most practical situations (see Part II) the analytic solution of (1.4) and (1.5) is intractable, one has to use approximations or suboptimal Bayesian algorithms. Conventional and more advanced approximations are described in Chapters 2 and 3.

A Priori Constraints

In some filtering problems one has to deal with a priori constraints imposed either on the state vector or the measurement process. In the case of target tracking, for example, there could be hard constraints on target position (flight corridors for commercial airplanes; a road network for ground vehicles), speed, or acceleration [3]. Sensors can have limitations too; for example, radars are usually characterized by certain blind velocity regions and the line-of-sight detection (hence, cannot "see" behind the mountains). These hard constraints introduce nonlinearities in the form of truncations on probability distributions, and further complicate the analytic derivation of the optimal nonlinear filter. In general, however, by incorporating the

---

[2] For clarity, the optimal Bayesian solution solves the problem of exact and complete characterization of the posterior density in a recursive manner. An optimal algorithm is a method for deducing this solution.

constraints (or any prior knowledge) into the filter, its performance should improve [4].

1.3 OPTIMAL ALGORITHMS

Optimal finite-dimensional algorithms for recursive Bayesian state estimation can be formulated in the following cases:

1. In a linear-Gaussian case, the functional recursion of (1.3) and (1.4) becomes the Kalman filter (see Section 1.3.1).

2. If the state space is discrete-valued with a finite number of states, the grid-based methods provide the optimal algorithm (see Section 1.3.2).

3. For certain subclasses of nonlinear problems, discovered by Beneš [5] and Daum [6, 7], it is also possible to formulate exact analytic solutions.

A brief description of these optimal algorithms is presented next.

1.3.1 The Kalman Filter

The Kalman filter [2] assumes that the posterior density at every time step is Gaussian and hence exactly and completely characterized by two parameters, its mean and covariance.

If $p(\mathbf{x}_{k-1}|\mathbf{Z}_{k-1})$ is Gaussian, it can be proved that $p(\mathbf{x}_k|\mathbf{Z}_k)$ is also Gaussian, provided that certain assumptions hold [8]:

- $\mathbf{v}_{k-1}$ and $\mathbf{w}_k$ are drawn from *Gaussian* densities of known parameters
- $\mathbf{f}_{k-1}(\mathbf{x}_{k-1}, \mathbf{v}_{k-1})$ is a known *linear* function of $\mathbf{x}_{k-1}$ and $\mathbf{v}_{k-1}$
- $\mathbf{h}_k(\mathbf{x}_k, \mathbf{w}_k)$ is a known *linear* function of $\mathbf{x}_k$ and $\mathbf{w}_k$

That is, suppose (1.1) and (1.2) can be rewritten as:

$$\mathbf{x}_k = \mathbf{F}_{k-1}\mathbf{x}_{k-1} + \mathbf{v}_{k-1} \quad (1.8)$$

$$\mathbf{z}_k = \mathbf{H}_k\mathbf{x}_k + \mathbf{w}_k \quad (1.9)$$

where $\mathbf{F}_{k-1}$ (of dimension $n_x \times n_x$) and $\mathbf{H}_k$ (of dimension $n_z \times n_x$) are known matrices defining the linear functions. Random sequences $\mathbf{v}_{k-1}$ and $\mathbf{w}_k$ are mutually independent zero-mean white Gaussian, with covariances $\mathbf{Q}_{k-1}$ and $\mathbf{R}_k$ respectively. Note that the system and measurement matrices $\mathbf{F}_{k-1}$ and $\mathbf{H}_k$, as well as noise covariances $\mathbf{Q}_{k-1}$ and $\mathbf{R}_k$, are allowed to be time-variant.

The Kalman filter algorithm, derived using (1.3) and (1.4), can then be viewed as the following recursive relationship:

$$p(\mathbf{x}_{k-1}|\mathbf{Z}_{k-1}) = \mathcal{N}(\mathbf{x}_{k-1}; \hat{\mathbf{x}}_{k-1|k-1}, \mathbf{P}_{k-1|k-1}) \quad (1.10)$$
$$p(\mathbf{x}_k|\mathbf{Z}_{k-1}) = \mathcal{N}(\mathbf{x}_k; \hat{\mathbf{x}}_{k|k-1}, \mathbf{P}_{k|k-1}) \quad (1.11)$$
$$p(\mathbf{x}_k|\mathbf{Z}_k) = \mathcal{N}(\mathbf{x}_k; \hat{\mathbf{x}}_{k|k}, \mathbf{P}_{k|k}) \quad (1.12)$$

where $\mathcal{N}(\mathbf{x}; \mathbf{m}, \mathbf{P})$ is a Gaussian density with argument $\mathbf{x}$, mean $\mathbf{m}$, and covariance $\mathbf{P}$; that is:

$$\mathcal{N}(\mathbf{x}; \mathbf{m}, \mathbf{P}) \triangleq |2\pi \mathbf{P}|^{-1/2} \exp\{-\frac{1}{2}(\mathbf{x} - \mathbf{m})^T \mathbf{P}^{-1}(\mathbf{x} - \mathbf{m})\}. \quad (1.13)$$

Notation $\mathbf{M}^T$ stands for the transpose of a matrix $\mathbf{M}$. The appropriate means and covariances of the Kalman filter are computed as follows [2, 9, 10]:

$$\hat{\mathbf{x}}_{k|k-1} = \mathbf{F}_{k-1}\hat{\mathbf{x}}_{k-1|k-1} \quad (1.14)$$
$$\mathbf{P}_{k|k-1} = \mathbf{Q}_{k-1} + \mathbf{F}_{k-1}\mathbf{P}_{k-1|k-1}\mathbf{F}_{k-1}^T \quad (1.15)$$
$$\hat{\mathbf{x}}_{k|k} = \hat{\mathbf{x}}_{k|k-1} + \mathbf{K}_k(\mathbf{z}_k - \mathbf{H}_k\hat{\mathbf{x}}_{k|k-1}) \quad (1.16)$$
$$\mathbf{P}_{k|k} = \mathbf{P}_{k|k-1} - \mathbf{K}_k \mathbf{S}_k \mathbf{K}_k^T \quad (1.17)$$

where

$$\mathbf{S}_k = \mathbf{H}_k \mathbf{P}_{k|k-1} \mathbf{H}_k^T + \mathbf{R}_k \quad (1.18)$$

is the covariance of the innovation term $\nu_k = \mathbf{z}_k - \mathbf{H}_k \hat{\mathbf{x}}_{k|k-1}$, and

$$\mathbf{K}_k = \mathbf{P}_{k|k-1} \mathbf{H}_k^T \mathbf{S}_k^{-1} \quad (1.19)$$

is the Kalman gain. Note that with (1.18) and (1.19), covariance update (1.17) can be written as:

$$\mathbf{P}_{k|k} = [\mathbf{I} - \mathbf{K}_k \mathbf{H}_k]\mathbf{P}_{k|k-1} \quad (1.20)$$

where $\mathbf{I}$ is the identity matrix of dimension $n_x \times n_x$.

The Kalman filter recursively computes the mean and covariance of the Gaussian posterior $p(\mathbf{x}_k|\mathbf{Z}_k)$. This is the optimal solution to the tracking problem – if the (highly restrictive) assumptions hold. The implication is that no algorithm can ever do better than the Kalman filter in this linear Gaussian environment. It should be noted that it is possible to derive the same results using a least squares (LS) argument [1], and that though the filter then optimally derives the mean and covariance of the posterior, this posterior is not necessarily Gaussian and so the filter is not certain to be optimal.

Similarly, if smoothed estimates of the states are required, that is, estimates of $p(\mathbf{x}_k|\mathbf{Z}_{k+\ell})$ where $\ell \geq 0$,[3] then the Kalman smoother is the optimal estimator. This holds if ℓ is fixed (*fixed-lag smoothing*), or if a batch of K data is considered and $0 \leq \ell \leq K$ (*fixed-interval smoothing*), or if the state at a particular time k is of interest and $\ell = 1, 2, \ldots$ (*fixed-point smoothing*). The problem of calculating smoothed densities is of interest because the densities at time k are then conditional not only on measurements up to and including time index k, but also on future measurements up to time $k + \ell$. Since there is more information on which to base the estimation, these smoothed densities are typically of lower variance than the filtered densities.

1.3.2 Grid-Based Methods

Grid-based methods provide the optimal recursion of the filtered density, $p(\mathbf{x}_k|\mathbf{Z}_k)$, if the state space is discrete and consists of a finite number of states. Suppose the state space at time $k-1$ consists of discrete states $\mathbf{x}_{k-1}^i$, $i = 1, \ldots, N$. For each state $\mathbf{x}_{k-1}^i$, let the conditional probability of that state, given measurements up to time $k-1$, be denoted by $w_{k-1|k-1}^i$; that is, $\mathrm{P}\{\mathbf{x}_{k-1} = \mathbf{x}_{k-1}^i|\mathbf{Z}_{k-1}\} \triangleq w_{k-1|k-1}^i$. Then, the posterior pdf at $k-1$ can be written as

$$p(\mathbf{x}_{k-1}|\mathbf{Z}_{k-1}) = \sum_{i=1}^{N} w_{k-1|k-1}^i \delta(\mathbf{x}_{k-1} - \mathbf{x}_{k-1}^i) \tag{1.21}$$

where $\delta(\cdot)$ is the Dirac delta measure, with defining properties: (1) $\delta(\mathbf{x} - \mathbf{a}) = 0$ for $\mathbf{x} \neq \mathbf{a}$, and (2) $\int_{-\infty}^{\infty} \delta(\mathbf{x} - \mathbf{a}) d\mathbf{x} = 1$. Substitution of (1.21) into (1.3) and (1.4) yields the prediction and update equations, respectively:

$$p(\mathbf{x}_k|\mathbf{Z}_{k-1}) = \sum_{i=1}^{N} w_{k|k-1}^i \delta(\mathbf{x}_k - \mathbf{x}_k^i) \tag{1.22}$$

$$p(\mathbf{x}_k|\mathbf{Z}_k) = \sum_{i=1}^{N} w_{k|k}^i \delta(\mathbf{x}_k - \mathbf{x}_k^i) \tag{1.23}$$

where

$$w_{k|k-1}^i \triangleq \sum_{j=1}^{N} w_{k-1|k-1}^j p(\mathbf{x}_k^i|\mathbf{x}_{k-1}^j), \tag{1.24}$$

$$w_{k|k}^i \triangleq \frac{w_{k|k-1}^i p(\mathbf{z}_k|\mathbf{x}_k^i)}{\sum_{j=1}^{N} w_{k|k-1}^j p(\mathbf{z}_k|\mathbf{x}_k^j)}. \tag{1.25}$$

---

[3] If $\ell = 0$ then the problem reduces to the estimation of $p(\mathbf{x}_k|\mathbf{Z}_k)$ considered up to this point.

The preceding assumes that transitional densities $p(x_k^i|x_{k-1}^j)$ and likelihood functions $p(z_k|x_k^i)$ are known, but does not constrain the particular form of these discrete densities. Again, this is the optimal solution if the assumptions made hold.

1.3.3 Beneš and Daum Filters

Beneš was the first to discover a class of nonlinear dynamic systems (assuming a linear measurement equation) for which the posterior pdf admits a "sufficient statistic" of a constant finite dimension [5]. This class is formulated in the continuous time with state vector $\mathbf{x}_t$ satisfying the following differential equation:

$$\frac{d\mathbf{x}_t}{dt} = \mathbf{f}(\mathbf{x}_t) + \mathbf{v}_t \qquad (1.26)$$

where $\mathbf{v}_t$ is Gaussian zero-mean white noise and function $\mathbf{f}(\mathbf{x})$ obeys:

$$\text{tr}\,[\nabla_\mathbf{x} \mathbf{f}] + \mathbf{f}^T \mathbf{f} = \mathbf{x}^T \mathbf{A} \mathbf{x}_t + \mathbf{b}^T \mathbf{x}_t + c. \qquad (1.27)$$

The explanation of terms in condition (1.27) is as follows: tr[ ] denotes the trace of a matrix; $\nabla_\mathbf{x}$ is the gradient operator with respect to $\mathbf{x}$; $\mathbf{A}$, $\mathbf{b}$, and c are constants independent of $\mathbf{x}_t$. In the scalar case, condition (1.27) simplifies to:

$$f'(x) + f^2(x) = ax^2 + bx + c \qquad (1.28)$$

where $f'(x)$ denotes the first derivative of f with respect to x. An example of a function that satisfies condition (1.28) is

$$f(x) = \frac{Ae^x - Be^{-x}}{Ae^x + Be^{-x}}, \qquad (1.29)$$

since then $f'(x) + f^2(x) = 1$ and thus $a = b = 0$, $c = 1$. For a special case where $A = B$, (1.29) simplifies to $f(x) = \tanh(x)$, which can be used as a model of a saturation phenomenon.

The measurement equation for the Beneš filtering problem can be specified in a discrete-time and is linear as in (1.9).

We have seen that the sufficient statistic for the Kalman filter consists of a pair: mean and covariance. The sufficient statistic of the Beneš filter also consists of a pair $(\mathbf{m}_t, \mathbf{P}_t)$, but $\mathbf{m}_t$ and $\mathbf{P}_t$ are not necessarily the mean and the covariance. The propagation of sufficient statistics between the measurement times can be described by the following ordinary differential equations [11]:

$$\dot{\mathbf{m}}_t = -\mathbf{P}_t \mathbf{A} \mathbf{m}_t - \frac{1}{2}\mathbf{P}_t \mathbf{b} \qquad (1.30)$$

$$\dot{\mathbf{P}}_t = \mathbf{I} - \mathbf{P}_t \mathbf{A} \mathbf{P}_t \qquad (1.31)$$

which can be solved numerically by integration over the interval $t_{k-1} \leq t \leq t_k$. Let $(\bar{\mathbf{m}}_k, \bar{\mathbf{P}}_k)$ denote the value of the sufficient statistic immediately before the measurement time t_k, and $(\mathbf{m}_k, \mathbf{P}_k)$ the value immediately after t_k. Then the filter update is performed at time t_k using the standard Kalman filter equations (1.16) and (1.17). After some manipulations (see, for example, [2, p. 133]), the update equations of the Beneš filter become:

$$\mathbf{P}_k = \bar{\mathbf{P}}_k - \bar{\mathbf{P}}_k \mathbf{H}_k^T (\mathbf{H}_k \bar{\mathbf{P}}_k \mathbf{H}_k^T + \mathbf{R}_k)^{-1} \mathbf{H}_k \bar{\mathbf{P}}_k \qquad (1.32)$$

$$\mathbf{m}_k = \bar{\mathbf{m}}_k + \mathbf{P}_k \mathbf{H}_k^T \mathbf{R}_k^{-1} (\mathbf{z}_k - \mathbf{H}_k \bar{\mathbf{m}}_k). \qquad (1.33)$$

Daum has further extended the class of nonlinear dynamic systems that admit a sufficient statistic of a constant finite dimension. The details can be found in [6, 7]. A study of the Beneš filter and its comparison with the particle filter is given in [12]. An application of the Daum filter is described in [13].

1.4 MULTIPLE SWITCHING DYNAMIC MODELS

In many engineering applications one deals with nonlinear dynamic systems characterized by a few possible modes (or regimes) of operation. In tracking, for example, this applies to maneuvering targets, where system behavior patterns (nonmaneuvering motion, various maneuvers) are referred to as system modes [2, 14]. These types of problems are often referred to as jump Markov or hybrid-state estimation problems [15] involving both continuous-valued target state and a discrete-valued regime (mode) variable.

A discrete-time hybrid system is described by the following dynamic and measurement equations:

$$\mathbf{x}_k = \mathbf{f}_{k-1}(\mathbf{x}_{k-1}, r_k, \mathbf{v}_{k-1}) \qquad (1.34)$$

$$\mathbf{z}_k = \mathbf{h}_k(\mathbf{x}_k, r_k, \mathbf{w}_k) \qquad (1.35)$$

where r_k is the regime (model) variable in effect during the sampling period $(t_{k-1}, t_k]$, that is with the impact of the new model, r_{k+1}, starting at t_k^+. The regime variable is commonly modeled by a time-homogeneous s-state first-order Markov chain with transitional probabilities

$$\pi_{ij} \triangleq P\{r_k = j | r_{k-1} = i\} \quad (i, j \in S) \qquad (1.36)$$

where $S \triangleq \{1, 2, \ldots, s\}$. The transitional probability matrix (TPM) $\Pi = [\pi_{ij}]$ is thus an $s \times s$ matrix with elements satisfying

$$\pi_{ij} \geq 0 \quad \text{and} \quad \sum_{j=1}^{s} \pi_{ij} = 1, \tag{1.37}$$

for each $i, j \in S$. The initial regime probabilities are denoted as

$$\mu_i \triangleq P\{r_1 = i\}, \tag{1.38}$$

for $i \in S$, such that

$$\mu_i \geq 0 \quad \text{and} \quad \sum_{i=1}^{s} \mu_i = 1. \tag{1.39}$$

The remaining terms in (1.34) and (1.35) have been defined before. Note that if $s = 1$, (1.34) is identical to (1.1) and (1.35) is identical to (1.2). Note also that the conceptual solution to the nonlinear filtering problem defined by (1.34) and (1.35) again is given by (1.3) and (1.4), the only difference being that the state vector now is an augmented hybrid state vector: $\mathbf{y}_k = [\mathbf{x}_k^T, r_k]^T$. For completeness we state here the conceptual recursive solution to the hybrid state estimation problem:

Prediction:
$$p(\mathbf{x}_k, r_k = j | \mathbf{Z}_{k-1}) = \sum_i \pi_{ij} \int p(\mathbf{x}_k | \mathbf{x}_{k-1}, r_k = j) p(\mathbf{x}_{k-1}, r_{k-1} = i | \mathbf{Z}_{k-1}) d\mathbf{x}_{k-1}$$
$$\tag{1.40}$$

Update:
$$p(\mathbf{x}_k, r_k = j | \mathbf{Z}_k) = \frac{p(\mathbf{z}_k | \mathbf{x}_k, r_k = j) p(\mathbf{x}_k, r_k = j | \mathbf{Z}_{k-1})}{\sum_j \int p(\mathbf{z}_k | \mathbf{x}_k, r_k = j) p(\mathbf{x}_k, r_k = j | \mathbf{Z}_{k-1}) d\mathbf{x}_k} \tag{1.41}$$

Jump Markov Linear System

A special case of a general hybrid system defined by (1.34) and (1.35) is the so-called jump Markov linear system (JMLS), defined as:

$$\mathbf{x}_k = \mathbf{F}_{k-1}(r_k)\mathbf{x}_{k-1} + \mathbf{v}_{k-1}(r_k) \tag{1.42}$$
$$\mathbf{z}_k = \mathbf{H}_k(r_k)\mathbf{x}_k + \mathbf{w}_k(r_k) \tag{1.43}$$

where $\mathbf{F}_{k-1}(r_k)$ and $\mathbf{H}_k(r_k)$ are linear functions and $\mathbf{v}_{k-1}(r_k)$ and $\mathbf{w}_k$ are zero-mean white Gaussian noise processes with covariances $\mathbf{Q}_{k-1}(r_k)$ and $\mathbf{R}_k(r_k)$

respectively. If $s = 1$, the JMLS reduces to (1.8) and (1.9), and we have seen that the optimal algorithm in this case is the Kalman filter.

The JMLS is a *nonlinear* system because $\mathbf{x}_k$ or $\mathbf{z}_k$ do not depend on the hybrid state vector $\mathbf{y}_k$ of the system in a linear fashion. The system becomes linear only if the regime variable r_k is given (fixed). We will see next that the number of parameters required to characterize the posterior pdf $p(\mathbf{x}_k|\mathbf{Z}_k)$ of JMLS grows exponentially with time. For this reason it is impossible to formulate a sufficient statistic for JMLS with a fixed finite dimension.

Suppose the sequence of regimes (a regime history) up to time index k is denoted as

$$R_k^\ell \triangleq \{r_1^\ell, \ldots, r_k^\ell\}, \quad \ell = 1, \ldots, s^k \tag{1.44}$$

where r_κ^ℓ, $(1 \leq \kappa \leq k)$, denotes the particular value that the regime variable takes during the sampling period $(t_{k-1}, t_k]$ in the ℓth regime sequence. The posterior density in the case of the JMLS can be expressed as a Gaussian mixture [2]:

$$p(\mathbf{x}_k|\mathbf{Z}_k) = \sum_{\ell=1}^{s^k} p(\mathbf{x}_k|R_k^\ell, \mathbf{Z}_k) \mathrm{P}\{R_k^\ell|\mathbf{Z}_k\} \tag{1.45}$$

where

- $\mathrm{P}\{R_k^\ell|\mathbf{Z}_k\}$ is the probability of a particular regime sequence given measurements $\mathbf{Z}_k$. Using the Bayes rule this probability can be expressed as [2]:

$$\begin{aligned}
\mathrm{P}\{R_k^\ell|\mathbf{Z}_k\} &= \mathrm{P}\{R_k^\ell|\mathbf{z}_k, \mathbf{Z}_{k-1}\} \\
&= \frac{1}{c} p(\mathbf{z}_k|R_k^\ell, \mathbf{Z}_{k-1}) \cdot \mathrm{P}\{R_k^\ell|\mathbf{Z}_{k-1}\} \\
&= \frac{1}{c} p(\mathbf{z}_k|R_k^\ell, \mathbf{Z}_{k-1}) \cdot \pi_{ij} \mathrm{P}\{R_{k-1}^\ell|\mathbf{Z}_{k-1}\}
\end{aligned} \tag{1.46}$$

where c is a normalization constant, R_{k-1}^ℓ is the parent regime sequence, $i = r_{k-1}^\ell$, and $j = r_k^\ell$.

- $p(\mathbf{x}_k|R_k^\ell, \mathbf{Z}_k)$ is the posterior density for a given regime sequence R_k^ℓ; this density is Gaussian and can be computed using a Kalman filter with parameters that correspond to the sequence of dynamic modes.

The problem with (1.45) is that the number of mixture components (i.e., regime histories) in the Gaussian sum grows exponentially with time k. Hence, a practical filter for JMLS has to be suboptimal, based on some approximation. Typical approximations are discussed in Chapters 2 and 3.

1.5 BASICS OF TARGET TRACKING

Target tracking is an element of a wider system that performs surveillance, guidance, obstacle avoidance, or a similar function. The top-level structure of such a system is usually of the form shown in Figure 1.1. Sensors provide signals that are channeled into the signal processing subsystem that outputs measurements. Tracking is part of the data-processing subsystem[4] whose role is to process measurements in order to form and maintain tracks. A track is a sequence of target state estimates up to the current time. The amount of data flow in Figure 1.1 reduces from left to right. A well-designed system, in this process, aims to preserve the information content.

The target state typically consists of kinematic components (position, velocity, acceleration, and so forth) and attributes (target signal-to-noise ratio (SNR), radar cross-section (RCS), spectral characteristics, class, allegiance, and so forth). The measurements are noise-corrupted observations, related to the target state. They are typically a result of an estimation/detection algorithm in the signal-processing subsystem. Detection is often performed by thresholding the data[5] and is characterized by the probability of detection P_D and the probability of false alarm P_{FA}. The kinematic measurements include target range, azimuth, elevation, and range-rate (extracted from Doppler frequency). Attribute measurements are typically the received signal strength, a low-resolution target image, radiated frequency, declared allegiance, and so forth. Measurements are collected by a single or possibly multiple sensors. In addition to target-originated measurements, the sensors often report false detections due to clutter, noise, or countermeasures. The tracking system forms and maintains a track for each target from a sequence of measurements that have been associated with the target over time. False detections and multiple target scenarios form an additional level of complexity in this process, due to the uncertainty in the measurement origin. A typical multiple target tracking system consists of following common blocks [19, 20, 21]:

- **Tracking filters.** The role of a tracking filter is to carry out recursive target state estimation given: (1) target dynamic equation of the form (1.1); (2) sensor measurement equation of the form (1.2); (3) target-originated measurements z_k. The most commonly used tracking filters are the fixed coefficient filters (alpha-beta), the Kalman filter, and the extended Kalman filter. The choice of the tracking filter often depends on the choice of the coordinate system for tracking. Suppose a target is moving with a *constant velocity* and

---

4 Note that future trends are towards the integration of signal processing with data processing. In addition, a feedback loop is used in some systems that allows the output of a tracker to control sensor scheduling or sensor pointing.

5 A modern trend is to use unthresholded data [16, 17, 18], although this dramatically increases the computational requirements.

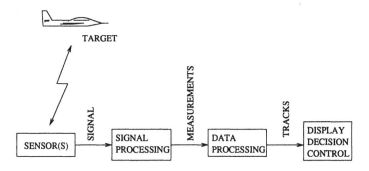

Figure 1.1 Top-level structure of a typical surveillance system.

the measurements are in polar coordinates (range, azimuth, elevation). Then if the state vector is formulated in Cartesian coordinates, the dynamic equation is linear and the measurement equation is nonlinear. However, if one adopts polar coordinates for the state vector, the dynamic equation becomes nonlinear and the measurement equation is linear.

- **Maneuver handling logic.** A maneuver can be described as a sudden change in the dynamic motion equation. A common model for describing the dynamics of a maneuvering target is given by (1.34) and (1.35). The nonmaneuvering segments of a target trajectory are modeled by a (nearly) constant velocity motion [2]. For the maneuvering segments there are many possible choices, such as coordinated turn motion, constant acceleration model, and constant jerk model [22].

- **Gating and data association.** These two components of a tracking system have a role to identify the origin of measurements. They perform an association of a discrete set of collected measurements at time k to a set of existing tracks. A typical measure of similarity used for data association is the distance in kinematic and/or the attribute domain. Common algorithms for data association are the joint probabilistic data association (JPDA) filter, multi-hypotheses tracker (MHT), S-dimensional assignment algorithms, and many others [21].

- **Track life management.** Track status is typically defined in terms of three life stages: tentative, confirmed, and deleted track. A tentative track is typically initiated on a measurement that has not been associated to any of the existing tracks. Track confirmation is based on the number of updates performed on a tentative track over a period of time or number of scans. Track deletion is based on the length of time or a number of scans without a track update.

Typically the trackers introduce the notion of track quality as a criterion for switching between the track life stages.

As we can see from this short review, filtering, which is the main focus of this book, is a fundamental component of a typical tracking system. Our presentation will often ignore the issue of data association and track life management; these two functions can be performed separately and independently from filtering. Thus, for most of the book, we adopt a framework where we consider a single target in the absence of false measurements. Only Chapter 12 departs from this framework and deals with multiple targets in clutter.

1.6 SUMMARY

This chapter introduced (1) the framework this book will use for nonlinear/non-Gaussian filtering, and (2) the role of filtering in a typical target tracking system. It turns out that the optimal recursive Bayesian estimator requires the entire posterior density of the target state to be computed sequentially, as the measurements arrive. The problem is difficult in the general case because the posterior density does not admit a sufficient statistic with a finite and constant dimension. Only for a very limited class of dynamic stochastic systems can one formulate and apply optimal algorithms. In most practical cases various approximate solutions need to be explored.

References

[1] A. H. Jazwinski, *Stochastic Processes and Filtering Theory*. New York: Academic Press, 1970.

[2] Y. Bar-Shalom, X. R. Li, and T. Kirubarajan, *Estimation with Applications to Tracking and Navigation*. New York: John Wiley & Sons, 2001.

[3] A. T. Alouani and W. D. Blair, "Use of kinematic constraint in tracking constant speed, maneuvering target," *IEEE Trans. Automatic Control*, vol. 38, pp. 1107–1111, July 1993.

[4] L.-S. Wang, Y.-T. Chiang, and F.-R. Chang, "Filtering method for nonlinear systems with constraints," *IEE Proc. - Control Theory and Appl.*, vol. 149, pp. 525–531, November 2002.

[5] V. E. Benes, "Exact finite-dimensional filters with certain diffusion non linear drift," *Stochastics*, vol. 5, pp. 65–92, 1981.

[6] F. E. Daum, "Exact finite dimensional nonlinear filters," *IEEE Trans. Automatic Control*, vol. 31, no. 7, pp. 616–622, 1986.

[7] F. E. Daum, "Beyond Kalman filters: practical design of nonlinear filters," in *Proc. SPIE*, vol. 2561, pp. 252–262, 1995.

[8] Y. C. Ho and R. C. K. Lee, "A Bayesian approach to problems in stochastic estimation and control," *IEEE Trans. Automatic Control*, vol. 9, pp. 333–339, 1964.

[9] B. D. O. Anderson and J. B. Moore, *Optimal Filtering*. Englewood Cliffs, NJ: Prentice-Hall, 1979.

[10] C. K. Chui and G. Chen, *Kalman Filtering with Real-Time Applications*. New York: Springer-Verlag, 2nd ed., 1991.

[11] F. E. Daum, "New exact nonlinear filters," in *Bayesian Analysis of Time Series and Dynamic Models* (J. C. Spall, ed.), ch. 8, pp. 199–226, New York: Marcel-Dekker, 1988.

[12] A. Farina, D. Benvenuti, and B. Ristic, "A comparative study of the Benes filtering problem," *Signal Processing*, vol. 82, pp. 133–147, 2002.

[13] G. C. Schmidt, "Designing nonlinear filters based on Daum's theory," *Journal of Guidance, Control and Dynamics*, vol. 16, pp. 371–376, March-April 1993.

[14] X. R. Li, "Engineer's guide to variable-structure multiple-model estimation for tracking," in *Multitarget-Multisensor Tracking: Applications and Advances* (Y. Bar-Shalom and W. D. Blair, eds.), vol. III, ch. 10, Norwood, MA: Artech House, 2000.

[15] D. D. Sworder and J. E. Boyd, *Estimation Problems in Hybrid Systems*. Cambridge, U.K.: Cambridge University Press, 1999.

[16] Y. Barniv, "Dynamic programming algorithm for detecting dim moving targets," in *Multitarget Multisensor Tracking: Advanced Applications* (Y. Bar-Shalom, ed.), ch. 4, Norwood, MA: Artech House, 1990.

[17] K. Kastella, "Finite difference methods for nonlinear filtering and automatic target recognition," in *Multitarget-Multisensor Tracking* (Y. Bar-Shalom and W. D. Blair, eds.), vol. III, ch. 5, Norwood, MA: Artech House, 2000.

[18] R. L. Streit, M. L. Graham, and M. J. Walsh, "Multi-target tracking of distributed targets using histogram-PMHT," in *Proc. 4th Int. Conf. Information Fusion (Fusion 2001)*, (Montreal, Canada), August 2001.

[19] A. Farina and F. A. Studer, *Radar Data Processing*. New York: John Wiley, 1985.

[20] Y. Bar-Shalom and T. E. Fortmann, *Tracking and Data Association*. Boston, MA: Academic Press, 1988.

[21] S. Blackman and R. Popoli, *Design and Analysis of Modern Tracking Systems*. Norwood, MA: Artech House, 1999.

[22] X. R. Li and V. P. Jilkov, "A survey of maneuvering target tracking, Part I: Dynamic models," *IEEE Trans. Aerospace and Electronic Systems*, 2002. In Review.

Chapter 2

Suboptimal Nonlinear Filters

Reality often manifests itself as being very complex: nonlinear, non-Gaussian, non-stationary, and with continuous-valued target states. Therefore, in most practical situations, the optimal nonlinear filters of Section 1.3 cannot be applied. Instead, one is forced to use approximations or suboptimal solutions.

Over the years a huge number of approximate nonlinear filters have been proposed [1, 2, 3]. Some are fairly general while others are more tailored to a particular application. Since it would take more than one book to review all of them in detail, in this chapter we restrict ourselves to those nonlinear filters that have been adopted by a wider scientific and engineering community. We group them into four broad classes: (1) analytic approximations; (2) numerical approximations; (3) Gaussian sum or multiple model filters; and (4) sampling approaches. The sampling approaches approximate the posterior density by a set of samples: the unscented Kalman filter (UKF) uses a small number of deterministically chosen samples, while the particle filter uses a large number of random (Monte Carlo) samples. We give a brief account of such suboptimal nonlinear filters in this chapter and the next.

2.1 ANALYTIC APPROXIMATIONS

In this category of nonlinear filters we include the extended Kalman filter (EKF) and its cousins, the higher-order EKF and iterated EKF. The main feature of these filters is that they approximate (linearize) the nonlinear functions in the state dynamic and measurement models [4, 5].

The extended Kalman filter is derived for nonlinear systems with additive noise; that is, for a special case of (1.1) and (1.2) given here:

$$\mathbf{x}_k = \mathbf{f}_{k-1}(\mathbf{x}_{k-1}) + \mathbf{v}_{k-1} \qquad (2.1)$$
$$\mathbf{z}_k = \mathbf{h}_k(\mathbf{x}_k) + \mathbf{w}_k. \qquad (2.2)$$

Random sequences $\mathbf{v}_{k-1}$ and $\mathbf{w}_k$ are mutually independent, zero-mean white Gaussian with covariances $\mathbf{Q}_{k-1}$ and $\mathbf{R}_k$, respectively. The nonlinear functions in (2.1) and (2.2) are approximated by the first term in their Taylor series expansion. The EKF is based on the assumption that local linearization of the above equations may be a sufficient description of nonlinearity. The posterior pdf $p(\mathbf{x}_k|\mathbf{Z}_k)$ is approximated by a Gaussian density and relationships (1.10)–(1.12) are assumed to hold. The mean and the covariance of the underlying Gaussian density are computed recursively as follows (for derivation of the EKF, see, for example, Sec. 10.3.2 of [5]):

$$\hat{\mathbf{x}}_{k|k-1} = \mathbf{f}_{k-1}(\hat{\mathbf{x}}_{k-1|k-1}) \tag{2.3}$$

$$\mathbf{P}_{k|k-1} = \mathbf{Q}_{k-1} + \hat{\mathbf{F}}_{k-1}\mathbf{P}_{k-1|k-1}\hat{\mathbf{F}}_{k-1}^T \tag{2.4}$$

$$\hat{\mathbf{x}}_{k|k} = \hat{\mathbf{x}}_{k|k-1} + \mathbf{K}_k(\mathbf{z}_k - \mathbf{h}_k(\hat{\mathbf{x}}_{k|k-1})) \tag{2.5}$$

$$\mathbf{P}_{k|k} = \mathbf{P}_{k|k-1} - \mathbf{K}_k \mathbf{S}_k \mathbf{K}_k^T \tag{2.6}$$

where

$$\mathbf{S}_k = \hat{\mathbf{H}}_k \mathbf{P}_{k|k-1} \hat{\mathbf{H}}_k^T + \mathbf{R}_k \tag{2.7}$$

$$\mathbf{K}_k = \mathbf{P}_{k|k-1} \hat{\mathbf{H}}_k^T \mathbf{S}_k^{-1} \tag{2.8}$$

and $\hat{\mathbf{F}}_{k-1}$ and $\hat{\mathbf{H}}_k$ are the local linearization of nonlinear functions $\mathbf{f}_{k-1}$ and $\mathbf{h}_k$, respectively. They are defined as Jacobians [5] evaluated at $\hat{\mathbf{x}}_{k-1|k-1}$ and $\hat{\mathbf{x}}_{k|k-1}$, respectively; that is:

$$\hat{\mathbf{F}}_{k-1} = [\nabla_{\mathbf{x}_{k-1}} \mathbf{f}_{k-1}^T(\mathbf{x}_{k-1})]^T \big|_{\mathbf{x}_{k-1}=\hat{\mathbf{x}}_{k-1|k-1}} \tag{2.9}$$

$$\hat{\mathbf{H}}_k = [\nabla_{\mathbf{x}_k} \mathbf{h}_k^T(\mathbf{x}_k)]^T \big|_{\mathbf{x}_k=\hat{\mathbf{x}}_{k|k-1}} \tag{2.10}$$

where

$$\nabla_{\mathbf{x}_k} = \left[\frac{\partial}{\partial \mathbf{x}_k[1]} \cdots \frac{\partial}{\partial \mathbf{x}_k[n_x]}\right]^T, \tag{2.11}$$

with $\mathbf{x}_k[i]$, $i = 1, \ldots, n_x$, being the ith component of vector $\mathbf{x}_k$. An element of, say, $\hat{\mathbf{H}}_k$ is then given by:

$$\hat{\mathbf{H}}_k[i,j] = \frac{\partial \mathbf{h}_k[i]}{\partial \mathbf{x}_k[j]} \bigg|_{\mathbf{x}_k=\hat{\mathbf{x}}_{k|k-1}} \tag{2.12}$$

where $\mathbf{h}_k[i]$ denotes the ith component of vector $\mathbf{h}_k(\mathbf{x}_k)$. We emphasize that according to (2.9) and (2.10), Jacobians $\hat{\mathbf{F}}_{k-1}$ and $\hat{\mathbf{H}}_k$ are evaluated at (and hence depend on) the estimates of the state vector.

The EKF and its cousins are referred to as *analytic* approximations because the Jacobians $\hat{\mathbf{F}}_{k-1}$ and $\hat{\mathbf{H}}_k$ have to be worked out analytically. If the nonlinear functions $\mathbf{f}_k$ or $\mathbf{h}_k$ are discontinuous, this class of filters cannot be applied.

Note that the EKF always approximates $p(\mathbf{x}_k|\mathbf{Z}_k)$ to be Gaussian. If the nonlinearity in models (2.1) and (2.2) is very severe, the non-Gaussianity of the true posterior density will be more pronounced (e.g., it can be bimodal or heavily skewed). In such cases the performance of the EKF will be degraded significantly. This situation is illustrated by the following example.

Example

Consider a Gaussian random variable in the polar coordinates, $\mathbf{a} \sim \mathcal{N}(\bar{\mathbf{a}}, \mathbf{P}_a)$. The first component of this random variable corresponds to the "range" and the second to the "angle." Let the mean range be 80 units and the mean angle 0.61 rad, with $\mathbf{P}_a[1,1] = 23$, $\mathbf{P}_a[2,2] = 0.0027$, and $\mathbf{P}_a[1,2] = \mathbf{P}_a[2,1] = 0.25$. Then we convert random variable $\mathbf{a}$ from polar to Cartesian coordinates, using the standard nonlinear transformation $\mathbf{g}$:

$$\mathbf{b} = \mathbf{g}(\mathbf{a}) = \begin{bmatrix} \mathbf{a}[1] \cdot \cos \mathbf{a}[2] \\ \mathbf{a}[1] \cdot \sin \mathbf{a}[2] \end{bmatrix} \quad (2.13)$$

Figure 2.1 shows the result of this transformation (uneven axes are used to emphasize the effect). The cloud of random samples represents the true density of random variable $\mathbf{b}$ (clearly non-Gaussian). This cloud is obtained by the Monte Carlo approach: 5000 samples were drawn from the Gaussian density $\mathcal{N}(\bar{\mathbf{a}}, \mathbf{P}_a)$ and then each sample was propagated through the nonlinear transformation $\mathbf{g}$. The true mean and covariance of $\mathbf{b}$ are computed from this sample cloud. The linearization approach, built into the EKF, would model the density of $\mathbf{b}$ by a Gaussian pdf with mean $\mathbf{g}(\bar{\mathbf{a}})$ and covariance $\mathbf{G}\mathbf{P}_a\mathbf{G}^T$, where

$$\mathbf{G} = [\nabla_\mathbf{a}\mathbf{g}(\mathbf{a})^T]^T = \begin{bmatrix} \cos \mathbf{a}[2] & \sin \mathbf{a}[2] \\ -\mathbf{a}[1] \sin \mathbf{a}[2] & \mathbf{a}[1] \cos \mathbf{a}[2] \end{bmatrix} \quad (2.14)$$

Figure 2.1 illustrates the true mean and covariance (2σ ellipse) of $\mathbf{b}$ together with the linearization approximation. In this two-dimensional example, the true 2σ ellipse should contain approximately 86% of random samples [6, p. 96]. Observe that linearization produces errors both in the mean and the covariance of $\mathbf{b}$.

The above example demonstrates that the EKF is an *inconsistent* estimator. A filter is consistent if the estimation error is zero-mean with the covariance matching the filter calculated covariance.[1]

---

[1] Note that despite inconsistency, the EKF may perform quite well if the nonlinearities in dynamics or measurement model are mild. Moreover, for the type of nonlinearity examined in above example, consistent estimators have been devised [7, 8]; these methods are problem-specific, but work well.

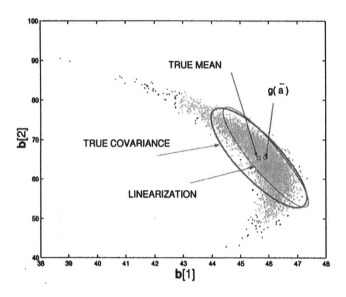

Figure 2.1 The probability density function of a nonlinearly transformed Gaussian random variable (observe the effect of linearization).

In order to overcome the described problems with the EKF, it has been proposed [2] to use higher-order EKFs, which retain further terms (in addition to the first one) in the Taylor expansion. However, the additional complexity of these filters have prohibited their widespread use. The iterated EKF (IEKF) is an attempt to alleviate the linearization errors in the measurement model (the IEKF differs from the EKF only in the update stage). The basic idea is to linearize the nonlinear measurement model around the *updated* state rather than the predicted state. This is achieved iteratively and involves the use of the current measurement z_k; details are given in [5]. It has been shown that the measurement update stage of the IEKF is very accurate only when the measurement model fully observes the state [9]. This is rarely the case in practice, certainly not in any of the applications that we consider in Part II of this book.

2.2 NUMERICAL METHODS

This class of nonlinear filters performs numerical integration in order to solve the multidimensional integrals in (1.3) and (1.5). Numerical methods are also referred to as approximate grid-based methods and essentially they replace the integral by summation involving a discretization of the integration variable [1, 10, 11, 12].

Although the numerical methods will not be used throughout this book, some basics are presented here for completeness. Let us decompose the continuous state space into N "cells," $\{\mathbf{x}_k^i : i = 1, \ldots, N\}$. Then the grid-based method of Section 1.3.2 can be applied to approximate the posterior density. Specifically, suppose the approximation to the posterior pdf at $k-1$ is given by

$$p(\mathbf{x}_{k-1}|\mathbf{Z}_{k-1}) \approx \sum_{i=1}^{N} w_{k-1|k-1}^i \delta(\mathbf{x}_{k-1} - \mathbf{x}_{k-1}^i). \qquad (2.15)$$

Then, the prediction and update equations can be written as

$$p(\mathbf{x}_k|\mathbf{Z}_{k-1}) \approx \sum_{i=1}^{N} w_{k|k-1}^i \delta(\mathbf{x}_k - \mathbf{x}_k^i) \qquad (2.16)$$

$$p(\mathbf{x}_k|\mathbf{Z}_k) \approx \sum_{i=1}^{N} w_{k|k}^i \delta(\mathbf{x}_k - \mathbf{x}_k^i) \qquad (2.17)$$

where

$$w_{k|k-1}^i \triangleq \sum_{j=1}^{N} w_{k-1|k-1}^j \int_{\mathbf{x} \in \mathbf{x}_k^i} p(\mathbf{x}|\bar{\mathbf{x}}_{k-1}^j) d\mathbf{x}, \qquad (2.18)$$

$$w_{k|k}^i \triangleq \frac{w_{k|k-1}^i \int_{\mathbf{x} \in \mathbf{x}_k^i} p(\mathbf{z}_k|\mathbf{x}) d\mathbf{x}}{\sum_{j=1}^{N} w_{k|k-1}^j \int_{\mathbf{x} \in \mathbf{x}_k^j} p(\mathbf{z}_k|\mathbf{x}) d\mathbf{x}} \qquad (2.19)$$

Here $\bar{\mathbf{x}}_{k-1}^j$ denotes the center of the jth "cell" at time index $k-1$. The integrals in (2.18) and (2.19) arise due to the fact that the grid points $\mathbf{x}_k^i$, $i = 1, \ldots, N$, represent regions of continuous state space, and thus the probabilities must be integrated over these regions. In practice, to simplify computation, a further approximation is made in the evaluation of $w_{k|k}^i$. Specifically, these weights are computed at the center of the "cell" corresponding to $\mathbf{x}_k^i$:

$$w_{k|k-1}^i \approx \sum_{j=1}^{N} w_{k-1|k-1}^j p(\bar{\mathbf{x}}_k^i|\bar{\mathbf{x}}_{k-1}^j) \qquad (2.20)$$

$$w_{k|k}^i \approx \frac{w_{k|k-1}^i p(\mathbf{z}_k|\bar{\mathbf{x}}_k^i)}{\sum_{j=1}^{N} w_{k|k-1}^j p(\mathbf{z}_k|\bar{\mathbf{x}}_k^j)} \qquad (2.21)$$

The grid must be sufficiently dense to get a good approximation to the continuous state space. As the dimensionality of the state space increases, the computational cost of the approach therefore increases dramatically. If the state space is not finite in extent, then using a grid-based approach necessitates some truncation of the state space. Another disadvantage of grid-based methods is that the state space must be predefined, and therefore cannot be partitioned unevenly to give greater resolution in high probability density regions unless prior knowledge is used.

Hidden Markov model (HMM) filters [13, 14] are an application of such approximate grid-based methods in a fixed-interval smoothing context and have been used extensively in speech processing. In HMM-based tracking, a common approach is to use the Viterbi algorithm [15] to calculate the maximum a posteriori estimate of the path through the trellis, that is, the sequence of discrete states that maximizes the probability of the state sequence given the data. Another approach, due to Baum-Welch [14], is to calculate the probability of each discrete state at each time epoch given the entire data sequence.[2]

2.3 GAUSSIAN SUM FILTERS

The key idea of the Gaussian sum filter [16] is to approximate the required posterior density $p(\mathbf{x}_k|\mathbf{Z}_k)$ by a Gaussian mixture (a weighted sum of Gaussian density functions); that is:

$$p(\mathbf{x}_k|\mathbf{Z}_k) \approx p_A(\mathbf{x}_k|\mathbf{Z}_k) = \sum_{i=1}^{q_k} w_k^i \mathcal{N}\left(\mathbf{x}_k^i; \hat{\mathbf{x}}_{k|k}^i, \mathbf{P}_{k|k}^i\right) \qquad (2.22)$$

where w_k^i are weights such that[3] $\sum_{i=1}^{q_k} w_k^i = 1$. The approximation p_A can be made as accurate as desirable through the choice of q_k, the number of mixture components. This type of approximation is quite reasonable especially when the posterior density is multimodal.[4] The problem is, however, how to formulate an algorithmic procedure for on-line computation of weights w_k^i, means $\hat{\mathbf{x}}_{k|k}^i$, and covariances $\mathbf{P}_{k|k}^i$. The problem is not trivial, since the number of components q_k, in general, can grow exponentially with k. We review briefly here two types of

---

2   The Viterbi and Baum-Welch algorithms are frequently applied when the state space is approximated to be discrete. The algorithms are optimal if and only if the underlying state space is truly discrete in nature.

3   Sorenson and Alspach [1, 17] state an additional condition here, that $w_k^i \geq 0$ for $i = 1, \ldots, q_k$. Koch [18] allows weights to be negative so that mixture densities with sharp edges can be constructed.

4   Note that filters such as the EKF or the unscented filter are inappropriate for multimodal densities.

Gaussian sum filters that have become very popular in the tracking community: the *static multiple-model* (MM) estimator (formulated originally in [17]) and the *dynamic MM* estimator [5].

2.3.1 Static MM Estimator

The static MM estimator (filter) is designed to solve the problem defined by (1.34) and (1.35), where the regime variable $r_k \in S = \{1, 2, \ldots, s\}$ is constant (i.e., $r_k = r$) and unknown. Consequently the number of components in the mixture density is fixed ($q_k = s$), as the target follows only one of s possible (nonlinear) models. The problem is that we do not know which model is correct.

The static MM filter consists of a bank of s nonlinear filters, each matched to a particular model. A convenient choice for a model-matched nonlinear filter is, for example, the EKF, or the unscented Kalman filter, or indeed any other nonlinear filter that adopts relationships (1.10)–(1.12). The output of model-matched filter i is the model-conditioned state estimate $\hat{\mathbf{x}}^i_{k|k}$ and its associated covariance $\mathbf{P}^i_{k|k}$. The probability of model i being correct (the mixture weight) is obtained from the Bayes formula:

$$\begin{aligned} w^i_k &\triangleq \mathrm{P}\{r = i | \mathbf{Z}_k\} \\ &= \mathrm{P}\{r = i | \mathbf{z}_k, \mathbf{Z}_{k-1}\} \\ &= \frac{p(\mathbf{z}_k | \mathbf{Z}_{k-1}, r = i) \cdot \mathrm{P}\{r = i | \mathbf{Z}_{k-1}\}}{p(\mathbf{z}_k | \mathbf{Z}_{k-1})} \\ &= \frac{p(\mathbf{z}_k | \mathbf{Z}_{k-1}, r = i) \cdot w^i_{k-1}}{\sum_{j=1}^{s} p(\mathbf{z}_k | \mathbf{Z}_{k-1}, j) \cdot w^j_{k-1}} \end{aligned} \quad (2.23)$$

where $p(\mathbf{z}_k | \mathbf{Z}_{k-1}, r = i) = \Lambda^i_k$ is the model conditioned likelihood function. Under the linear/Gaussian assumptions, this likelihood function is Gaussian:

$$\Lambda^i_k = p(\mathbf{z}_k | \mathbf{Z}_{k-1}, r = i) = \mathcal{N}(\nu^i_k; 0, \mathbf{S}^i_k) \quad (2.24)$$

where ν^i_k and $\mathbf{S}^i_k$ are the innovation and its covariance from the model-matched filter i. In this case:

$$\lim_{k \to \infty} w^i_k = \begin{cases} 0 & \text{if } r \neq i \\ 1 & \text{if } r = i. \end{cases} \quad (2.25)$$

If the dynamic or the measurement equation corresponding to model i is nonlinear or non-Gaussian, the likelihood function Λ^i_k is not Gaussian anymore. Nevertheless, even then, (2.24) is used in practice as an approximation.

The output of an MM filter (static or dynamic) is a Gaussian mixture (2.22). However, the MM filter often reports only the mean and the covariance of this Gaussian mixture, calculated as follows [5]:

$$\hat{\mathbf{x}}_{k|k} = \sum_{i=1}^{s} w_k^i \hat{\mathbf{x}}_{k|k}^i \qquad (2.26)$$

$$\mathbf{P}_{k|k} = \sum_{i=1}^{s} w_k^i \left[\mathbf{P}_{k|k}^i + \left(\hat{\mathbf{x}}_{k|k}^i - \hat{\mathbf{x}}_{k|k}\right)\left(\hat{\mathbf{x}}_{k|k}^i - \hat{\mathbf{x}}_{k|k}\right)^T \right]. \qquad (2.27)$$

2.3.2 Dynamic MM Filter

The dynamic MM filter was derived in Section 1.4 for the jump Markov linear system. The posterior density, given by (1.45), is a Gaussian mixture with an exponentially growing number of mixture components. A practical implementation of the MM filter has to limit this growth in some way. The simplest method would be to preserve only a fixed number of most probable components of the mixture and to "prune" the rest [19]. This is the basis of the multiple model pruning (MMP) algorithm. An alternative to pruning is "merging" of regime mixture components (or hypotheses). A popular merging type algorithm is the interactive MM (IMM) estimator [20].

MMP Algorithm

According to Section 1.4, R_{k-1}^ℓ denotes a regime sequence (history) up to time $k-1$ with index ℓ. Let there be M such sequences, each assigned a weight:

$$w_{k-1}^\ell = \mathrm{P}\{R_{k-1}^\ell | \mathbf{Z}_{k-1}\}, \qquad (\ell = 1, \ldots, M)$$

After receiving measurement $\mathbf{z}_k$ at time k, the MMP algorithm forms $M \times s$ hypotheses (s is the number of regimes or models). Each hypothesis is assigned a weight:

$$\begin{aligned}
w_k^{\ell,i} &\propto \mathrm{P}\{r_k = i, R_{k-1}^\ell | \mathbf{Z}_k\} \qquad (i=1,\ldots,s;\ \ell = 1,\ldots,M) \\
&\propto p(\mathbf{z}_k | \mathbf{Z}_{k-1}, r_k = i, R_{k-1}^\ell) p(r_k = i | R_{k-1}^\ell) \mathrm{P}\{R_{k-1}^\ell | \mathbf{Z}_{k-1}\} \\
&\propto p(\mathbf{z}_k | \mathbf{Z}_{k-1}, r_k = i, r_{k-1}^\ell) p(r_k = i | r_{k-1}^\ell) w_{k-1}^\ell \qquad (2.28)
\end{aligned}$$

where $p(r_k = i | r_{k-1}^\ell) = \pi_{r_{k-1}^\ell, i}$ is the transitional probability defined in (1.36) and $p(\mathbf{z}_k | \mathbf{Z}_{k-1}, r_k = i, r_{k-1}^\ell) = \Lambda_k^{i,\ell}$ is the conditional measurement likelihood. The next step of the MMP is pruning: only M best (those with highest weights) among $M \times s$ hypotheses are preserved and their weights are normalized.

IMM Algorithm

Similar to the static MM, the interactive MM estimator [5, 20] uses a bank of s model-matched filters. The IMM exploits the following feature of the dynamic MM filter: the input to the model-matched filter j is an interaction (mixture) of all s model-matched filters. Thus, before the model-matched filtering step, the IMM performs the "mixing" of models: the posterior density at $k-1$ for model j is represented by $\mathcal{N}\left(\mathbf{x}_{k-1}^j; \hat{\mathbf{x}}_{k-1|k-1}^{(j)}, \mathbf{P}_{k-1|k-1}^{(j)}\right)$, where

$$\hat{\mathbf{x}}_{k-1|k-1}^{(j)} = \sum_{i=1}^{s} \mu_{k-1}^{i|j} \hat{\mathbf{x}}_{k-1|k-1}^{i} \qquad (2.29)$$

$$\mathbf{P}_{k-1|k-1}^{(j)} = \sum_{i=1}^{s} \mu_{k-1}^{i|j} \left[\mathbf{P}_{k-1|k-1}^{i} + \left(\hat{\mathbf{x}}_{k-1|k-1}^{i} - \hat{\mathbf{x}}_{k-1|k-1}^{(j)}\right)\left(\hat{\mathbf{x}}_{k-1|k-1}^{i} - \hat{\mathbf{x}}_{k-1|k-1}^{(j)}\right)^T \right] \qquad (2.30)$$

Equations (2.29) and (2.30) have the same form as (2.26) and (2.27). The weights in (2.29) and (2.30), however, are the mixing probabilities $\mu_k^{i|j}$, defined as:

$$\mu_{k-1}^{i|j} \triangleq P\{r_{k-1} = i | r_k = j, \mathbf{Z}_{k-1}\}. \qquad (2.31)$$

Using Bayes' formula, it follows that:

$$\mu_{k-1}^{i|j} = \frac{\pi_{ij} w_{k-1}^i}{\sum_{i=1}^{s} \pi_{ij} w_{k-1}^i} \qquad (2.32)$$

where π_{ij} are transitional probabilities defined in (1.36) and w_{k-1}^i are the mode probabilities as before. After the model-matched filtering, the posterior density of mode j at k is represented by $\mathcal{N}\left(\mathbf{x}_k^j; \hat{\mathbf{x}}_{k|k}^j, \mathbf{P}_{k|k}^j\right)$. Using again the Bayes rule, the mode probabilities are updated as follows:

$$w_k^j = \frac{\Lambda_k^j \sum_{i=1}^{s} \pi_{ij} w_{k-1}^i}{\sum_{j=1}^{s} \Lambda_k^j \sum_{i=1}^{s} \pi_{ij} w_{k-1}^i} \qquad (2.33)$$

where $\Lambda_k^j = p(\mathbf{z}_k | \mathbf{Z}_{k-1}, r_k = j)$ is the model conditioned likelihood function of (2.24). The output of the IMM is computed again via (2.26) and (2.27). A schematic representation of IMM processing steps with $s = 2$ models is shown in Figure 2.2; M1 and M2 stand for model-matched filters 1 and 2, respectively.

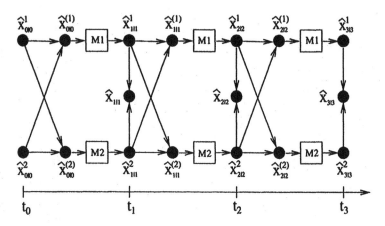

Figure 2.2 IMM with $s = 2$ models.

2.4 UNSCENTED KALMAN FILTER

Recently, the unscented transform has been used in the EKF framework [21, 22, 23] and the resulting filter is referred to as the unscented Kalman filter (UKF). Unlike the EKF, the UKF does not approximate nonlinear functions $\mathbf{f}_{k-1}$ and $\mathbf{h}_k$ in (2.1) and (2.2), respectively. Instead, it approximates the posterior $p(\mathbf{x}_k|\mathbf{Z}_k)$ by a Gaussian density [hence the relationships (1.10)–(1.12) hold], which is represented by a set of deterministically chosen sample points. These sample points completely capture the true mean and covariance of the Gaussian density. When propagated through a nonlinear transform, they capture the true mean and covariance up to the second order of nonlinearity (with errors introduced in the third and higher orders).

The idea of nonlinear filtering using the Gaussian representation of the posterior density via a set of deterministically chosen sample points has been proposed by several other authors [24, 25, 26]. This whole class of nonlinear filters (including the UKF) is referred to as linear regression Kalman filters in [9, 27], because they are all based on statistical linearization [28] rather than analytical linearization (as the EKF). The statistical linearization is performed via the linear regression through the regression (sample) points. The members of this class of filters differ only in the method of selecting the sample points (their number, their values, and weights), while the filtering equations are identical. We now present these filtering equations.

2.4.1 Filtering Equations

Consider the nonlinear filtering problem defined by (2.1) and (2.2). The assumption is that the posterior density at time $k-1$ is Gaussian: $p(\mathbf{x}_{k-1}|\mathbf{Z}_{k-1}) = \mathcal{N}(\mathbf{x}_{k-1}; \hat{\mathbf{x}}_{k-1|k-1}, \mathbf{P}_{k-1|k-1})$. The first step is to represent this density by a set of N sample points $\mathcal{X}_{k-1}^i$ and their weights W_{k-1}^i, $i = 0, \ldots, N-1$. The prediction step is then performed as follows:

$$\hat{\mathbf{x}}_{k|k-1} = \sum_{i=0}^{N-1} W_{k-1}^i \cdot \mathbf{f}_{k-1}(\mathcal{X}_{k-1}^i) \tag{2.34}$$

$$\mathbf{P}_{k|k-1} = \mathbf{Q}_{k-1} + \sum_{i=0}^{N-1} W_{k-1}^i \left[\mathbf{f}_{k-1}(\mathcal{X}_{k-1}^i) - \hat{\mathbf{x}}_{k|k-1}\right] \left[\mathbf{f}_{k-1}(\mathcal{X}_{k-1}^i) - \hat{\mathbf{x}}_{k|k-1}\right]^T. \tag{2.35}$$

The predicted density $p(\mathbf{x}_k|\mathbf{Z}_{k-1}) \approx \mathcal{N}(\mathbf{x}_k; \hat{\mathbf{x}}_{k|k-1}, \mathbf{P}_{k|k-1})$ is represented by a set of N sample points:

$$\mathcal{X}_{k|k-1}^i = \mathbf{f}_{k-1}(\mathcal{X}_{k-1}^i) \tag{2.36}$$

The predicted measurement is then:

$$\hat{\mathbf{z}}_{k|k-1} = \sum_{i=0}^{N-1} W_{k-1}^i \, \mathbf{h}_k(\mathcal{X}_{k|k-1}^i). \tag{2.37}$$

The update step is as follows:

$$\hat{\mathbf{x}}_{k|k} = \hat{\mathbf{x}}_{k|k-1} + \mathbf{K}_k(\mathbf{z}_k - \hat{\mathbf{z}}_{k|k-1}) \tag{2.38}$$

$$\mathbf{P}_{k|k} = \mathbf{P}_{k|k-1} - \mathbf{K}_k \mathbf{S}_k \mathbf{K}_k^T \tag{2.39}$$

where

$$\mathbf{K}_k = \mathbf{P}_{xz} \mathbf{S}_k^{-1} \tag{2.40}$$

$$\mathbf{S}_k = \mathbf{R}_k + \mathbf{P}_{zz} \tag{2.41}$$

$$\mathbf{P}_{xz} = \sum_{i=0}^{N-1} W_{k-1}^i \left(\mathcal{X}_{k|k-1}^i - \hat{\mathbf{x}}_{k|k-1}\right) \left(\mathbf{h}_k(\mathcal{X}_{k|k-1}^i) - \hat{\mathbf{z}}_{k|k-1}\right)^T \tag{2.42}$$

$$\mathbf{P}_{zz} = \sum_{i=0}^{N-1} W_{k-1}^i \left(\mathbf{h}_k(\mathcal{X}_{k|k-1}^i) - \hat{\mathbf{z}}_{k|k-1}\right) \left(\mathbf{h}_k(\mathcal{X}_{k|k-1}^i) - \hat{\mathbf{z}}_{k|k-1}\right)^T \tag{2.43}$$

Note that no explicit calculation of Jacobians is necessary to implement this algorithm. As a consequence, this class of filters is applicable even when there is a discontinuity in nonlinear functions f or h. The selection of sample points $\mathcal{X}_{k-1}^i$ and weights W_{k-1}^i of the UKF is presented next.

2.4.2 The Unscented Transform

The UT is a method for calculating the statistics of a random variable that undergoes a nonlinear transformation. Consider propagating a random variable a, with mean $\bar{a}$ and covariance $\mathbf{P}_a$, through an arbitrary nonlinear function $\mathbf{g} : \mathbb{R}^{n_a} \to \mathbb{R}^{n_b}$, to produce a random variable b:

$$\mathbf{b} = \mathbf{g}(\mathbf{a}). \tag{2.44}$$

The first two moments of b are computed using the UT as follows. First, $2n_a + 1$ weighted sample points $(\mathcal{A}_i, W_i)$ are deterministically chosen so that they completely describe (capture) the true mean $\bar{a}$ and covariance $\mathbf{P}_a$ of a. A scheme that satisfies this requirement is:

$$\mathcal{A}_0 = \bar{a} \qquad W_0 = \frac{\kappa}{(n_a + \kappa)} \qquad i = 0$$

$$\mathcal{A}_i = \bar{a} + \left(\sqrt{(n_a + \kappa)\mathbf{P}_a}\right)_i \qquad W_i = \frac{1}{2(n_a + \kappa)} \qquad i = 1, \ldots, n_a$$

$$\mathcal{A}_i = \bar{a} - \left(\sqrt{(n_a + \kappa)\mathbf{P}_a}\right)_i \qquad W_i = \frac{1}{2(n_a + \kappa)} \qquad i = n_1 + 1, \ldots, 2n_a$$

where κ is a scaling parameter (such that $\kappa + n_a \neq 0$) and $\left(\sqrt{(n_a + \kappa)\mathbf{P}_a}\right)_i$ is the ith row of the matrix square root $\mathbf{L}$ of $(n_a + \kappa)\mathbf{P}_a$, such that $(n_a + \kappa)\mathbf{P}_a = \mathbf{L}^T\mathbf{L}$. The weights are normalized; that is, satisfy $\sum_{i=0}^{2n_a} W_i = 1$. Now each sample point is propagated through the nonlinear function g:

$$\mathcal{B}_i = \mathbf{g}(\mathcal{A}_i) \quad (i = 0, 1, \ldots, 2n_a) \tag{2.45}$$

and the first two moments of b are computed as follows:

$$\bar{\mathbf{b}} = \sum_{i=0}^{2n_a} W_i \mathcal{B}_i \tag{2.46}$$

$$\mathbf{P}_b = \sum_{i=0}^{n_a} W_i (\mathcal{B}_i - \bar{\mathbf{b}})(\mathcal{B}_i - \bar{\mathbf{b}})^T. \tag{2.47}$$

It can be shown [21] that these estimates are accurate to the second order (third order if a is Gaussian) of the Taylor series expansion of $\mathbf{g}(\mathbf{a})$. Note that the distance of the sample points $\mathcal{A}_i$, $i = 1, \ldots, n_a$ from $\bar{\mathbf{a}}$ increases with dimension n_a. This can be controlled to a certain extent by the choice of κ. However when κ is negative, it is possible that $\mathbf{P}_b$ is not positive semidefinite. In order to add an extra degree of freedom in scaling the sample points away or towards $\bar{\mathbf{a}}$, the *scaled* UT was recently introduced [29].

Example

Here we revisit the example from Section 2.1, this time using the UT. The random variable **a** consists of two components (range, angle), hence it is represented by five sample points. The locations of the UT sample points $\mathcal{B}_i$ (after the nonlinear transformation **g**) are indicated by circles in Figure 2.3. The figure also displays the mean and covariance (2σ ellipse) of random variable $\mathbf{b} = \mathbf{g}(\mathbf{a})$ computed via (2.46) and (2.47). They appear almost identical to their true values, which illustrates the accuracy of the UT. One should not forget, however, the basic limitation of the UT (and the UKF), which can be seen from Figure 2.1. The actual pdf of **b** is distinctly non-Gaussian and therefore its first two moments, although accurately computed by the UT, are insufficient to characterize it completely. Julier [30] recently attempted to extend the sample point set in order to propagate information about the third-order moments.

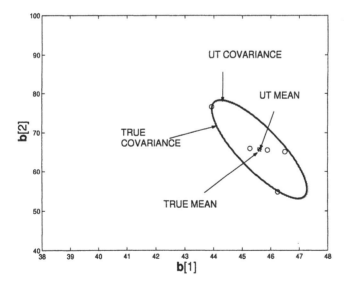

Figure 2.3 The approximation of mean and covariance via the unscented transform.

2.5 SUMMARY

This chapter gave a brief account of some common approaches to nonlinear filtering. The extended KF and the unscented KF operate in the framework of Gaussian approximation for the posterior density $p(\mathbf{x}_k|\mathbf{Z}_k)$. While this makes them simple to implement and fast to execute, they suffer from an inherent inability to model higher-order moments of truly non-Gaussian posterior densities. The approximate grid-based methods and the Gaussian sum filters (GSFs) overcome this limitation by approximating the entire posterior density. The grid-based methods, however, have a serious drawback: they are characterized by an exponentially growing computational complexity (with respect to the state vector dimension) that prevents their widespread use in practice. The Gaussian sum filters, on the other hand, have been used extensively in some specific applications (e.g., tracking maneuvering targets) where posterior densities are multimodal. There is, however, a lack of an automatic procedure for the computation of weights, means, and covariance of the GSF for a general nonlinear filtering problem. In the next chapter we review the remaining class of suboptimal nonlinear filters, the particle filters.

References

[1] H. W. Sorenson, "On the development of practical nonlinear filters," *Information Sciences*, vol. 7, pp. 253–270, 1974.

[2] P. S. Maybeck, *Stochastic Models, Estimation and Control*, vol. 2. New York: Academic Press, 1982.

[3] H. Tanizaki, *Nonlinear Filters: Estimation and Applications*. New York: Springer-Verlag, 2nd ed., 1996.

[4] B. D. O. Anderson and J. B. Moore, *Optimal Filtering*. Englewood Cliffs, NJ: Prentice-Hall, 1979.

[5] Y. Bar-Shalom, X. R. Li, and T. Kirubarajan, *Estimation with Applications to Tracking and Navigation*. New York: John Wiley & Sons, 2001.

[6] Y. Bar-Shalom and X. R. Li, *Multitarget-Multisensor Tracking: Principles and Techniques*. Storrs, CT: YBS Publishing, 1995.

[7] D. Lerro and Y. Bar-Shalom, "Tracking with debiased consistent converted measurements versus EKF," *IEEE Trans. Aerospace and Electronic Systems*, vol. 29, pp. 1015–1022, July 1993.

[8] M. Longbin, S. Xiaquan, Z. Yiyu, S. Z. Kang, and Y. Bar-Shalom, "Unbiased converted measurements for tracking," *IEEE Trans. Aerospace and Electronic Systems*, vol. 34, pp. 1023–1027, July 1998.

[9] T. Lefebvre, H. Bruyninckx, and J. D. Schutter, "Kalman Filters for Nonlinear Systems: A Comparison of Performance," Tech. Rep. 01R033, Katholieke Universiteit Leuven, 2001.

[10] S. C. Kramer and H. W. Sorenson, "Recursive Bayesian estimation using piece-wise constant approximations," *Automatica*, vol. 24, no. 6, pp. 789–801, 1988.

[11] K. Kastella, "Finite difference methods for nonlinear filtering and automatic target recognition," in *Multitarget-Multisensor Tracking* (Y. Bar-Shalom and W. D. Blair, eds.), vol. III, ch. 5, Norwood, MA: Artech House, 2000.

[12] L. D. Stone, "A Bayesian approach to multiple-target tracking," in *Handbook of Multisensor Data Fusion* (D. L. Hall and J. Llinas, eds.), ch. 10, Boca Raton, FL: CRC Press, 2001.

[13] F. Martinerie and P. Forster, "Data association and tracking using hidden Markov models and dynamic programming," in *Proc. Conf. ICASSP 92*, 1992.

[14] L. R. Rabiner, "A tutorial on hidden Markov models and selected applications in speech recognition," *Proc. of the IEEE*, vol. 77, pp. 257–285, February 1989.

[15] G. D. Forney, "The Viterbi algorithm," *Proc. of the IEEE*, vol. 61, pp. 268–278, March 1973.

[16] H. W. Sorenson and D. L. Alspach, "Recursive Bayesian estimation using Gaussian sums," *Automatica*, vol. 7, pp. 465–479, 1971.

[17] D. L. Alspach and H. W. Sorenson, "Nonlinear Bayesian estimation using Gaussian sum approximations," *IEEE Trans. Automatic Control*, vol. 17, no. 4, pp. 439–448, 1972.

[18] W. Koch and R. Klemm, "Ground target tracking with STAP radar," *IEEE Proc. - Radar, Sonar Navig.*, vol. 148, no. 3, pp. 173–185, 2001.

[19] F. Gustafsson, *Adaptive Filtering and Change Detection*. New York: John Wiley, 2000.

[20] H. A. P. Blom and Y. Bar-Shalom, "The interacting multiple model algorithm for systems with Markovian switching coefficients," *IEEE Trans. Automatic Control*, vol. 33, pp. 780–783, August 1988.

[21] S. Julier, J. Uhlmann, and H. F. Durrant-White, "A new method for nonlinear transformation of means and covariances in filters and estimators," *IEEE Trans. Automatic Control*, vol. 45, pp. 477–482, March 2000.

[22] S. Julier and J. Uhlmann, "Data fusion in nonlinear systems," in *Handbook of Multisensor Data Fusion* (D. L. Hall and J. Llinas, eds.), ch. 13, Boca Raton, FL: CRC, 2001.

[23] E. A. Wan and R. van der Merwe, "The unscented Kalman filter for nonlinear estimation," in *Proc. IEEE Symp. Adaptive Systems for Signal Proc., Comm. and Control (AS-SPCC)*, (Lake Louise, Alberta, Canada), pp. 153–158, 2000.

[24] T. S. Schei, "A finite-difference method for linearisation in nonlinear estimation algorithms," *Automatica*, vol. 33, no. 11, pp. 2053–2058, 1997.

[25] M. Norgaard, N. K. Poulsen, and O. Ravn, "New developments in state estimation of nonlinear systems," *Automatica*, vol. 36, pp. 1627–1638, 2000.

[26] K. Ito and K. Xiong, "Gaussian filters for nonlinear filtering problems," *IEEE Trans. Automatic Control*, vol. 45, pp. 910–927, May 2000.

[27] T. Lefebvre, H. Bruyninckx, and J. D. Schutter, "Comment on 'a new method for nonlinear transformation of means and covariances in filters and estimators'," *IEEE Trans. Automatic Control*, vol. 47, pp. 1406–1408, August 2002. See also Authors' Reply, same issue, p.1408.

[28] A. Gelb, *Applied Optimal Estimation*. Cambridge, MA: MIT Press, 1974.

[29] S. J. Julier, "The scaled unscented transform," in *Proc. American Control Conf.*, pp. 4555–4559, May 2002.

[30] S. J. Julier, "A skewed approach to filtering," in *Proc. SPIE*, vol. 3373, pp. 271–282, 1998.

Chapter 3

A Tutorial on Particle Filters

Particle filters (PFs) are also suboptimal filters. They perform sequential Monte Carlo (SMC) estimation based on point mass (or "particle") representation of probability densities. The basic SMC ideas in the form of sequential importance sampling had been introduced in statistics back in the 1950s [1]. Although these ideas continued to be explored sporadically during the 1960s and 1970s [2, 3], they were largely overlooked and ignored. Most likely the reason for this was the modest computational power available at the time. In addition, all these early implementations were based on plain sequential importance sampling, which, as we shall describe later, degenerates over time. The major contribution to the development of the SMC method was the inclusion of the resampling step [4], which, coupled with ever faster computers, made the particle filters useful in practice for the first time. Since then research activity in the field has dramatically increased [5, 6], resulting in many improvements of particle filters and their numerous applications. This chapter reviews the theoretical basis and the recent developments in a design of efficient particle filters, mainly following [7, 8].

3.1 MONTE CARLO INTEGRATION

Let us start with Monte Carlo integration, which is the basis of SMC methods. Suppose we want to numerically evaluate a multidimensional integral:

$$I = \int \mathbf{g}(\mathbf{x})d\mathbf{x} \tag{3.1}$$

where $\mathbf{x} \in \mathbb{R}^{n_x}$. Monte Carlo (MC) methods for numerical integration [9] factorize $\mathbf{g}(\mathbf{x}) = \mathbf{f}(\mathbf{x}) \cdot \pi(\mathbf{x})$ in such a way that $\pi(\mathbf{x})$ is interpreted as a probability density satisfying $\pi(\mathbf{x}) \geq 0$ and $\int \pi(\mathbf{x})d\mathbf{x} = 1$. The assumption is that it is possible to draw $N \gg 1$ samples $\{\mathbf{x}^i; i = 1,\ldots,N\}$ distributed according to $\pi(\mathbf{x})$. The MC estimate of integral

$$I = \int \mathbf{f}(\mathbf{x})\pi(\mathbf{x})d\mathbf{x} \qquad (3.2)$$

is the sample mean:

$$I_N = \frac{1}{N}\sum_{i=1}^{N}\mathbf{f}(\mathbf{x}^i). \qquad (3.3)$$

If the samples $\mathbf{x}^i$ are independent then I_N is an unbiased estimate and according to the law of large numbers I_N will almost surely converge to I. If the variance of $\mathbf{f}(\mathbf{x})$,

$$\sigma^2 = \int (\mathbf{f}(\mathbf{x}) - I)^2 \pi(\mathbf{x})d\mathbf{x}$$

is finite, then the central limit theorem holds and the estimation error converges in distribution:

$$\lim_{N\to\infty} \sqrt{N}(I_N - I) \sim \mathcal{N}(0,\sigma^2).$$

The error of the MC estimate, $e = I_N - I$, is of order $O(N^{-1/2})$, meaning that the rate of convergence of the estimate is independent of the dimension of the integrand (n_x). In contrast, any deterministic numerical integration (Section 2.2) has a rate of convergence that decreases as the dimension n_x increases. This useful and important property of MC integration is due to the choice of samples $\{\mathbf{x}^i; i = 1,\ldots,N\}$, as they automatically come from regions of the state space that are important for the integration result. In the Bayesian estimation context, density $\pi(\mathbf{x})$ is the posterior density. Unfortunately, usually it is not possible to sample effectively from the posterior distribution, being multivariate, nonstandard, and only known up to a proportionality constant. A possible solution is to apply the importance sampling method.

Importance Sampling

Ideally we want to generate samples directly from $\pi(\mathbf{x})$ and estimate I using (3.3). Suppose we can only generate samples from a density $q(\mathbf{x})$, which is similar to $\pi(\mathbf{x})$. Then a correct weighting of the sample set still makes the MC estimation possible. The pdf $q(\mathbf{x})$ is referred to as the *importance* or *proposal* density. Its "similarity" to $\pi(\mathbf{x})$ can be expressed by the following condition:

$$\pi(\mathbf{x}) > 0 \Rightarrow q(\mathbf{x}) > 0 \quad \text{for all } \mathbf{x} \in \mathbb{R}^{n_x} \qquad (3.4)$$

which means that $q(\mathbf{x})$ and $\pi(\mathbf{x})$ have the same support. Condition (3.4) is necessary for the importance sampling theory to hold and, if valid, any integral of the form

(3.2) can be rewritten as:

$$I = \int \mathbf{f}(\mathbf{x})\pi(\mathbf{x})d\mathbf{x} = \int \mathbf{f}(\mathbf{x})\frac{\pi(\mathbf{x})}{q(\mathbf{x})}q(\mathbf{x})d\mathbf{x} \qquad (3.5)$$

provided that $\pi(\mathbf{x})/q(\mathbf{x})$ is upper bounded. A Monte Carlo estimate of I is computed by generating $N \gg 1$ independent samples $\{\mathbf{x}^i; i = 1, \ldots, N\}$ distributed according to $q(\mathbf{x})$ and forming the weighted sum:

$$I_N = \frac{1}{N}\sum_{i=1}^{N} \mathbf{f}(\mathbf{x}^i)\tilde{w}(\mathbf{x}^i) \qquad (3.6)$$

where

$$\tilde{w}(\mathbf{x}^i) = \frac{\pi(\mathbf{x}^i)}{q(\mathbf{x}^i)} \qquad (3.7)$$

are the importance weights. If the normalizing factor of the desired density $\pi(\mathbf{x})$ is unknown, we need to perform normalization of the importance weights. Then we estimate I_N as follows:

$$I_N = \frac{\frac{1}{N}\sum_{i=1}^{N}\mathbf{f}(\mathbf{x}^i)\tilde{w}(\mathbf{x}^i)}{\frac{1}{N}\sum_{j=1}^{N}\tilde{w}(\mathbf{x}^j)} = \sum_{i=1}^{N} \mathbf{f}(\mathbf{x}^i)w(\mathbf{x}^i), \qquad (3.8)$$

where the normalized importance weights are given by:

$$w(\mathbf{x}^i) = \frac{\tilde{w}(\mathbf{x}^i)}{\sum_{j=1}^{N}\tilde{w}(\mathbf{x}^j)}. \qquad (3.9)$$

This technique is applied in the Bayesian framework, where $\pi(\mathbf{x})$ is the posterior density.

3.2 SEQUENTIAL IMPORTANCE SAMPLING

Importance sampling is a general MC integration method that we now apply to perform nonlinear filtering specified by the conceptual solution in Section 1.2. The resulting sequential importance sampling (SIS) algorithm is a Monte Carlo method that forms the basis for most sequential MC filters developed over the past decades; see [8, 10]. This sequential Monte Carlo approach is known variously as bootstrap filtering [4], the condensation algorithm [11], particle filtering [12], interacting particle approximations [13], and survival of the fittest [14]. It is a technique for implementing a recursive Bayesian filter by Monte Carlo simulations. The key idea is to represent the required posterior density function by a set of random samples with associated weights and to compute estimates based on these samples and weights. As the number of samples becomes very large, this Monte Carlo characterization becomes an equivalent representation to the usual functional description of the posterior pdf, and the SIS filter approaches the optimal Bayesian estimator.

Before we develop the details of the algorithm, let us introduce $\mathbf{X}_k = \{\mathbf{x}_j, j = 0, \ldots, k\}$, which represents the sequence of all target states up to time k. The joint posterior density at time k is denoted by $p(\mathbf{X}_k|\mathbf{Z}_k)$, and its marginal is $p(\mathbf{x}_k|\mathbf{Z}_k)$. Let $\{\mathbf{X}_k^i, w_k^i\}_{i=1}^N$ denote a random measure that characterizes the joint posterior $p(\mathbf{X}_k|\mathbf{Z}_k)$, where $\{\mathbf{X}_k^i, i = 1, \ldots, N\}$ is a set of support points with associated weights $\{w_k^i, i = 1, \ldots, N\}$. The weights are normalized such that $\sum_i w_k^i = 1$. Then, the joint posterior density at k can be approximated as follows [8]:

$$p(\mathbf{X}_k|\mathbf{Z}_k) \approx \sum_{i=1}^{N} w_k^i \delta(\mathbf{X}_k - \mathbf{X}_k^i). \qquad (3.10)$$

We therefore have a discrete weighted approximation of the true posterior, $p(\mathbf{X}_k|\mathbf{Z}_k)$. The normalized weights w_k^i are chosen using the principle of importance sampling described earlier. If the samples $\mathbf{X}_k^i$ were drawn from an importance density $q(\mathbf{X}_k|\mathbf{Z}_k)$, then according to (3.7)

$$w_k^i \propto \frac{p(\mathbf{X}_k^i|\mathbf{Z}_k)}{q(\mathbf{X}_k^i|\mathbf{Z}_k)}. \qquad (3.11)$$

Suppose at time step $k-1$ we have samples constituting an approximation to $p(\mathbf{X}_{k-1}|\mathbf{Z}_{k-1})$. With the reception of measurement $\mathbf{z}_k$ at time k, we wish to approximate $p(\mathbf{X}_k|\mathbf{Z}_k)$ with a new set of samples. If the importance density is chosen to factorize such that

$$q(\mathbf{X}_k|\mathbf{Z}_k) \triangleq q(\mathbf{x}_k|\mathbf{X}_{k-1}, \mathbf{Z}_k) \, q(\mathbf{X}_{k-1}|\mathbf{Z}_{k-1}) \qquad (3.12)$$

then one can obtain samples $\mathbf{X}_k^i \sim q(\mathbf{X}_k|\mathbf{Z}_k)$ by augmenting each of the existing samples $\mathbf{X}_{k-1}^i \sim q(\mathbf{X}_{k-1}|\mathbf{Z}_{k-1})$ with the new state $\mathbf{x}_k^i \sim q(\mathbf{x}_k|\mathbf{X}_{k-1}, \mathbf{Z}_k)$. To derive the weight update equation, the pdf $p(\mathbf{X}_k|\mathbf{Z}_k)$ is first expressed in terms of

$p(\mathbf{X}_{k-1}|\mathbf{Z}_{k-1})$, $p(\mathbf{z}_k|\mathbf{x}_k)$ and $p(\mathbf{x}_k|\mathbf{x}_{k-1})$:

$$\begin{aligned}p(\mathbf{X}_k|\mathbf{Z}_k) &= \frac{p(\mathbf{z}_k|\mathbf{X}_k,\mathbf{Z}_{k-1})p(\mathbf{X}_k|\mathbf{Z}_{k-1})}{p(\mathbf{z}_k|\mathbf{Z}_{k-1})}\\ &= \frac{p(\mathbf{z}_k|\mathbf{X}_k,\mathbf{Z}_{k-1})p(\mathbf{x}_k|\mathbf{X}_{k-1},\mathbf{Z}_{k-1})p(\mathbf{X}_{k-1}|\mathbf{Z}_{k-1})}{p(\mathbf{z}_k|\mathbf{Z}_{k-1})}\\ &= \frac{p(\mathbf{z}_k|\mathbf{x}_k)p(\mathbf{x}_k|\mathbf{x}_{k-1})}{p(\mathbf{z}_k|\mathbf{Z}_{k-1})}p(\mathbf{X}_{k-1}|\mathbf{Z}_{k-1}) \quad (3.13)\\ &\propto p(\mathbf{z}_k|\mathbf{x}_k)p(\mathbf{x}_k|\mathbf{x}_{k-1})p(\mathbf{X}_{k-1}|\mathbf{Z}_{k-1}). \quad (3.14)\end{aligned}$$

By substituting (3.12) and (3.14) into (3.11), the weight update equation can then be shown to be

$$\begin{aligned}w_k^i &\propto \frac{p(\mathbf{z}_k|\mathbf{x}_k^i)p(\mathbf{x}_k^i|\mathbf{x}_{k-1}^i)p(\mathbf{X}_{k-1}^i|\mathbf{Z}_{k-1})}{q(\mathbf{x}_k^i|\mathbf{X}_{k-1}^i,\mathbf{Z}_k)q(\mathbf{X}_{k-1}^i|\mathbf{Z}_{k-1})}\\ &= w_{k-1}^i \frac{p(\mathbf{z}_k|\mathbf{x}_k^i)p(\mathbf{x}_k^i|\mathbf{x}_{k-1}^i)}{q(\mathbf{x}_k^i|\mathbf{X}_{k-1}^i,\mathbf{Z}_k)}. \quad (3.15)\end{aligned}$$

Furthermore, if $q(\mathbf{x}_k|\mathbf{X}_{k-1},\mathbf{Z}_k) = q(\mathbf{x}_k|\mathbf{x}_{k-1},\mathbf{z}_k)$, then the importance density becomes only dependent on the $\mathbf{x}_{k-1}$ and $\mathbf{z}_k$. This is particularly useful in the common case when only a filtered estimate of posterior $p(\mathbf{x}_k|\mathbf{Z}_k)$ is required at each time step. From this point on, we shall assume such a case, except when explicitly stated otherwise. In such scenarios, only $\mathbf{x}_k^i$ need be stored, and so one can discard the path, $\mathbf{X}_{k-1}^i$, and the history of observations, $\mathbf{Z}_{k-1}$. The modified weight is then

$$w_k^i \propto w_{k-1}^i \frac{p(\mathbf{z}_k|\mathbf{x}_k^i)p(\mathbf{x}_k^i|\mathbf{x}_{k-1}^i)}{q(\mathbf{x}_k^i|\mathbf{x}_{k-1}^i,\mathbf{z}_k)} \quad (3.16)$$

and the posterior filtered density $p(\mathbf{x}_k|\mathbf{Z}_k)$ can be approximated as

$$p(\mathbf{x}_k|\mathbf{Z}_k) \approx \sum_{i=1}^N w_k^i \delta(\mathbf{x}_k - \mathbf{x}_k^i) \quad (3.17)$$

where the weights are defined in (3.16). It can be shown that as $N \to \infty$ the approximation (3.17) approaches the true posterior density $p(\mathbf{x}_k|\mathbf{Z}_k)$.

Filtering via SIS thus consists of recursive propagation of importance weights w_k^i and support points $\mathbf{x}_k^i$ as each measurement is received sequentially. A pseudo-code description of this algorithm is given in Table 3.1 [7, 15].

This simple and general algorithm forms the basis of most particle filters. The choice of the importance density function plays a crucial role in the design and this will be discussed in Section 3.4.

Table 3.1
Filtering via SIS

$[\{x_k^i, w_k^i\}_{i=1}^N] = \text{SIS}\,[\{x_{k-1}^i, w_{k-1}^i\}_{i=1}^N, z_k]$

- FOR $i = 1 : N$
  - Draw $x_k^i \sim q(x_k|x_{k-1}^i, z_k)$
  - Evaluate the importance weights up to a normalizing constant according to (3.16)

$$\tilde{w}_k^i = w_{k-1}^i \frac{p(z_k|x_k^i)\,p(x_k^i|x_{k-1}^i)}{q(x_k^i|x_{k-1}^i, z_k)}$$

- END FOR
- Calculate total weight: $t = \text{SUM}\,[\{\tilde{w}_k^i\}_{i=1}^N]$
- FOR $i = 1 : N$
  - Normalize: $w_k^i = t^{-1}\tilde{w}_k^i$
- END FOR

Degeneracy Problem

Ideally the importance density function should be the posterior distribution itself, $p(x_k|Z_k)$. For the importance function of the form (3.12), it has been shown [8] that the variance of importance weights can only increase over time. The variance increase has a harmful effect on the accuracy and leads to a common problem with the SIS particle filter: the degeneracy phenomenon. In practical terms this means that after a certain number of recursive steps, all but one particle will have negligible normalized weights. The degeneracy is impossible to avoid in the SIS framework and hence it was a major stumbling block in the development of sequential MC methods. Effectively a large computational effort is devoted to updating particles whose contribution to the approximation of $p(x_k|Z_k)$ is almost zero. A suitable measure of degeneracy of an algorithm is the effective sample size N_{eff} introduced in [16] and estimated as follows:

$$\hat{N}_{eff} = \frac{1}{\sum_{i=1}^N (w_k^i)^2}, \qquad (3.18)$$

where w_k^i is the normalized weight obtained using (3.15). It is straightforward to verify that $1 \leq N_{eff} \leq N$ with the following two extreme cases: (1) if the weights are uniform (i.e., $w_k^i = \frac{1}{N}$ for $i = 1, \ldots, N$) then $N_{eff} = N$; and (2) if $\exists j \in \{1, \ldots, N\}$ such that $w_k^j = 1$, and $w_k^i = 0$ for all $i \neq j$, then $N_{eff} = 1$. Hence, small N_{eff} indicates a severe degeneracy and vice versa. The next section presents a strategy to overcome degeneracy of samples in SIS.

3.3 RESAMPLING

Whenever a significant degeneracy is observed (i.e., when N_{eff} falls below some threshold N_{thr}), resampling is required in the SIS algorithm. Resampling eliminates samples with low importance weights and multiplies samples with high importance weights (in apparent analogy to genetic algorithms). It involves a mapping of random measure $\{x_k^i, w_k^i\}$ into a random measure $\{x_k^{i*}, 1/N\}$ with uniform weights. The new set of random samples $\{x_k^{i*}\}_{i=1}^N$ is generated by resampling (with replacement) N times from an approximate discrete representation of $p(x_k|Z_k)$ given by

$$p(x_k|Z_k) \approx \sum_{i=1}^{N} w_k^i \delta(x_k - x_k^i) \qquad (3.19)$$

so that $P\{x_k^{i*} = x_k^j\} = w_k^j$. The resulting sample is an i.i.d. sample from the discrete density (3.19), and hence the new weights are uniform. The selection $x_k^{i*} = x_k^j$ is schematically shown in Figure 3.1 [17], where acronym CSW stands for the cumulative sum of weights of random measure $\{x_k^i, w_k^i\}$, and random variable u_i, $i = 1, \ldots, N$ is uniformly distributed on the interval $[0, 1]$.

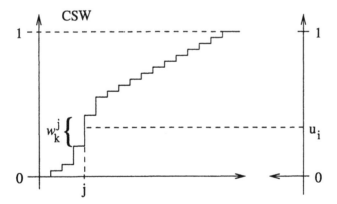

Figure 3.1 The process of resampling: $u_i \sim \mathcal{U}[0, 1]$ maps into index j; the corresponding particle x_k^j has a good chance of being selected and multiplied because of its high value of w_k^j.

A direct implementation of resampling would consist of generating N i.i.d. variables from the uniform distribution, sorting them in ascending order and comparing them with the cumulative sum of normalized weights. The best sorting algorithm has a complexity of $O(N \log N)$ and this is a major limitation in practical applications. However, it is possible to implement the resampling procedure in $O(N)$ operations by sampling N ordered uniforms using an algorithm based on order statistics [12, 18]. Note that other efficient (in terms of reduced MC variation) resampling schemes such as stratified sampling and residual sampling [15] may be applied as alternatives to this algorithm. Systematic resampling [19] is another efficient scheme: it is simple to implement, its computational complexity is $O(N)$ and it minimizes the MC variation. The pseudocode of the systematic resampling algorithm is described in Table 3.2. For each resampled particle $\mathbf{x}_k^{j\,*}$, this resampling algorithm also stores the index of its parent, denoted by i^j. In general this is unnecessary, but may be required as in Section 3.5.4.

Table 3.2

Resampling Algorithm

$[\{\mathbf{x}_k^{j\,*}, w_k^j, i^j\}_{j=1}^N]$ = RESAMPLE $[\{\mathbf{x}_k^i, w_k^i\}_{i=1}^N]$

- Initialize the CSW: $c_1 = w_k^1$
- FOR $i = 2 : N$
    - Construct CSW: $c_i = c_{i-1} + w_k^i$
- END FOR
- Start at the bottom of the CSW: $i = 1$
- Draw a starting point: $u_1 \sim \mathcal{U}\left[0, N^{-1}\right]$
- FOR $j = 1 : N$
    - Move along the CSW: $u_j = u_1 + N^{-1}(j - 1)$
    - WHILE $u_j > c_i$
        * $i = i + 1$
    - END WHILE
    - Assign sample: $\mathbf{x}_k^{j\,*} = \mathbf{x}_k^i$
    - Assign weight: $w_k^j = N^{-1}$
    - Assign parent: $i^j = i$
- END FOR

We have now defined the main steps of a generic particle filter. Its pseudocode is given in Table 3.3. A graphical representation of a PF (with only $N = 10$ samples), which uses the transitional density as the importance function, that is, $q(\mathbf{x}_k|\mathbf{x}_{k-1}^i, \mathbf{z}_k) = p(\mathbf{x}_k|\mathbf{x}_{k-1}^i)$, is shown in Figure 3.2 [17]. At the top we start with a uniformly weighted random measure $\{\mathbf{x}_k^i, N^{-1}\}$, which approximates prediction density $p(\mathbf{x}_k|\mathbf{Z}_{k-1})$. Then we use the received measurement $\mathbf{z}_k$ to compute for each particle its importance weight. This involves the likelihood function $p(\mathbf{z}_k|\mathbf{x}_k)$ via (3.15). The result is a random measure $\{\mathbf{x}_k^i, w_k^i\}$, which is an approximation of $p(\mathbf{x}_k|\mathbf{Z}_k)$. If $\hat{N}_{eff} < N_{thr}$, the resampling step is executed – it selects the "important" particles to obtain the uniformly weighted measure $\{\mathbf{x}_k^{i*}, N^{-1}\}$, which still approximates $p(\mathbf{x}_k|\mathbf{Z}_k)$. The last step is the prediction, which introduces a variety (due to process noise) and results in the measure $\{\mathbf{x}_{k+1}^i, N^{-1}\}$ that approximates $p(\mathbf{x}_{k+1}|\mathbf{Z}_k)$. Note that the SIS algorithm performs both the prediction and the importance weights computation.

Table 3.3
Generic Particle Filter

$[\{\mathbf{x}_k^i, w_k^i\}_{i=1}^N] = \text{PF}\,[\{\mathbf{x}_{k-1}^i, w_{k-1}^i\}_{i=1}^N, \mathbf{z}_k]$

- Filtering via SIS (Table 3.1):
  $[\{\mathbf{x}_k^i, w_k^i\}_{i=1}^N] = \text{SIS}\,[\{\mathbf{x}_{k-1}^i, w_{k-1}^i\}_{i=1}^N, \mathbf{z}_k]$
- Calculate $\hat{N}_{eff}$ using (3.18)
- IF $\hat{N}_{eff} < N_{thr}$
  - Resample using the algorithm in Table 3.2:
    $[\{\mathbf{x}_k^i, w_k^i, -\}_{i=1}^N] = \text{RESAMPLE}\,[\{\mathbf{x}_k^i, w_k^i\}_{i=1}^N]$
- END IF

Although the resampling step reduces the effects of degeneracy, it introduces other practical problems. First, it limits the opportunity to parallelize the implementation since all the particles must be combined. Second, the particles that have high weights w_k^i are statistically selected many times. This leads to a loss of diversity among the particles as the resultant sample will contain many repeated points. This problem, known as *sample impoverishment*, is severe in the case where process noise in state dynamics is very small. It leads to the situation where all particles will collapse to a single point within a few iterations. Third, since the diversity of the paths of the particles is reduced, any smoothed estimates based on the particles'

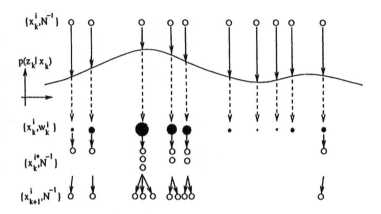

Figure 3.2 A single cycle of a particle filter.

paths degenerate.[1] Schemes exist to counteract this effect. One approach considers the states for the particles to be predetermined by the forward filter and then obtains the smoothed estimates by recalculating the particles' weights via a recursion from the final to the first time step [21]. Another approach is to use a Markov chain Monte Carlo (MCMC) move step [22].

There have been some systematic techniques proposed recently to solve the problem of sample impoverishment. One such technique is the *resample-move* algorithm [23]. Although this technique draws conceptually on the same technologies of importance sampling-resampling and MCMC sampling, it avoids sample impoverishment. It does this in a rigorous manner that ensures the particles to asymptotically approximate the samples from the posterior. An alternative solution to the same problem is based on the *regularization* step [24]. The latter approach is frequently found to improve performance despite a less rigorous derivation and is included here for completeness. Both regularization and MCMC move step are discussed in Section 3.5.3. In general, when process noise in the state dynamics equation is zero, then using a particle filter is not entirely appropriate. Particle filtering is a method well suited to the estimation of dynamic stochastic states. If the state dynamics is deterministic, the problem can be regarded as the one of parameter estimation, and alternative approaches are more appropriate [25, 26, 27].

As a general rule, the accuracy of any estimate of a function of the distribution can only decrease as a result of resampling. Therefore if quantities such as the mean and covariance of the samples are to be reported by the particle filter, these should be calculated prior to resampling.

---

1   Since the particles actually represent paths through the state space, smoothed estimates of the state can be obtained by storing the trajectory taken by each particle [20].

3.4 SELECTION OF IMPORTANCE DENSITY

3.4.1 The Optimal Choice

The choice of importance density $q(\mathbf{x}_k|\mathbf{x}_{k-1}^i, \mathbf{z}_k)$ is one of the most critical issues in the design of a particle filter. The optimal importance density function that minimizes the variance of importance weights, conditioned upon $\mathbf{x}_{k-1}^i$ and $\mathbf{z}_k$ has been shown [8] to be

$$q(\mathbf{x}_k|\mathbf{x}_{k-1}^i, \mathbf{z}_k)_{opt} = p(\mathbf{x}_k|\mathbf{x}_{k-1}^i, \mathbf{z}_k)$$
$$= \frac{p(\mathbf{z}_k|\mathbf{x}_k, \mathbf{x}_{k-1}^i) p(\mathbf{x}_k|\mathbf{x}_{k-1}^i)}{p(\mathbf{z}_k|\mathbf{x}_{k-1}^i)}. \quad (3.20)$$

Substitution of (3.20) into (3.16) yields

$$w_k^i \propto w_{k-1}^i p(\mathbf{z}_k|\mathbf{x}_{k-1}^i), \quad (3.21)$$

which states that importance weights at time k can be computed (and resampling, if necessary, can be carried out) *before* the particles are propagated to time k. In order to use the optimal importance function, one has to be able to: (1) sample from $p(\mathbf{x}_k|\mathbf{x}_{k-1}^i, \mathbf{z}_k)$, and (2) evaluate

$$p(\mathbf{z}_k|\mathbf{x}_{k-1}^i) = \int p(\mathbf{z}_k|\mathbf{x}_k) p(\mathbf{x}_k|\mathbf{x}_{k-1}^i) d\mathbf{x}_k \quad (3.22)$$

up to a normalizing constant. In the general case either of these two may not be straightforward.

However, there are some special cases where the use of the optimal importance density is possible. The first case is when $\mathbf{x}_k$ is a member of a finite set. In such a case, the integral in (3.22) becomes a sum and sampling from $p(\mathbf{x}_k|\mathbf{x}_{k-1}^i, \mathbf{z}_k)$ is possible. An example of an application where $\mathbf{x}_k$ is a member of a finite set is a jump-Markov linear system for tracking maneuvering targets [28]. The second case is a class of models for which $p(\mathbf{x}_k|\mathbf{x}_{k-1}^i, \mathbf{z}_k)$ is Gaussian [8].

Gaussian Optimal Importance Function

Consider the case where the state dynamics is nonlinear, the measurement equation is linear, and all the random elements in the model are additive Gaussian. Such a system is given by

$$\mathbf{x}_k = \mathbf{f}_{k-1}(\mathbf{x}_{k-1}) + \mathbf{v}_{k-1} \quad (3.23)$$
$$\mathbf{z}_k = \mathbf{H}_k \mathbf{x}_k + \mathbf{w}_k. \quad (3.24)$$

where $\mathbf{v}_{k-1}$ and $\mathbf{w}_k$ are mutually independent zero-mean white Gaussian sequences with covariances $\mathbf{Q}_{k-1}$ and $\mathbf{R}_k$, respectively. It can be shown that in this case, both the optimal importance density and $p(\mathbf{z}_k|\mathbf{x}_{k-1})$ are Gaussian, that is:

$$p(\mathbf{x}_k|\mathbf{x}_{k-1},\mathbf{z}_k) = \mathcal{N}(\mathbf{x}_k;\mathbf{a}_k,\boldsymbol{\Sigma}_k) \qquad (3.25)$$
$$p(\mathbf{z}_k|\mathbf{x}_{k-1}) = \mathcal{N}(\mathbf{z}_k;\mathbf{b}_k,\mathbf{S}_k) \qquad (3.26)$$

where

$$\mathbf{a}_k = \mathbf{f}_{k-1}(\mathbf{x}_{k-1}) + \boldsymbol{\Sigma}_k \mathbf{H}_k^T \mathbf{R}_k^{-1}(\mathbf{z}_k - \mathbf{b}_k) \qquad (3.27)$$
$$\boldsymbol{\Sigma}_k = \mathbf{Q}_{k-1} - \mathbf{Q}_{k-1}\mathbf{H}_k^T \mathbf{S}_k^{-1} \mathbf{H}_k \mathbf{Q}_{k-1} \qquad (3.28)$$
$$\mathbf{S}_k = \mathbf{H}_k \mathbf{Q}_{k-1} \mathbf{H}_k^T + \mathbf{R}_k \qquad (3.29)$$
$$\mathbf{b}_k = \mathbf{H}_k \mathbf{f}_{k-1}(\mathbf{x}_{k-1}). \qquad (3.30)$$

Next we provide the basic steps of the proof. According to the assumptions related to (3.23) and (3.24), observe that:

$$p(\mathbf{x}_k|\mathbf{x}_{k-1}) = \mathcal{N}(\mathbf{x}_k;\mathbf{f}_{k-1}(\mathbf{x}_{k-1}),\mathbf{Q}_{k-1}) \qquad (3.31)$$
$$p(\mathbf{z}_k|\mathbf{x}_k) = \mathcal{N}(\mathbf{z}_k;\mathbf{H}_k\mathbf{x}_k,\mathbf{R}_k). \qquad (3.32)$$

From the Bayes formula it follows that:

$$p(\mathbf{x}_k|\mathbf{x}_{k-1},\mathbf{z}_k) = \frac{p(\mathbf{z}_k|\mathbf{x}_k)\,p(\mathbf{x}_k|\mathbf{x}_{k-1})}{p(\mathbf{z}_k|\mathbf{x}_{k-1})} \qquad (3.33)$$

which can be rearranged as:

$$p(\mathbf{z}_k|\mathbf{x}_k)\,p(\mathbf{x}_k|\mathbf{x}_{k-1}) = p(\mathbf{x}_k|\mathbf{x}_{k-1},\mathbf{z}_k)\,p(\mathbf{z}_k|\mathbf{x}_{k-1}). \qquad (3.34)$$

The exponents on both sides of (3.34) can be equated to yield:

$$(\mathbf{z}_k - \mathbf{H}_k\mathbf{x}_k)^T \mathbf{R}_k^{-1}(\mathbf{z}_k - \mathbf{H}_k\mathbf{x}_k)$$
$$+ (\mathbf{x}_k - \mathbf{f}_{k-1}(\mathbf{x}_{k-1}))^T \mathbf{Q}_{k-1}^{-1}(\mathbf{x}_k - \mathbf{f}_{k-1}(\mathbf{x}_{k-1}))$$
$$= (\mathbf{x}_k - \mathbf{a}_k)^T \boldsymbol{\Sigma}_k^{-1}(\mathbf{x}_k - \mathbf{a}_k) + (\mathbf{z}_k - \mathbf{b}_k)^T \mathbf{S}_k^{-1}(\mathbf{z}_k - \mathbf{b}_k). \qquad (3.35)$$

Now one needs to expand both sides of (3.35) and to equate the similar terms. For example, the quadratic form terms yield the following equality:

$$\mathbf{x}_k^T \mathbf{H}_k^T \mathbf{R}_k^{-1} \mathbf{H}_k \mathbf{x}_k + \mathbf{x}_k^T \mathbf{Q}_{k-1}^{-1} \mathbf{x}_k = \mathbf{x}_k^T \boldsymbol{\Sigma}_k^{-1} \mathbf{x}_k \qquad (3.36)$$

which leads to $\Sigma_k^{-1} = \mathbf{Q}_{k-1}^{-1} + \mathbf{H}_k^T \mathbf{R}_k^{-1} \mathbf{H}_k$. The application of the matrix inversion lemma[2] then proves (3.28). The proof of (3.27), (3.29), and (3.30) is given in Section 3.8.

The described analytic evaluation of $p(\mathbf{x}_k|\mathbf{x}_{k-1}, \mathbf{z}_k)$ and $p(\mathbf{z}_k|\mathbf{x}_{k-1})$ is difficult for most other cases. The next section therefore describes the suboptimal methods that approximate the optimal importance density.

3.4.2 Suboptimal Choices

The most popular suboptimal choice is the transitional prior,

$$q(\mathbf{x}_k|\mathbf{x}_{k-1}^i, \mathbf{z}_k) = p(\mathbf{x}_k|\mathbf{x}_{k-1}^i). \qquad (3.37)$$

If an additive zero-mean Gaussian process noise model is used as in (3.23), the transitional prior is simply:

$$p(\mathbf{x}_k|\mathbf{x}_{k-1}^i) = \mathcal{N}(\mathbf{x}_k; \mathbf{f}_{k-1}(\mathbf{x}_{k-1}^i), \mathbf{Q}_{k-1}). \qquad (3.38)$$

Substitution of (3.37) into (3.16) then yields

$$w_k^i \propto w_{k-1}^i p(\mathbf{z}_k|\mathbf{x}_k^i). \qquad (3.39)$$

Note that when the optimal importance function is used, the importance weights, according to (3.21), can be computed *before* the particles are propagated to time k. Equation (3.39) states that this is not possible with the transitional prior.

If the transitional prior $p(\mathbf{x}_k|\mathbf{x}_{k-1})$ is used as the importance density and is a much broader distribution than the likelihood, $p(\mathbf{z}_k|\mathbf{x}_k)$, then only a few particles will be assigned a high weight. Consequently, the particles will degenerate rapidly and the filter will not work. Methods exist for encouraging the particles to be in the right place (in the regions of high likelihood) by incorporating the current observation. One such a method is used in the auxiliary particle filter [29], described in detail in Section 3.5.2. Another possibility is to use bridging densities [30] or progressive correction [31]; both introduce intermediate distributions between the prior and likelihood. The particles are then reweighted according to these intermediate distributions and resampled. This "herds" the particles into the right part of the state space. Partitioned sampling [11] is useful if the likelihood is very peaked, but can be factorized into a number of broader distributions. Typically, this occurs because each of the partitioned distributions are functions of some (not all) of the states. By treating each of these partitioned distributions in turn and resampling

2 $(\mathbf{P}^{-1} + \mathbf{H}^T \mathbf{R}^{-1} \mathbf{H})^{-1} = \mathbf{P} - \mathbf{P} \mathbf{H}^T (\mathbf{H} \mathbf{P} \mathbf{H}^T + \mathbf{R})^{-1} \mathbf{H} \mathbf{P}$

on the basis of each such partitioned distribution, the particles are again herded towards the peaked likelihood.

It is also possible to construct suboptimal approximations to the optimal importance density by using local linearization techniques [8]. Such linearizations use an importance density that is a Gaussian approximation to $p(\mathbf{x}_k|\mathbf{x}_{k-1},\mathbf{z}_k)$ and are based on the extended KF or the unscented KF (see Section 3.5.4).

3.5 VERSIONS OF PARTICLE FILTERS

The sequential importance sampling algorithm presented in Section 3.2 forms the basis for most particle filters that have been developed so far. There is a plethora of particle filters on the menu, but in this section we present only five. Their selection is based on their widespread use in practice as well as throughout Part II of this book. All five particle filters are derived from the SIS algorithm by an appropriate choice of the importance sampling density and/or modification of the resampling step. The following types of filters are described: (1) sampling importance resampling (SIR) filter; (2) auxiliary sampling importance resampling (ASIR) filter; (3) particle filters with improved sample diversity; (4) local linearization particle filters; and (5) multiple-model particle filter. Note that these filters can be (and often are) mutually combined.

3.5.1 SIR Filter

The SIR filter was proposed in [4] under the name "bootstrap" filter. The assumptions required to use the SIR filter are very weak: (1) the state dynamics and measurement functions, $\mathbf{f}_{k-1}(\cdot,\cdot)$ and $\mathbf{h}_k(\cdot,\cdot)$ in (1.1) and (1.2), respectively, need to be known; and (2) it is required to be able to sample realizations from the process noise distribution of $\mathbf{v}_{k-1}$ and from the prior. Finally, the likelihood function $p(\mathbf{z}_k|\mathbf{x}_k)$ needs to be available for pointwise evaluation (at least up to proportionality). The SIR algorithm is derived from the SIS algorithm by choosing the importance density to be the transitional prior and by performing the resampling step at every time index.

The above choice of importance density implies that we need samples from $p(\mathbf{x}_k|\mathbf{x}_{k-1}^i)$. A sample $\mathbf{x}_k^i \sim p(\mathbf{x}_k|\mathbf{x}_{k-1}^i)$ can be generated by first generating a process noise sample $\mathbf{v}_{k-1}^i \sim p_v(\mathbf{v}_{k-1})$ and setting $\mathbf{x}_k^i = \mathbf{f}_{k-1}(\mathbf{x}_{k-1}^i, \mathbf{v}_{k-1}^i)$, where $p_v(\cdot)$ is the pdf of $\mathbf{v}_{k-1}$. For this particular choice of importance density, it is evident that the weights are given by (3.39). Note, however, that resampling is applied at every time index, and therefore $w_{k-1}^i = 1/N$ for all $i = 1, \ldots, N$. This means two things: (1) there is no need to pass on the importance weights (being known) from one time step of the algorithm to the next; and (2) relationship (3.39)

simplifies to:
$$w_k^i \propto p(\mathbf{z}_k|\mathbf{x}_k^i). \tag{3.40}$$

The weights given by the proportionality in (3.40) are normalized before the resampling stage. An iteration of the algorithm is then described in Table 3.4.

Table 3.4
SIR Particle Filter

$[\{\mathbf{x}_k^i\}_{i=1}^N] = \text{SIR}[\{\mathbf{x}_{k-1}^i\}_{i=1}^N, \mathbf{z}_k]$

- FOR $i = 1 : N$
    - Draw $\mathbf{x}_k^i \sim p(\mathbf{x}_k|\mathbf{x}_{k-1}^i)$
    - Calculate $\tilde{w}_k^i = p(\mathbf{z}_k|\mathbf{x}_k^i)$
- END FOR
- Calculate total weight: $t = \text{SUM}\,[\{\tilde{w}_k^i\}_{i=1}^N]$
- FOR $i = 1 : N$
    - Normalize: $w_k^i = t^{-1}\tilde{w}_k^i$
- END FOR
- Resample using the algorithm in Table 3.2:
  $[\{\mathbf{x}_k^i, -, -\}_{i=1}^N] = \text{RESAMPLE}\,[\{\mathbf{x}_k^i, w_k^i\}_{i=1}^N]$

As the importance sampling density for the SIR filter is independent of measurement $\mathbf{z}_k$, the state space is explored without any knowledge of the observations. Therefore, this filter can be inefficient and is sensitive to outliers. Furthermore, as resampling is applied at every iteration, this can result in a rapid loss of diversity in particles. The SIR method, however, has the advantage that the importance weights are easily evaluated and the importance density can be easily sampled.

3.5.2 Auxiliary SIR Filter

The auxiliary SIR (ASIR) filter was introduced by Pitt and Shephard [29] as a variant of the standard SIR. The basic idea is to perform the resampling step at time $k - 1$ (using the available measurement at time k), before the particles are propagated to time k. In this way the ASIR filter attempts to mimic the sequence of steps carried out when the optimal importance density is available. The derivation of ASIR filter follows from the SIS framework by introducing an importance density $q(\mathbf{x}_k, i|\mathbf{Z}_k)$, which samples the pair $\{\mathbf{x}_k^j, i^j\}_{j=1}^N$, where i^j refers to the index of the particle at $k - 1$.

By applying Bayes' rule, a proportionality can be derived for $p(\mathbf{x}_k, i|\mathbf{Z}_k)$ as:

$$\begin{aligned} p(\mathbf{x}_k, i|\mathbf{Z}_k) &\propto p(\mathbf{z}_k|\mathbf{x}_k) p(\mathbf{x}_k, i|\mathbf{Z}_{k-1}) \\ &= p(\mathbf{z}_k|\mathbf{x}_k) p(\mathbf{x}_k|i, \mathbf{Z}_{k-1}) p(i|\mathbf{Z}_{k-1}) \\ &= p(\mathbf{z}_k|\mathbf{x}_k) p(\mathbf{x}_k|\mathbf{x}_{k-1}^i) w_{k-1}^i. \end{aligned} \quad (3.41)$$

The ASIR filter operates by obtaining a sample from the joint density $p(\mathbf{x}_k, i|\mathbf{Z}_k)$ and then omitting the indices i in the pair $(\mathbf{x}_k, i)$ to produce a sample $\{\mathbf{x}_k^j\}_{j=1}^N$ from the marginalized density $p(\mathbf{x}_k|\mathbf{Z}_k)$. The importance density used to draw the sample $\{\mathbf{x}_k^j, i^j\}_{j=1}^N$ is defined to satisfy the proportionality

$$q(\mathbf{x}_k, i|\mathbf{Z}_k) \propto p(\mathbf{z}_k|\boldsymbol{\mu}_k^i) p(\mathbf{x}_k|\mathbf{x}_{k-1}^i) w_{k-1}^i \quad (3.42)$$

where $\boldsymbol{\mu}_k^i$ is some characterization of $\mathbf{x}_k$ given $\mathbf{x}_{k-1}^i$. This could be the mean, in which case $\boldsymbol{\mu}_k^i = \mathbb{E}[\mathbf{x}_k|\mathbf{x}_{k-1}^i]$ or a sample $\boldsymbol{\mu}_k^i \sim p(\mathbf{x}_k|\mathbf{x}_{k-1}^i)$. By writing

$$q(\mathbf{x}_k, i|\mathbf{Z}_k) = q(i|\mathbf{Z}_k) q(\mathbf{x}_k|i, \mathbf{Z}_k) \quad (3.43)$$

and defining

$$q(\mathbf{x}_k|i, \mathbf{Z}_k) \triangleq p(\mathbf{x}_k|\mathbf{x}_{k-1}^i) \quad (3.44)$$

it follows from (3.42) that

$$q(i|\mathbf{Z}_k) \propto p(\mathbf{z}_k|\boldsymbol{\mu}_k^i) w_{k-1}^i. \quad (3.45)$$

According to (3.16), the sample $\{\mathbf{x}_k^j, i^j\}_{j=1}^N$ is then assigned a weight proportional to the ratio of the right-hand side of (3.41) and (3.42):

$$w_k^j \propto w_{k-1}^{i^j} \frac{p(\mathbf{z}_k|\mathbf{x}_k^j) p\left(\mathbf{x}_k^j|\mathbf{x}_{k-1}^{i^j}\right)}{q(\mathbf{x}_k^j, i^j|\mathbf{Z}_k)} = \frac{p(\mathbf{z}_k|\mathbf{x}_k^j)}{p\left(\mathbf{z}_k|\boldsymbol{\mu}_k^{i^j}\right)}. \quad (3.46)$$

A pseudocode description of a single cycle of the ASIR filter is presented in Table 3.5. The original ASIR filter as proposed in [29] included an additional resampling step at the end of each cycle to produce an equally weighted set of particles. This additional resampling step was shown to be unnecessary [32] and is therefore omitted from Table 3.5.

Compared to SIR filter, the advantage of the ASIR filter is that it naturally generates points from the sample at $k-1$, which, conditioned on the current measurement, are most likely to be in the region of high likelihood. ASIR can be viewed as resampling at the "previous" time step based on some point estimates

Table 3.5
Auxiliary SIR Particle Filter

$[\{\mathbf{x}_k^i, w_k^i\}_{i=1}^N] = \text{ASIR}[\{\mathbf{x}_{k-1}^i, w_{k-1}^i\}_{i=1}^N, \mathbf{z}_k]$

- FOR $i = 1 : N$
    - Calculate μ_k^i
    - Calculate $\tilde{w}_k^i = q(i|\mathbf{Z}_k) \propto p(\mathbf{z}_k|\mu_k^i) w_{k-1}^i$.
- END FOR
- Calculate total weight: t = SUM $[\{\tilde{w}_k^i\}_{i=1}^N]$
- FOR $i = 1 : N$
    - Normalize: $w_k^i = t^{-1} \tilde{w}_k^i$
- END FOR
- Resample using the algorithm in Table 3.2:
  $[\{-, -, i^j\}_{i=1}^N] = \text{RESAMPLE}\,[\{\mu_k^i, w_k^i\}_{i=1}^N]$
- FOR $i = 1 : N$
    - Draw $\mathbf{x}_k^j \sim q(\mathbf{x}_k|i^j, \mathbf{Z}_k) = p\left(\mathbf{x}_k|\mathbf{x}_{k-1}^{i^j}\right)$, as in the SIR filter
    - Assign weight $\tilde{w}_k^i$ using (3.46)
- END FOR
- Calculate total weight: t = SUM $[\{\tilde{w}_k^i\}_{i=1}^N]$
- FOR $i = 1 : N$
    - Normalize: $w_k^i = t^{-1} \tilde{w}_k^i$
- END FOR

μ_k^i that characterize $p(\mathbf{x}_k|\mathbf{x}_{k-1}^i)$. If process noise is small, so that $p(\mathbf{x}_k|\mathbf{x}_{k-1}^i)$ is well characterized by μ_k^i, then the ASIR filter is often less sensitive to outliers than SIR filter, and the weights w_k^i are more even. If process noise is large, however, a single point in the state space does not characterize $p(\mathbf{x}_k|\mathbf{x}_{k-1}^i)$ well and hence ASIR resamples based on a poor approximation of $p(\mathbf{x}_k|\mathbf{x}_{k-1}^i)$. In cases like this, the use of ASIR filter can even degrade the performance.

3.5.3 Particle Filters with an Improved Sample Diversity

Resampling was suggested in the SIR filter as a method to reduce the degeneracy problem. However, it has been pointed out that resampling in turn introduces other problems, in particular, the problem of loss of diversity among the particles. This arises due to the fact that in the resampling stage, samples are drawn from a discrete distribution rather than a continuous one. If this problem is not addressed properly, it may lead to "particle collapse," where all N particles occupy the same point in the state space, giving a poor representation of the posterior density. Here we present two techniques that attempt to improve the diversity among the particles. The first is based on the *regularization* step (a kind of jittering of particles) and hence the name regularized particle filter (RPF) [24]. The second is based on the MCMC move step [23]. The particle filter that improves the sample diversity either via regularization or by the MCMC move step is identical to the SIR filter except for the resampling stage.

Regularized PF

The RPF resamples from a continuous approximation of the posterior density $p(\mathbf{x}_k|\mathbf{Z}_k)$ while the SIR resamples from the discrete approximation (3.19). Specifically, in the RPF, samples are drawn from the approximation

$$p(\mathbf{x}_k|\mathbf{Z}_k) \approx \sum_{i=1}^{N} w_k^i K_h(\mathbf{x}_k - \mathbf{x}_k^i) \qquad (3.47)$$

where

$$K_h(\mathbf{x}) = \frac{1}{h^{n_x}} K\left(\frac{\mathbf{x}}{h}\right) \qquad (3.48)$$

is the rescaled kernel density $K(\cdot)$, $h > 0$ is the kernel bandwidth (a scalar parameter), n_x is the dimension of the state vector $\mathbf{x}$, and w_k^i, $i = 1, \ldots, N$, are normalized weights. The kernel density is a symmetric probability density function such that

$$\int \mathbf{x} K(\mathbf{x})d\mathbf{x} = 0, \quad \int \|\mathbf{x}\|^2 K(\mathbf{x})d\mathbf{x} < \infty.$$

The kernel $K(\cdot)$ and bandwidth h are chosen to minimize the mean integrated square error (MISE) between the true posterior density and the corresponding regularized empirical representation in (3.47), which is defined as

$$\text{MISE}(\hat{p}) = \mathbb{E}\left[\int [\hat{p}(\mathbf{x}_k|\mathbf{Z}_k) - p(\mathbf{x}_k|\mathbf{Z}_k)]^2 \, d\mathbf{x}_k\right] \quad (3.49)$$

where $\hat{p}(\cdot|\cdot)$ denotes the approximation to $p(\mathbf{x}_k|\mathbf{Z}_k)$ given by the right-hand side of (3.47). It is worth noting that the use of the kernel approximation becomes increasingly less appropriate as n_x (dimensionality of the state) increases. In the special case of an equally weighted sample, the optimal choice of the kernel is the Epanechnikov kernel [24],

$$K_{opt} = \begin{cases} \frac{n_x+2}{2c_{n_x}}(1 - \|\mathbf{x}\|^2) & \text{if } \|\mathbf{x}\| < 1 \\ 0 & \text{otherwise} \end{cases} \quad (3.50)$$

where c_{n_x} is the volume of the unit hypersphere in $\mathbb{R}^{n_x}$. Furthermore, when the underlying density is Gaussian with a unit covariance matrix, the optimal choice for the bandwidth is [24]

$$h_{opt} = A \cdot N^{-\frac{1}{n_x+4}} \text{ with } A = [8c_{n_x}^{-1}(n_x + 4)(2\sqrt{\pi})^{n_x}]^{\frac{1}{n_x+4}}. \quad (3.51)$$

Though the results of (3.50) and (3.51) are optimal only in the special case of equally weighted particles and underlying Gaussian density, these results can still be used in the general case to obtain a suboptimal filter. In order to reduce the computational cost, the samples can be generated from the Gaussian kernel instead of the Epanechnikov kernel. The optimal bandwidth in this case is [24]:

$$h_{opt} = A \cdot N^{-\frac{1}{n_x+4}} \text{ with } A = [4/(n_x + 2)]^{\frac{1}{n_x+4}}. \quad (3.52)$$

One cycle of the regularized PF is described by an algorithm given in Table 3.6. The RPF differs from the generic particle filter (Table 3.3) only in additional regularization steps when conducting the resampling. Note also that the calculation of the empirical covariance matrix $\mathbf{S}_k$ is carried out prior to resampling and hence is a function of both the $\mathbf{x}_k^i$ and w_k^i. The key step is

$$\mathbf{x}_k^{i*} = \mathbf{x}_k^i + h_{opt}\mathbf{D}_k \epsilon^i \quad (3.53)$$

which effectively jitters the resampled values. The theoretical disadvantage of the RPF is that its samples are no longer guaranteed to asymptotically approximate those from the posterior.

Table 3.6
Regularized Particle Filter

$[\{\mathbf{x}_k^{i\,*}, w_k^i\}_{i=1}^N]$ = RPF $[\{\mathbf{x}_{k-1}^i, w_{k-1}^i\}_{i=1}^N, \mathbf{z}_k]$

- FOR $i = 1 : N$
    - Draw $\mathbf{x}_k^i \sim p(\mathbf{x}_k|\mathbf{x}_{k-1}^i)$
    - Calculate $\tilde{w}_k^i = p(\mathbf{z}_k|\mathbf{x}_k^i)$
- END FOR
- Calculate total weight: t = SUM $[\{\tilde{w}_k^i\}_{i=1}^N]$
- FOR $i = 1 : N$
    - Normalize: $w_k^i = t^{-1}\tilde{w}_k^i$
- END FOR
- Calculate $\hat{N}_{eff}$ using (3.18)
- IF $\hat{N}_{eff} < N_{thr}$
    - Calculate the empirical covariance matrix $\mathbf{S}_k$ of $\{\mathbf{x}_k^i, w_k^i\}_{i=1}^N$
    - Compute $\mathbf{D}_k$ such that $\mathbf{D}_k\mathbf{D}_k^T = \mathbf{S}_k$.
    - Resample using the algorithm in Table 3.2:
      $[\{\mathbf{x}_k^i, w_k^i, -\}_{i=1}^N]$ = RESAMPLE $[\{\mathbf{x}_k^i, w_k^i\}_{i=1}^N]$
    - FOR $i = 1 : N$
        * Draw $\epsilon^i \sim K$ from the Epanechnikov/Gaussian kernel
        * $\mathbf{x}_k^{i\,*} = \mathbf{x}_k^i + h_{opt}\mathbf{D}_k\epsilon^i$
    - END FOR
- END IF

MCMC Move Step

The theoretical foundations of this step are given in [23]. We describe a particular implementation of the MCMC move step based on the Metropolis-Hastings algorithm [33]. The key idea is that a resampled particle x_k^i is moved to a new state $x_k^{i\,*}$ according to (3.53), only if $u \leq \alpha$, where $u \sim \mathcal{U}[0,1]$ and α is the acceptance probability. Otherwise, the move is rejected. The acceptance probability of the move is derived next. The objective (desired) density of the MCMC step is $p(\mathbf{X}_k|\mathbf{Z}_k)$, which has been expressed by (3.13) as:

$$p(\mathbf{X}_k|\mathbf{Z}_k) = \frac{p(\mathbf{z}_k|\mathbf{x}_k)\,p(\mathbf{x}_k|\mathbf{x}_{k-1})}{p(\mathbf{z}_k|\mathbf{Z}_{k-1})} p(\mathbf{X}_{k-1}|\mathbf{Z}_{k-1}). \qquad (3.54)$$

Consider the situation where you have generated

- Particle x_k^i by the resampling step, so that $\mathbf{X}_k^i = \{x_k^i, \mathbf{X}_{k-1}^i\}$;
- Particle $x_k^{i\,*}$ by sampling from instrumental (proposal) distribution $q(\cdot|x_k^i)$, so that $\mathbf{X}_k^{i\,*} = \{x_k^{i\,*}, \mathbf{X}_{k-1}^i\}$.

The Metropolis-Hastings acceptance probability [33] is given by:

$$\alpha = \min\left\{1, \frac{p(\mathbf{X}_k^{i\,*}|\mathbf{Z}_k)q(x_k^i|x_k^{i\,*})}{p(\mathbf{X}_k^i|\mathbf{Z}_k)q(x_k^{i\,*}|x_k^i)}\right\}. \qquad (3.55)$$

Let $q(\cdot|x_k^i)$ correspond to sampling according to (3.53). Since in this case $q(\cdot|x_k^i)$ is symmetric in its arguments, that is $q(x_k^{i\,*}|x_k^i) = q(x_k^i|x_k^{i\,*})$, the substitution of (3.54) into (3.55) yields:

$$\alpha = \min\left\{1, \frac{p(\mathbf{z}_k|x_k^{i\,*})\,p(x_k^{i\,*}|x_{k-1}^i)}{p(\mathbf{z}_k|x_k^i)\,p(x_k^i|x_{k-1}^i)}\right\}. \qquad (3.56)$$

The theoretical advantage of the MCMC move particle filter over the regularized PF is that its samples are guaranteed to asymptotically approximate those from the posterior. In terms of the computational complexity, both the RPF and the MCMC move PF are comparable to SIR, since the only requirement is N additional generations from the kernel $K(\cdot)$ at each time step. In practical scenarios, both the RPF and the MCMC move PF perform better than the SIR in cases where sample impoverishment is severe; for example when process noise is small.

3.5.4 Local Linearization Particle Filters

The optimal importance density can be approximated by incorporating the most current measurement z_k via a bank of extended or unscented Kalman filters [8, 17].

The idea is to use for each particle (index i) a separate EKF(i) or UKF(i) to generate and propagate a Gaussian importance distribution; that is,

$$q(\mathbf{x}_k^i|\mathbf{x}_{k-1}^i, \mathbf{z}_k) = \mathcal{N}\left(\mathbf{x}_k^i; \hat{\mathbf{x}}_k^i, \hat{\mathbf{P}}_k^i\right), \qquad (3.57)$$

where $\hat{\mathbf{x}}_k^i$ and $\hat{\mathbf{P}}_k^i$ are estimates of the mean and covariance computed by EKF(i) or UKF(i) at time k using measurement $\mathbf{z}_k$. We refer to the corresponding particle filter as the local linearization[3] particle filter (LLPF). A single cycle of this filter is given by the algorithm in Table 3.7. This routine performs resampling at every time step and therefore the importance weights are not passed on from one cycle to the next (the same as in the SIR filter).

Table 3.7
Local Linearization Particle Filter

$[\{\mathbf{x}_k^j, \mathbf{P}_k^j\}_{j=1}^N] = \text{LLPF}\,[\{\mathbf{x}_{k-1}^i, \mathbf{P}_{k-1}^i\}_{i=1}^N, \mathbf{z}_k]$

- FOR $i = 1 : N$
    - Run EKF or UKF (Chapter 2)
      $[\hat{\mathbf{x}}_k^i, \hat{\mathbf{P}}_k^i] = \text{EKF/UKF}\,[\mathbf{x}_{k-1}^i, \mathbf{P}_{k-1}^i, \mathbf{z}_k]$
    - Draw a sample from importance density given by (3.57):
      $\mathbf{x}_k^i \sim \mathcal{N}\left(\mathbf{x}_k^i; \hat{\mathbf{x}}_k^i, \hat{\mathbf{P}}_k^i\right)$
    - Calculate $\tilde{w}_k^i = \frac{p(\mathbf{z}_k|\mathbf{x}_k^i)p(\mathbf{x}_k^i|\mathbf{x}_{k-1}^i)}{q(\mathbf{x}_k^i|\mathbf{x}_{k-1}^i, \mathbf{z}_k)}$, where $q(\mathbf{x}_k^i|\mathbf{x}_{k-1}^i, \mathbf{z}_k)$ is given by (3.57).
- END FOR
- Calculate total weight: $t = \text{SUM}\,[\{\tilde{w}_k^i\}_{i=1}^N]$
- FOR $i = 1 : N$
    - Normalize: $w_k^i = t^{-1}\tilde{w}_k^i$
- END FOR
- Resample using the algorithm in Table 3.2:
  $[\{\mathbf{x}_k^j, -, i^j\}_{j=1}^N] = \text{RESAMPLE}\,[\{\mathbf{x}_k^i, w_k^i\}_{i=1}^N]$
- FOR $j = 1 : N$
    - Assign covariance: $\mathbf{P}_k^j = \hat{\mathbf{P}}_k^{i^j}$
- END FOR

---

[3] In Chapter 2 we have seen that the EKF is based on analytic local linearization, while the UKF performs statistical local linearization.

The local linearization method for approximation of the importance density propagates the particles towards the likelihood function and consequently the LLPF performs better than the SIR filter. The additional computational cost of using such an importance density is often more than offset by a reduction in the number of samples required to achieve a certain level of performance. Using UKF instead of the EKF in the local linearization PF is reported to improve the performance [17].

3.5.5 Multiple-Model Particle Filter

The multiple-model (MM) particle filter has been proposed by several authors [24, 34, 35] to perform nonlinear filtering with switching dynamic models. As we observed in Section 1.4 the problem belongs to a wider class of *hybrid state estimation* problems where the (augmented) state vector consists of both a continuous-valued part and a discrete-valued part. The components of the continuous-valued vector are usually target kinematic variables (position, velocity, and so forth) and possibly unknown parameters (e.g., ballistic coefficient in Chapter 5). The discrete-valued vector can be any combination of:

- The regime variable r_k, which determines which dynamic model is in effect from t_{k-1}^+ to t_k (see Section 1.4);

- Target nonkinematic attributes (class, allegiance), used in joint target and classification problems [36] (see Chapter 11);

- Data association vector, which determines the origin of measurements in the presence of clutter and/or multiple targets [37].

The MM particle filter is a sequential Monte Carlo approximation of the conceptual solution given by (1.40) and (1.41). Let the augmented state vector be defined as $\mathbf{y}_k = [\mathbf{x}_k^T \ r_k]^T$ where $r_k \in S = \{1, 2, \ldots, s\}$ is in accordance with definitions and notation of Section 1.4. The initial densities $p(\mathbf{x}_0)$ and $p(r_1) = \sum_{i=1}^{s} \mu_i \delta(r_1 - i)$ are assumed to be known. Furthermore, let $\{\mathbf{y}_k^n, w_k^n\}_{n=1}^{N}$ denote a random measure that characterizes the posterior density $p(\mathbf{y}_k|\mathbf{Z}_k)$, such that each particle $\mathbf{y}_k^n$ consists of two components: $\mathbf{x}_k^n$ and r_k^n. A pseudocode of a generic MM particle filter (MMPF) is shown in Table 3.8.

The first step in this algorithm is to generate a random set $\{r_k^n\}_{n=1}^{N}$ based on $\{r_{k-1}^n\}_{n=1}^{N}$ and the transitional probability matrix $\mathbf{\Pi} = [\pi_{ij}]$, where $i, j \in S$. This can be done using an algorithm shown in Table 3.9, which implements the rule that if $r_{k-1}^n = i$ then r_k^n should be set to j with probability π_{ij}. The implementation is as follows: if $r_{k-1}^n = i$ and $u_n \sim \mathcal{U}[0, 1]$, then r_k^n is set to $m \in S$ such that:

$$\sum_{j=1}^{m-1} \pi_{ij} < u_n \leq \sum_{j=1}^{m} \pi_{ij}. \tag{3.58}$$

Table 3.8
Generic MMPF

$[\{\mathbf{y}_k^n, w_k^n\}_{n=1}^N] = \text{MMPF}[\{\mathbf{y}_{k-1}^n, w_{k-1}^n\}_{n=1}^N, \mathbf{z}_k]$

- Regime transition (Table 3.9):
  $[\{r_k^n\}_{n=1}^N] = \text{RT}\,[\{r_{k-1}^n\}_{n=1}^N, \Pi]$

- Regime conditioned SIS (Table 3.10):
  $[\{\mathbf{x}_k^n, w_k^n\}_{n=1}^N] = \text{RC-SIS}\,[\{\mathbf{x}_{k-1}^n, r_k^n, w_{k-1}^n\}_{n=1}^N, \mathbf{z}_k]$

- Calculate $\hat{N}_{eff}$ using (3.18)

- IF $\hat{N}_{eff} < N_{thr}$

  - Resample using the algorithm in Table 3.2:
    $[\{\mathbf{y}_k^n, w_k^n, -\}_{n=1}^N] = \text{RESAMPLE}\,[\{\mathbf{y}_k^n, w_k^n\}_{n=1}^N]$

- END IF

The sum $\sum_{j=1}^m \pi_{ij}$ represents the cumulative distribution function of discrete random variable r_k given $r_{k-1} = i$.

The next step in the generic MMPF performs a *regime conditioned* SIS filtering, described in Table 3.10. The optimal regime conditioned importance density is given by

$$q(\mathbf{x}_k|\mathbf{x}_{k-1}^n, r_k^n, \mathbf{z}_k)_{opt} = p(\mathbf{x}_k|\mathbf{x}_{k-1}^n, r_k^n, \mathbf{z}_k), \qquad (3.59)$$

although the most popular choice [24, 34, 35] appears to be the transitional prior

$$q(\mathbf{x}_k|\mathbf{x}_{k-1}^n, r_k^n, \mathbf{z}_k) = p(\mathbf{x}_k|\mathbf{x}_{k-1}^n, r_k^n). \qquad (3.60)$$

Various forms of the MMPF will be developed in Part II of this book.

3.6 COMPUTATIONAL ASPECTS

We have seen in Section 3.1 that the error of a Monte Carlo estimate is of order $O(N^{-1/2})$ that is independent of the state vector dimension. This property of the estimation error, however, is based on the assumption that N is the number of *statistically independent* samples and that the function to be integrated is known exactly. In the case of sequential Monte Carlo estimation, unfortunately, none of these two assumptions is valid. Crisan and Doucet [38] have recently shown that an upper bound on the variance of estimation error has the form $c \cdot O(N^{-1})$, where c is

Table 3.9

Regime Transition

$[\{r_k^n\}_{n=1}^N] = \text{RT}\,[\{r_{k-1}^n\}_{n=1}^N, \Pi]$

- FOR $i = 1 : s$
    - $c_i(0) = 0$
    - FOR $j = 1 : s$
        * $c_i(j) = c_i(j-1) + \pi_{ij}$
    - END FOR
- END FOR
- FOR $n = 1 : N$
    - Draw $u_n \sim \mathcal{U}[0,1]$
    - Set $i = r_{k-1}^n$
    - $m = 1$
    - WHILE $(c_i(m) < u_n)$
        * $m = m + 1$
    - END WHILE
    - Set $r_k^n = m$
- END FOR

---

a constant. Daum [39], however, argues that constant c in this upper bound depends heavily on the state vector dimension. He then derives a formula which states that the variance of error for n_x dimensional sequential Monte Carlo integration, using importance sampling with a good importance density, is approximately linear in n_x. For a poorly chosen density, however, the variance of error becomes exponential in n_x, corresponding to what is commonly referred to as the "curse of dimensionality." Indeed numerous authors point out that the number of required particles N needs to be quite high for higher-dimensional systems [40, 41] and emphasize how crucial it is to choose a good importance density.

For some classes of state-space models one can further reduce the number of particles in the particle filter using a variance reduction method known as Rao-Blackwellization [42]. The idea is to partition the state vector so that one component of the partition is a conditionally linear Gaussian state-space model; for this component one can work out the solution analytically and use the Kalman

Table 3.10

Regime Conditioned SIS

$[\{x_k^n, w_k^n\}_{n=1}^N] = $ RC-SIS $[\{x_{k-1}^n, r_k^n, w_{k-1}^n\}_{n=1}^N, z_k]$

- FOR $n = 1 : N$
    - Draw $x_k^n \sim q(x_k | x_{k-1}^n, r_k^n, z_k)$
    - Evaluate the importance weights up to a normalizing constant
    $$\tilde{w}_k^n = w_{k-1}^n \frac{p(z_k | x_k^n, r_k^n) \, p\left(x_k^n | x_{k-1}^n, r_k^n\right)}{q\left(x_k^n | x_{k-1}^n, r_k^n, z_k\right)}$$
- END FOR
- Calculate total weight: $t = $ SUM $[\{\tilde{w}_k^n\}_{n=1}^N]$
- FOR $n = 1 : N$
    - Normalize: $w_k^n = t^{-1} \tilde{w}_k^n$
- END FOR

---

filter. The particle filter is then used only for the nonlinear non-Gaussian portion of the state-space. In this way the majority of the computational effort is devoted to the hard part of the problem rather than to the easy part [8]. To explain the rationale for this approach, consider first the definition of the conditional variance of random variable τ given u:

$$\text{var}(\tau | u) \triangleq \mathbb{E}\{[\tau - \mathbb{E}(\tau | u)]^2 | u\}. \tag{3.61}$$

The following property of the conditional variance can be proved:

$$\text{var}(\tau) = \mathbb{E}[\text{var}(\tau | u)] + \text{var}[\mathbb{E}(\tau | u)]. \tag{3.62}$$

If we think of τ as an estimator depending on variables u and v, then estimator $\tau' = \mathbb{E}[\tau | u]$ is referred to as the Rao-Blackwellized version of τ. Estimator τ' has the same mean as τ, but according to (3.62) its variance is smaller by $\mathbb{E}\{\text{var}(\tau | u)\}$. Therefore, from the statistical point of view, whenever τ' can be computed, it makes sense to use it. Gustafsson et al. [40] present a particularly convincing example of the Rao-Blackwellized particle filter for inertial navigation, where the state vector can have as many as 27 components, but the particle filter is applied only to the three-dimensional target position. Other efficient Rao-Blackwellized particle filters for target tracking are presented in [43, 44] and in Chapter 12 of this book.

3.7 SUMMARY

This chapter described the basic techniques for particle filter design. The two key steps of a generic particle filter are: (1) the sequential importance sampling and (2) resampling. The choice of a good importance density has been emphasized and some possible solutions reviewed. Finally the chapter has presented five relatively widespread versions of particle filters, backed up by their pseudocode for easier implementation. The class of particle filters, however, is much broader, and many other variations have been proposed (see, for example, [5] for further reading). New particle filtering schemes are being proposed constantly, in an attempt to improve their computational efficiency (the goal being to use as few particles as possible). Throughout the book, however, we will primarily be concerned with the *statistical* efficiency, sometimes ignoring the computational aspects.[4] In fact, the design of efficient particle filters is still more of an art than a science. At present, there are no clear theoretical rules as to when it is necessary to resample (e.g., at what level of $\hat{N}_{eff}$), at what level of process noise the particles must be diversified (e.g., MCMC move step applied), or even more importantly, what is the minimum number of particles that will enable a statistically efficient implementation. In practice, one often tries a few versions of particle filters before a suitable implementation (both statistically and computationally) is found. Recently an attempt to automatically optimize the parameters of a particle filter using training data sequences has been reported in [45].

Since the particle filters are very expensive in terms of computational requirements, as a general rule in practice one should use them only when problems are difficult and the conventional Kalman filter-based methods do not produce satisfactory results.

3.8 APPENDIX: COMBINATION OF QUADRATIC TERMS

We want to prove that if $\mathbf{R}$ and $\mathbf{Q}$ are symmetric and positive definite matrices then the following relationship holds:

$$(\mathbf{z} - \mathbf{Hx})^T \mathbf{R}^{-1}(\mathbf{z} - \mathbf{Hx}) + (\mathbf{x} - \mathbf{f})^T \mathbf{Q}^{-1}(\mathbf{x} - \mathbf{f}) = \\ (\mathbf{x} - \mathbf{a})^T \mathbf{\Sigma}^{-1}(\mathbf{x} - \mathbf{a}) + (\mathbf{z} - \mathbf{b}) \mathbf{S}^{-1}(\mathbf{z} - \mathbf{b}) \quad (3.63)$$

---

4 In order to measure the statistical performance we describe the Cramér-Rao lower bounds for nonlinear filtering in the next chapter.

with

$$\begin{align}
\mathbf{S} &= \mathbf{R} + \mathbf{HQH}^T \\
\mathbf{b} &= \mathbf{Hf} \\
\mathbf{a} &= \mathbf{f} + \mathbf{\Sigma H}^T \mathbf{R}^{-1}(\mathbf{z} - \mathbf{Hf}) \\
\mathbf{\Sigma}^{-1} &= \mathbf{H}^T \mathbf{R}^{-1} \mathbf{H} + \mathbf{Q}^{-1}.
\end{align}$$

The proof [46] starts by expansion of the left side of (3.63):

$$\begin{aligned}
&(\mathbf{z} - \mathbf{Hx})^T \mathbf{R}^{-1}(\mathbf{z} - \mathbf{Hx}) + (\mathbf{x} - \mathbf{f})^T \mathbf{Q}^{-1}(\mathbf{x} - \mathbf{f}) \\
&= \mathbf{x}^T(\mathbf{H}^T \mathbf{R}^{-1} \mathbf{H} + \mathbf{Q}^{-1})\mathbf{x} - \mathbf{x}^T(\mathbf{H}^T \mathbf{R}^{-1} \mathbf{z} + \mathbf{Q}^{-1}\mathbf{f}) \\
&\quad - (\mathbf{z}\mathbf{R}^{-1}\mathbf{H} + \mathbf{f}^T \mathbf{Q}^{-1})\mathbf{x} + \mathbf{z}^T \mathbf{R}^{-1} \mathbf{z} + \mathbf{f}^T \mathbf{Q}^{-1} \mathbf{f} \\
&= \mathbf{x}^T \mathbf{\Sigma}^{-1} \mathbf{x} - \mathbf{x}^T [\mathbf{H}^T \mathbf{R}^{-1}(\mathbf{z} - \mathbf{Hf}) + \mathbf{\Sigma}^{-1}\mathbf{f}] \\
&\quad - [(\mathbf{z} - \mathbf{Hf})^T \mathbf{R}^{-1}\mathbf{H} + \mathbf{f}^T \mathbf{\Sigma}^{-1}]\mathbf{x} + \mathbf{z}^T \mathbf{R}^{-1} \mathbf{z} + \mathbf{f}^T \mathbf{Q}^{-1}\mathbf{f} \\
&= \mathbf{x}^T \mathbf{\Sigma}^{-1} \mathbf{x} - \mathbf{x}^T \mathbf{\Sigma}^{-1}\mathbf{a} - \mathbf{a}^T \mathbf{\Sigma}^{-1}\mathbf{x} + \mathbf{z}^T \mathbf{R}^{-1}\mathbf{z} + \mathbf{f}^T \mathbf{Q}^{-1}\mathbf{f} \\
&= (\mathbf{x} - \mathbf{a})^T \mathbf{\Sigma}^{-1}(\mathbf{x} - \mathbf{a}) + m
\end{aligned}$$

which already has a form of (3.63). Next we evaluate m, a term independent of $\mathbf{x}$, as follows:

$$\begin{aligned}
m &= -\mathbf{a}^T \mathbf{\Sigma}^{-1}\mathbf{a} + \mathbf{z}^T \mathbf{R}^{-1}\mathbf{z} + \mathbf{f}^T \mathbf{Q}^{-1}\mathbf{f} \\
&= -(\mathbf{z} - \mathbf{Hf})^T \mathbf{R}^{-1}\mathbf{H}\mathbf{\Sigma}\mathbf{H}^T \mathbf{R}^{-1}(\mathbf{z} - \mathbf{Hf}) - \mathbf{z}^T \mathbf{R}^{-1}\mathbf{Hf} \\
&\quad + \mathbf{f}^T \mathbf{H}^T \mathbf{R}^{-1}\mathbf{Hf} - \mathbf{f}^T \mathbf{H}^T \mathbf{R}^{-1}\mathbf{z} + \mathbf{f}^T \mathbf{H}^T \mathbf{R}^{-1}\mathbf{Hf} \\
&\quad - \mathbf{f}^T(\mathbf{H}^T \mathbf{R}^{-1}\mathbf{H} + \mathbf{Q}^{-1})\mathbf{f} + \mathbf{z}^T \mathbf{R}^{-1}\mathbf{z} + \mathbf{f}^T \mathbf{Q}^{-1}\mathbf{f} \\
&= (\mathbf{z} - \mathbf{Hf})^T [\mathbf{R}^{-1} - \mathbf{R}^{-1}\mathbf{H}(\mathbf{H}^T \mathbf{R}^{-1}\mathbf{H} + \mathbf{Q}^{-1})^{-1}\mathbf{H}^T \mathbf{R}^{-1}](\mathbf{z} - \mathbf{Hf}).
\end{aligned}$$

Using matrix inversion lemma it follows that $m = (\mathbf{z} - \mathbf{b})^T \mathbf{S}^{-1}(\mathbf{z} - \mathbf{b})$, which completes the proof of (3.63).

References

[1] J. M. Hammersley and K. W. Morton, "Poor man's Monte Carlo," *Journal of the Royal Statistical Society B*, vol. 16, pp. 23–38, 1954.

[2] J. E. Handschin and D. Q. Mayne, "Monte Carlo techniques to estimate the conditional expectation in multi-stage non-linear filtering," *Intern. Journal of Control*, vol. 9, no. 5, pp. 547–559, 1969.

[3] H. Akashi and H. Kumamoto, "Random sampling approach to state estimation in switching environments," *Automatica*, vol. 13, pp. 429–434, 1977.

[4] N. J. Gordon, D. J. Salmond, and A. F. M. Smith, "Novel approach to nonlinear/non-Gaussian Bayesian state estimation," *IEE Proc.-F*, vol. 140, no. 2, pp. 107–113, 1993.

[5] A. Doucet, N. de Freitas, and N. J. Gordon, eds., *Sequential Monte Carlo Methods in Practice*. New York: Springer, 2001.

[6] "Special issue on Monte Carlo methods for statistical signal processing," *IEEE Trans. Signal Processing*, vol. 50, February 2002.

[7] M. S. Arulampalam, S. Maskell, N. Gordon, and T. Clapp, "A tutorial on particle filters for nonlinear/non-Gaussian Bayesian tracking," *IEEE Trans. Signal Processing*, vol. 50, pp. 174–188, February 2002.

[8] A. Doucet, S. Godsill, and C. Andrieu, "On sequential Monte Carlo sampling methods for Bayesian filtering," *Statistics and Computing*, vol. 10, no. 3, pp. 197–208, 2000.

[9] P. J. Davis and P. Rabinowitz, *Methods of Numerical Integration*. New York: Academic Press, 1984.

[10] A. Doucet, N. de Freitas, and N. J. Gordon, "An introduction to sequential Monte Carlo methods," in *Sequential Monte Carlo Methods in Practice* (A. Doucet, N. de Freitas, and N. J. Gordon, eds.), New York: Springer, 2001.

[11] J. MacCormick and A. Blake, "A probabilistic exclusion principle for tracking multiple objects," in *Proc Int. Conf. Computer Vision*, pp. 572–578, 1999.

[12] J. Carpenter, P. Clifford, and P. Fearnhead, "Improved particle filter for non-linear problems," *IEE Proc. Part F: Radar and Sonar Navigation*, vol. 146, pp. 2–7, February 1999.

[13] P. D. Moral, "Measure valued processes and interacting particle systems: Application to non-linear filtering problems," *Annals of Applied Probability*, vol. 8, no. 2, pp. 438–495, 1998.

[14] K. Kanazawa, D. Koller, and S. J. Russell, "Stochastic simulation algorithms for dynamic probabilistic networks," in *Proceedings of the Eleventh Annual Conference on Uncertainty in AI (UAI '95)*, pp. 346–351, 1995.

[15] J. S. Liu and R. Chen, "Sequential Monte Carlo methods for dynamical systems," *Journal of the American Statistical Association*, vol. 93, pp. 1032–1044, 1998.

[16] A. Kong, J. S. Liu, and W. H. Wong, "Sequential imputations and Bayesian missing data problems," *Journal of the American Statistical Association*, vol. 89, no. 425, pp. 278–288, 1994.

[17] R. van der Merwe, A. Doucet, N. de Freitas, and E. Wan, "The Unscented Particle Filter," Tech. Rep. CUED/F-INFENG/TR 380, Cambridge University Engineering Department, 2000.

[18] B. Ripley, *Stochastic Simulation*. New York: John Wiley, 1987.

[19] G. Kitagawa, "Monte Carlo filter and smoother for non-Gaussian non-linear state space models," *Journal Of Computational and Graphical Statistics*, vol. 5, no. 1, pp. 1–25, 1996.

[20] N. Bergman, A. Doucet, and N. Gordon, "Optimal estimation and Cramer-Rao bounds for partial non-Gaussian state space models," *Ann. Inst. Statist. Math.*, vol. 53, no. 1, pp. 97–112, 2001.

[21] S. Godsill, A. Doucet, and M. West, "Methodology for Monte Carlo smoothing with application to time-varying autoregressions," in *Proc. International Symposium on Frontiers of Time Series Modelling*, 2000.

[22] B. P. Carlin, N. G. Polson, and D. S. Stoffer, "A Monte Carlo approach to nonnormal and non-linear state-space modelling," *Journal of the American Statistical Association*, vol. 87, no. 418, pp. 493–500, 1992.

[23] W. R. Gilks and C. Berzuini, "Following a moving target – Monte Carlo inference for dynamic Baysian models," *Journal of the Royal Statistical Society, B*, vol. 63, pp. 127–146, 2001.

[24] C. Musso, N. Oudjane, and F. LeGland, "Improving regularised particle filters," in *Sequential Monte Carlo Methods in Practice* (A. Doucet, N. de Freitas, and N. J. Gordon, eds.), New York: Springer, 2001.

[25] J. Liu and M. West, "Combined parameter and state estimation in simulation-based filtering," in *Sequential Monte Carlo Methods in Practice* (A. Doucet, N. de Freitas, and N. J. Gordon, eds.), New York: Springer, 2001.

[26] G. Storvik, "Particle filters for state-space models with the presence of unknown state parameters," *IEEE Trans. Signal Processing*, vol. 50, no. 2, pp. 281–289, 2002.

[27] D. S. Lee and N. K. K. Chia, "A particle algorithm for sequential Bayesian parameter estimation and model selection," *IEEE Trans. Signal Processing*, vol. 50, pp. 326–336, February 2002.

[28] A. Doucet, N. Gordon, and V. Krishnamurthy, "Particle filters for state estimation of jump Markov linear systems," *IEEE Trans. Signal Processing*, vol. 49, pp. 613–624, March 2001.

[29] M. Pitt and N. Shephard, "Filtering via simulation: Auxiliary particle filters," *Journal of the American Statistical Association*, vol. 94, no. 446, pp. 590–599, 1999.

[30] T. Clapp and S. Godsill, "Improvement strategies for Monte Carlo particle filters," in *Sequential Monte Carlo Methods in Practice* (A. Doucet, N. de Freitas, and N. J. Gordon, eds.), New York: Springer, 2001.

[31] N. Oudjane and C. Musso, "Progressive correction for regularized particle filters," in *Proc. 3rd Int. Conf. Information Fusion*, (Paris, France), 2000.

[32] P. Fearnhead, *Sequential Monte Carlo Methods in Filter Theory*. PhD thesis, University of Oxford, 1998.

[33] C. P. Robert and G. Casella, *Monte Carlo Statistical Methods*. New York: Springer, 1999.

[34] S. McGinnity and G. W. Irwin, "Multiple model bootstrap filter for maneuvering target tracking," *IEEE Trans. Aerospace and Electronic Systems*, vol. 36, no. 3, pp. 1006–1012, 2000.

[35] D. S. Angelova, T. A. Semerdjiev, V. P. Jilkov, and E. A. Semerdjiev, "Application of Monte Carlo method for tracking maneuvering target in clutter," *Mathematics and Computers in Simulation*, vol. 1851, pp. 1–9, 2000.

[36] Y. Boers and H. Driessen, "Hybrid state estimation: a target tracking application," *Automatica*, vol. 38, pp. 2153–2158, December 2002.

[37] C. Hue, J. L. Cadre, and P. Pérez, "Sequential Monte Carlo methods for multiple target tracking and data fusion," *IEEE Trans. Signal Processing*, vol. 50, pp. 309–325, February 2002.

[38] D. Crisan and A. Doucet, "A survey of convergence results on particle filtering methods for practitioners," *IEEE Trans. Signal Processing*, vol. 50, pp. 736–746, March 2002.

[39] F. Daum and J. Huang, "Curse of dimensionality and particle filters," in *Proc. IEEE Aerospace Conf.*, (Big Sky, MT), 2003.

[40] F. Gustafsson, F. Gunnarsson, N. Bergman, U. Forssell, J. Jansson, R. Karlsson, and P.-J. Nordlund, "Particle filters for positioning, navigation and tracking," *IEEE Trans. Signal Processing*, vol. 50, pp. 425–437, February 2002.

[41] D. J. Ballantyne, H. Y. Chan, and M. A. Kouritzin, "Novel branching particle method for tracking," in *Proc. 4th Int. Conf. Information Fusion (Fusion 2001)*, vol. I, (Montreal, Canada), August 2001.

[42] G. Castella and C. P. Robert, "Rao-Blackwellisation of sampling schemes," *Biometrika*, vol. 83, pp. 81–94, 1996.

[43] S. Maskell, M. Rollason, N. Gordon, and D. Salmond, "Efficient particle filtering for multiple target tracking with application to tracking in structured images," in *Proc. SPIE, Signal and Data Processing of Small Targets*, vol. 4728, April 2002.

[44] M. R. Morelande and S. Challa, "An algorithm for tracking group targets," in *Proc. Workshop on Multiple Hypothesis Tracking: A Tribute to S. Blackman*, (San Diego, CA), May 2003.

[45] A. Doucet and V. B. Tadic, "On-line optimization of sequential Monte Carlo methods using stochastic approximation," in *American Control Conference*, vol. 4, pp. 2565–2570, 2002.

[46] D. J. Salmond, *Tracking in Uncertain Environments*. PhD thesis, University of Sussex, 1989.

Chapter 4

Cramér-Rao Bounds for Nonlinear Filtering

For a general nonlinear filtering problem, we have seen in Chapter 1 that the optimal recursive Bayesian estimator requires the complete posterior density of the state to be determined as a function of time. A closed form analytic solution to this problem cannot be formulated and in all practical applications, nonlinear filtering is performed by some form of approximation. Despite the absence of a closed form solution, the best achievable error performance for nonlinear filtering has been formulated in the form of a theoretical Cramér-Rao lower bound (CRLB).

We wish to emphasize that the CRLB provides a lower bound for second-order (mean-squared) error only. We have seen, however, that posterior densities, which result from nonlinear filtering, are in general non-Gaussian. A full statistical characterization of a non-Gaussian density requires higher-order moments (higher than two), in addition to the mean and the covariance. As a consequence, the CRLB for nonlinear filtering does not fully characterize the accuracy of filtering algorithms.

Nevertheless, the CRLB for nonlinear filtering is a useful tool that deserves attention for several reasons. Traditionally, the bound has been used as a benchmark for the comparison of implemented suboptimal filtering algorithms and the assessment of the effects of introduced approximations. Since the bound predicts the best achievable performance even before the system is built, it has been used as a system design tool [1]. Increasingly, the bound is also being used as a tool for sensor management: (1) to optimize schedules and configuration for the deployment of passive sonobuoys in submarine tracking [2]; (2) to optimize radar scheduling [3]; and (3) to determine the optimal observer trajectory for passive ranging [4, 5] and terrain navigation [6].

The formulation of the CRLB for nonlinear filtering has a long history. A comprehensive review of pre-1989 attempts is presented in [7], with some key references for the discrete-time case being [8, 9, 10, 11]. The key modern reference for recursive calculation of the information matrix (the inverse of the CRLB) is [12], with further extensions applicable to a larger class of nonlinear models presented in [13]. The bounds in [12, 13] are referred to as *posterior* CRLBs, to emphasize that they are applicable to the cases where the state dynamics is modeled as being stochastic (with nonzero process noise). The recursive computation of the posterior CRLBs for *prediction* and *smoothing* is presented in [14].

The chapter starts by defining the CRLB and providing some necessary background information in Section 4.1. The general recursive solution for nonlinear filtering, mainly following [12], is presented in Section 4.2. Some special cases of CRLBs are described in Section 4.3. The CRLB for multiple switching dynamic models is discussed in Section 4.4.

4.1 BACKGROUND

Consider the nonlinear filtering problem defined by (1.1) and (1.2), repeated here for convenience:

$$\mathbf{x}_k = \mathbf{f}_{k-1}(\mathbf{x}_{k-1}, \mathbf{v}_{k-1}),$$
$$\mathbf{z}_k = \mathbf{h}_k(\mathbf{x}_k, \mathbf{w}_k).$$

Let $\hat{\mathbf{x}}_{k|k}$ be an *unbiased* estimator of the state vector $\mathbf{x}_k$, based on measurement sequence $\mathbf{Z}_k = \{\mathbf{z}_1, \ldots, \mathbf{z}_k\}$ and prior knowledge of initial density $p(\mathbf{x}_0)$. The covariance matrix of $\hat{\mathbf{x}}_{k|k}$, denoted as $\mathbf{P}_{k|k}$, has a lower bound (referred to as the CRLB) expressed as follows [15]:

$$\mathbf{P}_{k|k} \triangleq \mathbb{E}\left\{ \left(\hat{\mathbf{x}}_{k|k} - \mathbf{x}_k\right) \left(\hat{\mathbf{x}}_{k|k} - \mathbf{x}_k\right)^T \right\} \geq \mathbf{J}_k^{-1}. \quad (4.1)$$

The inequality in (4.1) means that the difference $\mathbf{P}_{k|k} - \mathbf{J}_k^{-1}$ is a positive semi-definite matrix. Matrix $\mathbf{J}_k$ in (4.1), referred to as the *filtering* information matrix, is of dimension $(n_x \times n_x)$. Its inverse is the filtering CRLB that we want to derive. Before we go any further, however, we must introduce the *trajectory* information matrix $\mathbf{I}_k$, which will be the starting point of our derivation.

Consider a sequence of target states (referred to as a trajectory) $\mathbf{X}_k = \{\mathbf{x}_0, \mathbf{x}_1, \ldots, \mathbf{x}_k\}$, and its unbiased estimate $\hat{\mathbf{X}}_{k|k}$ based on $\mathbf{Z}_k = \{\mathbf{z}_1, \ldots, \mathbf{z}_k\}$ and $p(\mathbf{x}_0)$. The covariance matrix of $\hat{\mathbf{X}}_{k|k}$ has a Cramér-Rao lower bound, defined as the inverse of the information matrix $\mathbf{I}_k$,

$$\mathbb{E}\left\{ \left(\hat{\mathbf{X}}_{k|k} - \mathbf{X}_k\right) \left(\hat{\mathbf{X}}_{k|k} - \mathbf{X}_k\right)^T \right\} \geq \mathbf{I}_k^{-1}. \quad (4.2)$$

The trajectory information matrix $\mathbf{I}_k$ is defined as [12]:

$$\mathbf{I}_k \triangleq \mathbb{E}\left\{ [\nabla_{\mathbf{X}_k} \log p(\mathbf{X}_k, \mathbf{Z}_k)] \, [\nabla_{\mathbf{X}_k} \log p(\mathbf{X}_k, \mathbf{Z}_k)]^T \right\} \qquad (4.3)$$

or equivalently as [15, 16]

$$\mathbf{I}_k \triangleq -\mathbb{E}\left\{ \nabla_{\mathbf{X}_k} [\nabla_{\mathbf{X}_k} \log p(\mathbf{X}_k, \mathbf{Z}_k)]^T \right\}, \qquad (4.4)$$

where $\nabla_{\mathbf{X}_k}$ is the first-order partial derivative operator with respect to $\mathbf{X}_k$; see (2.11). The dimension of $\mathbf{I}_k$ is $(k+1)n_x \times (k+1)n_x$. We make a few important remarks in relation to definitions (4.3) or (4.4):

1. The joint probability distribution $p(\mathbf{X}_k, \mathbf{Z}_k)$ in definition (4.3) or (4.4) is used because of the assumption that the target state is stochastic (due to process noise in target dynamics). If one adopts a deterministic model of state dynamics (zero process noise), then $p(\mathbf{X}_k, \mathbf{Z}_k)$ in (4.3) and (4.4) is replaced with the likelihood function $p(\mathbf{Z}_k|\mathbf{X}_k)$ [15].

2. The partial derivatives in (4.3) and (4.4) are evaluated at the *true value* of $\mathbf{X}_k$, while the expectation operator is taken with respect to $\mathbf{Z}_k$ and $\mathbf{X}_k$.

3. The proof of the statement: "The posterior CRLB for estimation of $\mathbf{X}_k$ is defined as the inverse of $\mathbf{I}_k$" holds under some mild regularity conditions imposed on the density $p(\mathbf{X}_k, \mathbf{Z}_k)$, which imply that the density does not have infinite moments. Details are given in [15, 17].

4. An unbiased state estimator, with the corresponding covariance matrix *equal* to the CRLB, is said to be (statistically) *efficient*.

Computation of $\mathbf{J}_k$ from $\mathbf{I}_k$

The relationship between information matrices $\mathbf{J}_k$ and $\mathbf{I}_k$ is examined next. This relationship is the first step in formulating the recursive solution for the computation of the filtering information matrix $\mathbf{J}_k$, introduced by (4.1). Recall that the inverse of $\mathbf{J}_k$ is the CRLB we seek to derive. Suppose we decompose $\mathbf{X}_k = [\mathbf{X}_{k-1}^T \; \mathbf{x}_k^T]^T$. The information matrix $\mathbf{I}_k$ can be accordingly decomposed into blocks:

$$\mathbf{I}_k = \begin{bmatrix} \mathbf{A}_k & \mathbf{B}_k \\ \mathbf{B}_k^T & \mathbf{C}_k \end{bmatrix} \qquad (4.5)$$

where

$$\mathbf{A}_k = -\mathbb{E}\left\{\nabla_{\mathbf{x}_{k-1}}\left[\nabla_{\mathbf{x}_{k-1}}\log p(\mathbf{X}_k,\mathbf{Z}_k)\right]^T\right\} \quad (4.6)$$

$$\mathbf{B}_k = -\mathbb{E}\left\{\nabla_{\mathbf{x}_{k-1}}\left[\nabla_{\mathbf{x}_k}\log p(\mathbf{X}_k,\mathbf{Z}_k)\right]^T\right\} \quad (4.7)$$

$$\mathbf{C}_k = -\mathbb{E}\left\{\nabla_{\mathbf{x}_k}\left[\nabla_{\mathbf{x}_k}\log p(\mathbf{X}_k,\mathbf{Z}_k)\right]^T\right\}. \quad (4.8)$$

It can be easily shown[1] [12] that the covariance of the error in estimating $\mathbf{x}_k$ is lower bounded by the $(n_x \times n_x)$ right-lower block of $\mathbf{I}_k^{-1}$; that is:

$$\mathbf{J}_k = \mathbf{C}_k - \mathbf{B}_k^T \mathbf{A}_k^{-1} \mathbf{B}_k. \quad (4.9)$$

The next section describes how $\mathbf{J}_k$ can be computed recursively, without a need for inverting large matrices such as $\mathbf{A}_k$.

Nonlinear Transformation of the State Vector

In some cases it may be of interest to compute the CRLB of a quantity that is a nonlinear function of the state vector[2]; that is:

$$\mathbf{s}_k = \mathbf{g}(\mathbf{x}_k). \quad (4.10)$$

According to [15, p. 83], the CRLB of vector $\mathbf{s}_k$ is then:

$$\mathbf{CRLB}(\mathbf{s}_k) = \mathbf{G}_k \mathbf{J}_k^{-1} \mathbf{G}_k^T \quad (4.11)$$

where $\mathbf{J}_k^{-1} = \mathbf{CRLB}(\mathbf{x}_k)$ and $\mathbf{G}_k = \left[\nabla_{\mathbf{x}_k}\mathbf{g}^T(\mathbf{x}_k)\right]^T$ is the Jacobian with elements:

$$[\mathbf{G}_k]_{ij} = \frac{\partial g_k[i]}{\partial x_k[j]} \quad (4.12)$$

The partial derivatives in (4.12) are evaluated at true values of $\mathbf{x}_k$ and $g_k[i]$ denotes the ith component of $\mathbf{g}(\mathbf{x}_k)$. A word of caution is in order here: if $\hat{\mathbf{x}}_{k|k}$ is an efficient estimator of $\mathbf{x}_k$, then $\mathbf{g}(\hat{\mathbf{x}}_{k|k})$, in general, is not an efficient estimator of $\mathbf{s}_k$, because $\mathbf{g}(\cdot)$ is a nonlinear function [15, p. 84].

---

1   Follows from the inverse of a 2×2 block matrix; see, for example, [18, p. 13]. Matrix $\mathbf{J}_k$, in this context, is referred to as the Schur complement of $\mathbf{A}_k$ in the block matrix (4.5).
2   In the bearings-only tracking problem (see Chapter 6), one may wish to compute the CRLB for the estimation of target range, while the target state vector is in the Cartesian coordinates [19].

4.2 RECURSIVE COMPUTATION OF THE FILTERING INFORMATION MATRIX

Tichavsky et al. [12] provided an elegant method of computing the information matrix $\mathbf{J}_k$ recursively:

$$\mathbf{J}_{k+1} = \mathbf{D}_k^{22} - \mathbf{D}_k^{21}(\mathbf{J}_k + \mathbf{D}_k^{11})^{-1}\mathbf{D}_k^{12} \quad (k > 0) \quad (4.13)$$

where

$$\mathbf{D}_k^{11} = -\mathbb{E}\left\{\nabla_{\mathbf{x}_k}\left[\nabla_{\mathbf{x}_k}\log p(\mathbf{x}_{k+1}|\mathbf{x}_k)\right]^T\right\} \quad (4.14)$$

$$\mathbf{D}_k^{21} = -\mathbb{E}\left\{\nabla_{\mathbf{x}_k}\left[\nabla_{\mathbf{x}_{k+1}}\log p(\mathbf{x}_{k+1}|\mathbf{x}_k)\right]^T\right\} \quad (4.15)$$

$$\mathbf{D}_k^{12} = -\mathbb{E}\left\{\nabla_{\mathbf{x}_{k+1}}\left[\nabla_{\mathbf{x}_k}\log p(\mathbf{x}_{k+1}|\mathbf{x}_k)\right]^T\right\} = \left[\mathbf{D}_k^{21}\right]^T \quad (4.16)$$

$$\mathbf{D}_k^{22} = -\mathbb{E}\left\{\nabla_{\mathbf{x}_{k+1}}\left[\nabla_{\mathbf{x}_{k+1}}\log p(\mathbf{x}_{k+1}|\mathbf{x}_k)\right]^T\right\}$$
$$\quad -\mathbb{E}\left\{\nabla_{\mathbf{x}_{k+1}}\left[\nabla_{\mathbf{x}_{k+1}}\log p(\mathbf{z}_{k+1}|\mathbf{x}_{k+1})\right]^T\right\}. \quad (4.17)$$

The expectation $\mathbb{E}\{\cdot\}$ in (4.14), (4.16), and (4.15) is with respect to $\mathbf{x}_k$ and $\mathbf{x}_{k+1}$, whereas in (4.17) is with respect to $\mathbf{x}_k$, $\mathbf{x}_{k+1}$ and $\mathbf{z}_{k+1}$.

The proof requires us to establish the following recursive property of joint density of states and measurements:

$$\begin{aligned} p(\mathbf{X}_{k+1}, \mathbf{Z}_{k+1}) &= p(\mathbf{x}_{k+1}, \mathbf{X}_k, \mathbf{z}_{k+1}, \mathbf{Z}_k) \\ &= p(\mathbf{z}_{k+1}|\mathbf{x}_{k+1}, \mathbf{X}_k, \mathbf{Z}_k)\, p(\mathbf{x}_{k+1}|\mathbf{X}_k, \mathbf{Z}_k)\, p(\mathbf{X}_k, \mathbf{Z}_k) \\ &= p(\mathbf{z}_{k+1}|\mathbf{x}_{k+1})\, p(\mathbf{x}_{k+1}|\mathbf{x}_k)\, p(\mathbf{X}_k, \mathbf{Z}_k), \quad (4.18) \end{aligned}$$

which follows from the assumptions made in relation to (1.1) and (1.2). Adopt the following notation for brevity: $p_k = p(\mathbf{X}_k, \mathbf{Z}_k)$. Now let us decompose $\mathbf{X}_{k+1}$ as follows: $\mathbf{X}_{k+1} = \begin{bmatrix} \mathbf{X}_{k-1}^T & \mathbf{x}_k^T & \mathbf{x}_{k+1}^T \end{bmatrix}^T$. Then the information matrix $\mathbf{J}_{k+1}$ can be accordingly decomposed as:

$$\mathbf{I}_{k+1} = \begin{bmatrix} \mathbf{A}_{k+1} & \mathbf{B}_{k+1} & \mathbf{L}_{k+1} \\ \mathbf{B}_{k+1}^T & \mathbf{C}_{k+1} & \mathbf{E}_{k+1} \\ \mathbf{L}_{k+1}^T & \mathbf{E}_{k+1}^T & \mathbf{F}_{k+1} \end{bmatrix} \quad (4.19)$$

where the individual submatrices have to be worked out one by one. Start with

$$\begin{aligned}
\mathbf{A}_{k+1} &= -\mathbb{E}\left\{\nabla_{\mathbf{x}_{k-1}}[\nabla_{\mathbf{x}_{k-1}}\log p_{k+1}]^T\right\} \\
&= -\mathbb{E}\left\{\nabla_{\mathbf{x}_{k-1}}[\nabla_{\mathbf{x}_{k-1}}(\log p_k + \log p(\mathbf{x}_{k+1}|\mathbf{x}_k) + \log p(\mathbf{z}_{k+1}|\mathbf{x}_{k+1}))]^T\right\} \\
&= -\mathbb{E}\left\{\nabla_{\mathbf{x}_{k-1}}[\nabla_{\mathbf{x}_{k-1}}\log p_k]^T\right\} + 0 + 0 \\
&= \mathbf{A}_k,
\end{aligned} \quad (4.20)$$

where $\mathbf{A}_k$ was introduced by (4.6). Similarly,

$$\begin{aligned}
\mathbf{C}_{k+1} &= -\mathbb{E}\left\{\nabla_{\mathbf{x}_k}[\nabla_{\mathbf{x}_k}\log p_{k+1}]^T\right\} \\
&= -\mathbb{E}\left\{\nabla_{\mathbf{x}_k}[\nabla_{\mathbf{x}_k}(\log p_k + \log p(\mathbf{x}_{k+1}|\mathbf{x}_k) + \log p(\mathbf{z}_{k+1}|\mathbf{x}_{k+1}))]^T\right\} \\
&= -\mathbb{E}\left\{\nabla_{\mathbf{x}_k}[\nabla_{\mathbf{x}_k}\log p_k]^T\right\} - \mathbb{E}\left\{\nabla_{\mathbf{x}_k}[\nabla_{\mathbf{x}_k}\log p(\mathbf{x}_{k+1}|\mathbf{x}_k)]^T\right\} + 0 \\
&= \mathbf{C}_k + \mathbf{D}_k^{11}.
\end{aligned} \quad (4.21)$$

It is straightforward to show that the remaining terms in (4.19) are as follows:

$$\mathbf{B}_{k+1} = \mathbf{B}_k \quad \mathbf{L}_{k+1} = 0 \quad \mathbf{E}_{k+1} = \mathbf{D}_k^{12} \quad \mathbf{F}_{k+1} = \mathbf{D}_k^{22}.$$

Now (4.19) can be rewritten as:

$$\mathbf{I}_{k+1} = \begin{bmatrix} \mathbf{A}_k & \mathbf{B}_k & 0 \\ \mathbf{B}_k^T & \mathbf{C}_k + \mathbf{D}_k^{11} & \mathbf{D}_k^{12} \\ 0 & \mathbf{D}_k^{21} & \mathbf{D}_k^{22} \end{bmatrix}, \quad (4.22)$$

where $\mathbf{0}$ stands for a zero submatrix of appropriate dimension. Information matrix $\mathbf{J}_{k+1}$ is computed as the inverse of the right-lower $(n_x \times n_x)$ submatrix of $\mathbf{I}_{k+1}^{-1}$. Using the same approach as in (4.9) we have:

$$\begin{aligned}
\mathbf{J}_{k+1} &= \mathbf{D}_k^{22} - [0 \; \mathbf{D}_k^{21}] \begin{bmatrix} \mathbf{A}_k & \mathbf{B}_k \\ \mathbf{B}_k^T & \mathbf{C}_k + \mathbf{D}_k^{11} \end{bmatrix}^{-1} \begin{bmatrix} 0 \\ \mathbf{D}_k^{12} \end{bmatrix} \quad (4.23) \\
&= \mathbf{D}_k^{22} - \mathbf{D}_k^{21}\left[\mathbf{C}_k + \mathbf{D}_k^{11} - \mathbf{B}_k^T \mathbf{A}_k^{-1} \mathbf{B}_k\right]^{-1} \mathbf{D}_k^{12} \quad (4.24)
\end{aligned}$$

Using the definition of $\mathbf{J}_k$ given by (4.9) we prove the desired recursive formula (4.13). Note that the recursion of (4.13) involves computation with matrices of dimension $(n_x \times n_x)$ only. Hence, the computational complexity of the posterior CRLB is independent of the discrete-time index k.

The recursions start with the initial information matrix $\mathbf{J}_0$, which can be computed from the initial density $p(\mathbf{x}_0)$ as follows:

$$\mathbf{J}_0 = \mathbb{E}\left\{[\nabla_{\mathbf{x}_0} \log p(\mathbf{x}_0)][\nabla_{\mathbf{x}_0} \log p(\mathbf{x}_0)]^T\right\}, \qquad (4.25)$$

where the expectation is with respect to $\mathbf{x}_0$. The recursive method for computation of the posterior CRLB is applicable to the most general nonlinear/non-Gaussian filtering problem defined by (1.1) and (1.2). Next we consider some simpler forms that appear in special cases.

4.3 SPECIAL CASES

Consider first the case where the initial distribution is Gaussian, that is, $p(\mathbf{x}_0) = \mathcal{N}(\mathbf{x}_0; \bar{\mathbf{x}}_0, \mathbf{P}_0)$. Then

$$\nabla_{\mathbf{x}_0} \log p(\mathbf{x}_0) = \nabla_{\mathbf{x}_0}\left\{c - \frac{1}{2}\left[(\mathbf{x}_0 - \bar{\mathbf{x}}_0)^T \mathbf{P}_0^{-1}(\mathbf{x}_0 - \bar{\mathbf{x}}_0)\right]\right\} = -\mathbf{P}_0^{-1}(\mathbf{x}_0 - \bar{\mathbf{x}}_0), \qquad (4.26)$$

where c is a constant. Using the basic rules of matrix algebra and the fact that the covariance matrix (and its inverse) are symmetric matrices, it follows that:

$$\mathbf{J}_0 = \mathbb{E}\left\{\mathbf{P}_0^{-1}(\mathbf{x}_0 - \bar{\mathbf{x}}_0)(\mathbf{x}_0 - \bar{\mathbf{x}}_0)^T \left[\mathbf{P}_0^{-1}\right]^T\right\} \qquad (4.27)$$

$$= \mathbf{P}_0^{-1} \mathbb{E}\left\{(\mathbf{x}_0 - \bar{\mathbf{x}}_0)(\mathbf{x}_0 - \bar{\mathbf{x}}_0)^T\right\} \mathbf{P}_0^{-1} \qquad (4.28)$$

$$= \mathbf{P}_0^{-1} \mathbf{P}_0 \mathbf{P}_0^{-1} = \mathbf{P}_0^{-1}. \qquad (4.29)$$

Next we simplify expressions for $\mathbf{D}_k^{11}$, $\mathbf{D}_k^{12}$ and $\mathbf{D}_k^{22}$ for the additive Gaussian noise case.

4.3.1 Additive Gaussian Noise

Consider the nonlinear filtering problem defined by (2.1) and (2.2), repeated here for convenience:

$$\mathbf{x}_{k+1} = \mathbf{f}_k(\mathbf{x}_k) + \mathbf{v}_k \qquad (4.30)$$

$$\mathbf{z}_{k+1} = \mathbf{h}_{k+1}(\mathbf{x}_{k+1}) + \mathbf{w}_{k+1}. \qquad (4.31)$$

Random noise sequences $\mathbf{v}_k$ and $\mathbf{w}_{k+1}$ are mutually independent, zero-mean white Gaussian, with covariances $\mathbf{Q}_k$ and $\mathbf{R}_{k+1}$, respectively. Here we impose an additional condition on covariances: they have to be invertible (nonsingular).

For the stated assumptions we have:

$$\nabla_{\mathbf{x}_k} \log p(\mathbf{x}_{k+1}|\mathbf{x}_k) = \nabla_{\mathbf{x}_k} \left[-\frac{1}{2}[\mathbf{x}_{k+1} - \mathbf{f}_k(\mathbf{x}_k)]^T \mathbf{Q}_k^{-1} [\mathbf{x}_{k+1} - \mathbf{f}_k(\mathbf{x}_k)] \right]$$
$$= \left[\nabla_{\mathbf{x}_k} \mathbf{f}_k^T(\mathbf{x}_k) \right] \mathbf{Q}_k^{-1} [\mathbf{x}_{k+1} - \mathbf{f}_k(\mathbf{x}_k)] \quad (4.32)$$

and similarly

$$\nabla_{\mathbf{x}_k} \log p(\mathbf{z}_{k+1}|\mathbf{x}_{k+1}) = \left[\nabla_{\mathbf{x}_{k+1}} \mathbf{h}_{k+1}^T(\mathbf{x}_{k+1}) \right] \mathbf{R}_{k+1}^{-1} [\mathbf{z}_{k+1} - \mathbf{h}_{k+1}(\mathbf{x}_{k+1})]. \quad (4.33)$$

Matrix $\mathbf{D}_k^{11}$ defined by (4.14) now simplifies as follows:

$$\begin{aligned}
\mathbf{D}_k^{11} &= \mathbb{E}\left\{ [\nabla_{\mathbf{x}_k} \log p(\mathbf{x}_{k+1}|\mathbf{x}_k)] [\nabla_{\mathbf{x}_k} \log p(\mathbf{x}_{k+1}|\mathbf{x}_k)]^T \right\} \\
&= \mathbb{E}\left\{ [\nabla_{\mathbf{x}_k} \mathbf{f}_k^T(\mathbf{x}_k)] \cdot \mathbf{Q}_k^{-1} \cdot [\nabla_{\mathbf{x}_k} \mathbf{f}_k^T(\mathbf{x}_k)]^T \right\} \\
&= \mathbb{E}\left\{ \tilde{\mathbf{F}}_k^T \mathbf{Q}_k^{-1} \tilde{\mathbf{F}}_k \right\} \quad (4.34)
\end{aligned}$$

where

$$\tilde{\mathbf{F}}_k = \left[\nabla_{\mathbf{x}_k} \mathbf{f}_k^T(\mathbf{x}_k) \right]^T \quad (4.35)$$

is the Jacobian[3] of $\mathbf{f}_k(\mathbf{x}_k)$ evaluated at the true value of $\mathbf{x}_k$. Similarly one can show that:

$$\mathbf{D}_k^{12} = -\mathbb{E}\left\{ \tilde{\mathbf{F}}_k^T \right\} \mathbf{Q}_k^{-1} \quad (4.36)$$

$$\mathbf{D}_k^{22} = \mathbf{Q}_k^{-1} + \mathbb{E}\left\{ \tilde{\mathbf{H}}_{k+1}^T \mathbf{R}_{k+1}^{-1} \tilde{\mathbf{H}}_{k+1} \right\} \quad (4.37)$$

where

$$\tilde{\mathbf{H}}_{k+1} = \left[\nabla_{\mathbf{x}_{k+1}} \mathbf{h}_{k+1}^T(\mathbf{x}_{k+1}) \right]^T \quad (4.38)$$

is the Jacobian of $\mathbf{h}_{k+1}(\mathbf{x}_{k+1})$ evaluated at the true value of $\mathbf{x}_{k+1}$.

In a few chapters of Part II of the book we apply these equations to commutate the theoretical CRLBs. The most difficult problem in practical calculations appears to be the expectation operator $\mathbb{E}$, which features in (4.34), (4.36), and (4.37). The expectation is taken with respect to the state vector $\mathbf{x}_k$ only (the bound is independent of the actual measurement sequence). A Monte Carlo approximation can be applied to implement the theoretical posterior CRLB formulas. One first needs to create an ensemble of state vector realizations, the so-called target "trajectories." Then the appropriate terms in (4.34), (4.36), and (4.37) are computed as the average over this ensemble.

---

[3] Throughout the book we use the convention that $\hat{\mathbf{F}}_k$ and $\tilde{\mathbf{F}}_k$ denote the Jacobians of $\mathbf{f}_k(\mathbf{x}_k)$, evaluated at the estimated and the true state vector, respectively.

4.3.2 Linear/Gaussian Case

The linear Gaussian case is a special case of (4.30) and (4.31) with

$$f_k(\mathbf{x}_k) = \mathbf{F}_k \mathbf{x}_k \qquad (4.39)$$
$$h_{k+1}(\mathbf{x}_{k+1}) = \mathbf{H}_{k+1} \mathbf{x}_{k+1} \qquad (4.40)$$

(see also Section 1.3.1). The Jacobians $\tilde{\mathbf{F}}_k$ and $\tilde{\mathbf{H}}_{k+1}$, given by (4.35) and (4.38) respectively, now simplify as follows:

$$\tilde{\mathbf{F}}_k = \left[\nabla_{\mathbf{x}_k}\left(\mathbf{x}_k^T \mathbf{F}_k^T\right)\right]^T = \mathbf{F}_k \qquad (4.41)$$
$$\tilde{\mathbf{H}}_{k+1} = \left[\nabla_{\mathbf{x}_{k+1}}\left(\mathbf{x}_{k+1}^T \mathbf{H}_{k+1}^T\right)\right]^T = \mathbf{H}_{k+1} \qquad (4.42)$$

The most important observation to make from (4.41) and (4.42) is that in the linear-Gaussian case, the Jacobians are independent of the target state. Consequently $\mathbf{D}_k^{11}$, $\mathbf{D}_k^{12}$, and $\mathbf{D}_k^{22}$ are deterministic and the expectation operator can be dropped out. Thus we have:

$$\mathbf{D}_k^{11} = \mathbf{F}_k^T \mathbf{Q}_k^{-1} \mathbf{F}_k \qquad (4.43)$$
$$\mathbf{D}_k^{12} = -\mathbf{F}_k^T \mathbf{Q}_k^{-1} \qquad (4.44)$$
$$\mathbf{D}_k^{22} = \mathbf{Q}_k^{-1} + \mathbf{H}_{k+1}^T \mathbf{R}_{k+1}^{-1} \mathbf{H}_{k+1}. \qquad (4.45)$$

The substitution of (4.43)–(4.45) into the recursion (4.13) yields:

$$\mathbf{J}_{k+1} = \mathbf{Q}_k^{-1} + \mathbf{H}_{k+1}^T \mathbf{R}_{k+1}^{-1} \mathbf{H}_{k+1} - \mathbf{Q}_k^{-1} \mathbf{F}_k \left(\mathbf{J}_k + \mathbf{F}_k^T \mathbf{Q}_k^{-1} \mathbf{F}_k\right)^{-1} \mathbf{F}_k^T \mathbf{Q}_k^{-1}. \qquad (4.46)$$

Let us denote the information matrix $\mathbf{J}_k$ by $\mathbf{P}_{k|k}^{-1}$, the inverse of a filter error covariance matrix. The application of the matrix inversion lemma[4] to (4.46) then yields:

$$\mathbf{P}_{k+1|k+1}^{-1} = \left(\mathbf{Q}_k + \mathbf{F}_k \mathbf{P}_{k|k} \mathbf{F}_k^T\right)^{-1} + \mathbf{H}_{k+1}^T \mathbf{R}_{k+1}^{-1} \mathbf{H}_{k+1}$$
$$= \mathbf{P}_{k+1|k}^{-1} + \mathbf{H}_{k+1}^T \mathbf{R}_{k+1}^{-1} \mathbf{H}_{k+1}. \qquad (4.47)$$

Using again the matrix inversion lemma, it can be easily shown [16] that (4.47) is algebraically equivalent to (1.17).

The conclusion is that the CRLB for the linear Gaussian filtering problem is equivalent to the covariance matrix of the Kalman filter.[5] The filter using (4.47) rather than (1.17) is referred to as the *information matrix filter* [16].

---
[4] $(\mathbf{A} + \mathbf{B}\mathbf{C}\mathbf{B}^T)^{-1} = \mathbf{A}^{-1} - \mathbf{A}^{-1}\mathbf{B}(\mathbf{B}^T\mathbf{A}^{-1}\mathbf{B} + \mathbf{C}^{-1})^{-1}\mathbf{B}^T\mathbf{A}^{-1}$
[5] It is well known that the covariance recursion of the Kalman filter is independent of measurements and can be iterated forward off-line [16, 20].

4.3.3 Zero Process Noise

In the absence of process noise, the evolution of the state vector (the target trajectory) is purely deterministic. Hence, the expectation operator in (4.34), (4.36), and (4.37) can be dropped out. The recursion of (4.13) then can be written as:

$$\mathbf{J}_{k+1} = \mathbf{Q}_k^{-1} + \tilde{\mathbf{H}}_{k+1}^T \mathbf{R}_{k+1}^{-1} \tilde{\mathbf{H}}_{k+1} - \mathbf{Q}_k^{-1} \tilde{\mathbf{F}}_k \left(\mathbf{J}_k + \tilde{\mathbf{F}}_k^T \mathbf{Q}_k^{-1} \tilde{\mathbf{F}}_k \right)^{-1} \tilde{\mathbf{F}}_k^T \mathbf{Q}_k^{-1}. \tag{4.48}$$

Using the matrix inversion lemma this simplifies to:

$$\mathbf{J}_{k+1} = \left(\mathbf{Q}_k + \tilde{\mathbf{F}}_k \mathbf{J}_k^{-1} \tilde{\mathbf{F}}_k^T \right)^{-1} + \tilde{\mathbf{H}}_{k+1}^T \mathbf{R}_{k+1}^{-1} \tilde{\mathbf{H}}_{k+1}. \tag{4.49}$$

Due to the absence of process noise, $\mathbf{Q}_k = \mathbf{0}$ and it follows that:

$$\mathbf{J}_{k+1} = \left[\tilde{\mathbf{F}}_k^{-1} \right]^T \mathbf{J}_k \tilde{\mathbf{F}}_k^{-1} + \tilde{\mathbf{H}}_{k+1}^T \mathbf{R}_{k+1}^{-1} \tilde{\mathbf{H}}_{k+1}. \tag{4.50}$$

Compare (4.50) to the EKF covariance computation in (2.6). If we replace $\mathbf{J}_k$ by $\mathbf{P}_{k|k}^{-1}$ and apply the matrix inversion lemma, these two equations become identical in their form. The only difference is that the EKF equation features the Jacobians $\hat{\mathbf{F}}_k$ and $\hat{\mathbf{H}}_{k+1}$, while (4.50) is based on Jacobians $\tilde{\mathbf{F}}_k$ and $\tilde{\mathbf{H}}_{k+1}$. The difference between a Jacobian with a hat ˆ and with a tilde sign ˜ is that the latter is evaluated at the *true value* of the state vector (which obviously is not available to the EKF). The conclusion is that the CRLB recursion for nonlinear filtering, in the absence of process noise, is identical to the covariance matrix propagation of the EKF, where the Jacobians are evaluated at the true state vector $\mathbf{x}_k$.

The information matrix recursion (4.50) and its corresponding CRLB for nonlinear filtering was first reported by Taylor [9].

4.4 MULTIPLE-SWITCHING DYNAMIC MODELS

This section is devoted to the derivation of the CRLB for the hybrid state estimation problem defined in Section 1.4. The problem was defined by (1.34)–(1.36), and we seek to derive the posterior CRLB defined as the inverse of the filtering information matrix $\mathbf{J}_k$. Recall that the hybrid state estimation problem attempts to estimate jointly the state vector $\mathbf{x}_k$ and the regime variable r_k. The posterior CRLB for joint estimation of $\mathbf{x}_k$ and r_k is the right-lower submatrix of $[\mathbf{I}_k]^{-1}$ where

$$\mathbf{I}_k \triangleq \mathbb{E}\left\{ [\nabla_{\mathbf{Y}_k} \log p(\mathbf{Y}_k, \mathbf{Z}_k)] [\nabla_{\mathbf{Y}_k} \log p(\mathbf{Y}_k, \mathbf{Z}_k)]^T \right\}, \tag{4.51}$$

and $\mathbf{Y}_k = \begin{bmatrix} \mathbf{X}_k^T & R_k^T \end{bmatrix}^T$, with $R_k = \{r_1, \ldots, r_k\}$ being the random regime sequence.

The derivation of the analytical posterior CRLB for the hybrid state estimation problem, at first thought, may follow from the framework described in Sections 4.1 and 4.2. Unfortunately, this approach would require differentiation of terms such as $\log p(r_{k+1}|r_k)$, with respect to r_k. Clearly, for a discrete regime variable this is not possible.

As an alternative, we propose in Section 4.4.1 to approximate the desired CRLB as the expected value over the regime sequence [21]. A special case, often considered in the tracking literature, is to assume a deterministic target trajectory with a few segments of different dynamic motions. A bound for this case is considered in Section 4.4.2.

4.4.1 Enumeration Method

In order to simplify analysis, suppose that the process and measurement noise sequences are additive, zero-mean Gaussian. Equations (1.34) and (1.35) from Section 1.4 can now be rewritten as:

$$\mathbf{x}_k = \mathbf{f}_{k-1}(\mathbf{x}_{k-1}, r_k) + \mathbf{v}_{k-1}(r_k) \qquad (4.52)$$
$$\mathbf{z}_k = \mathbf{h}_k(\mathbf{x}_k, r_k) + \mathbf{w}_k, \qquad (4.53)$$

where r_k is the regime variable modeled by a time-homogeneous s-state first-order Markov chain with transitional probabilities defined in (1.36). The covariance matrices of process noise $\mathbf{v}_{k-1}$ and measurement noise $\mathbf{w}_k$ are $\mathbf{Q}_{k-1}(r_k)$ and $\mathbf{R}_k$, respectively.

Section 1.4 introduced the regime sequence R_k^ℓ in (1.44), $\ell = 1, \ldots, s^k$. The probability of a regime sequence R_k^ℓ can be computed using the (known or specified) transitional probability matrix $\mathbf{\Pi} = [\pi_{ij}]$ and the initial regime probabilities $\mu_i \triangleq \mathrm{P}\{r_1 = i\}$, as follows[6]:

$$\mathrm{P}(R_k^\ell) = \left[\prod_{j=1}^s \mu_j^{\delta(r_1^\ell, j)}\right] \prod_{i=1}^s \left[\prod_{j=1}^s \pi_{ij}^{n_{ij}(R_k^\ell)}\right] \qquad (4.54)$$

where $\delta(r_i, j) = 1$ if $r_i = j$ and zero otherwise and

$$n_{ij}(R_k) = \sum_{\eta=2}^k \delta(r_{\eta-1}, i)\, \delta(r_\eta, j) \qquad (4.55)$$

---

[6] Note that (4.54) represents the prior probability of a regime sequence, independent of measurements, as opposed to (1.46).

is the number of transitions from regime i to regime j in the regime sequence R_k. For example, if $s = 2$ and $k = 9$, the probability of regime sequence $\{1,1,1,2,2,2,2,1,2\}$ equals $\mu_1 \pi_{11}^2 \pi_{12}^2 \pi_{21} \pi_{22}^3$.

Given a particular regime sequence R_k^ℓ, the covariance matrix of an unbiased estimator $\hat{\mathbf{x}}_{k|k}$ has a lower bound as follows:

$$\mathbb{E}\left\{\left(\hat{\mathbf{x}}_{k|k} - \mathbf{x}_k\right)\left(\hat{\mathbf{x}}_{k|k} - \mathbf{x}_k\right)^T \Big| R_k^\ell\right\} \geq [\mathbf{J}_k^\ell]^{-1}. \tag{4.56}$$

It follows from (4.13) that the conditional information matrix $\mathbf{J}_k^\ell$ can be computed as:

$$\mathbf{J}_k^\ell = \mathbf{D}_{k-1}^{\ell,22} - \left[\mathbf{D}_{k-1}^{\ell,12}\right]^T \left(\mathbf{J}_{k-1}^\ell + \mathbf{D}_{k-1}^{\ell,11}\right)^{-1} \mathbf{D}_{k-1}^{\ell,12} \tag{4.57}$$

where

$$\mathbf{D}_{k-1}^{\ell,11} = \mathbb{E}\left\{\left[\tilde{\mathbf{F}}_{k-1}^\ell\right]^T [\mathbf{Q}_{k-1}^\ell]^{-1} \tilde{\mathbf{F}}_{k-1}^\ell\right\} \tag{4.58}$$

$$\mathbf{D}_{k-1}^{\ell,12} = -\mathbb{E}\left\{\left[\tilde{\mathbf{F}}_{k-1}^\ell\right]^T\right\} [\mathbf{Q}_{k-1}^\ell]^{-1} \tag{4.59}$$

$$\mathbf{D}_k^{\ell,22} = [\mathbf{Q}_{k-1}^\ell]^{-1} + \mathbb{E}\left\{\left[\tilde{\mathbf{H}}_k^\ell\right]^T \mathbf{R}_k^{-1} \tilde{\mathbf{H}}_k^\ell\right\} \tag{4.60}$$

with $\mathbf{Q}_{k-1}^\ell = \mathbf{Q}_{k-1}(r_k^\ell)$ and Jacobians

$$\tilde{\mathbf{F}}_{k-1}^\ell = \left[\nabla_{\mathbf{x}_{k-1}} \mathbf{f}_{k-1}^T(\mathbf{x}_{k-1}, r_k^\ell)\right]^T \tag{4.61}$$

$$\tilde{\mathbf{H}}_k^\ell = \left[\nabla_{\mathbf{x}_k} \mathbf{h}_k^T(\mathbf{x}_k, r_k^\ell)\right]^T. \tag{4.62}$$

In summary, the computation of the conditional information matrix $\mathbf{J}_k^\ell$ can be done recursively, using the given regime sequence. The corresponding conditional CRLB is then defined as

$$\text{CRLB}^\ell(\mathbf{x}_k) \triangleq [\mathbf{J}_k^\ell]^{-1}. \tag{4.63}$$

Remark. The implementation of recursion (4.57) may require us to change the dimension of the information matrix whenever there is a regime switch. This possibility exists if we deal with different state vector sizes for different dynamic models (e.g., the dynamic models are constant velocity and constant acceleration models). In this case, the augmentation or reduction of the information matrix *must* be done in the inverse matrix domain. For example, suppose an $n \times n$ information matrix $\mathbf{J}_k^\ell(i)$, corresponding to model i, has to be augmented to $(n+1) \times (n+1)$ matrix $\mathbf{J}_k^\ell(j)$, corresponding to model j, where $i, j \in \{1, 2, \ldots, r\}$. Let the extra component in the state vector of model j be the $(n+1)$th component. Then

$$\left[\mathbf{J}_k^\ell(j)\right]^{-1} = \begin{bmatrix} \left[\mathbf{J}_k^\ell(i)\right]^{-1} & 0 \\ 0 & \sigma_{n+1}^2 \end{bmatrix} \qquad (4.64)$$

where σ_{n+1} denotes the standard deviation of the additional component in model j.

The unconditional posterior CRLB for the hybrid state estimation problem is defined as the expected value of conditional CRLBs; that is

$$\text{CRLB}(\mathbf{x}_k) \triangleq \mathbb{E}\left\{\text{CRLB}^\ell(\mathbf{x}_k)\right\} \qquad (4.65)$$

where $\text{CRLB}^\ell(\mathbf{x}_k)$ is given by (4.63) and the expectation is taken over the regime sequences. It follows then

$$\text{CRLB}(\mathbf{x}_k) = \sum_{\ell=1}^{s^k} \text{P}(R_k^\ell) \cdot \left[\mathbf{J}_k^\ell\right]^{-1} \qquad (4.66)$$

where $\text{P}(R_k^\ell)$ is given by (4.54) and $\mathbf{J}_k^\ell$ by (4.57). Equation (4.66) represents a type of the posterior CRLB for the general nonlinear filtering problem with switching dynamics. It does not follow from the basic principles, and therefore it is not exact. In addition, its computational complexity grows exponentially with time, and in practice it can be computed only for a small value of k. In doing so, one can exploit the structure of the TPM matrix to approximate the weighted sum in (4.66) and thus simplify the computation[7] [21]. Despite the proposed simplification, however, with the current computer technology it becomes practically impossible to work out the bound for $k > 50$.

4.4.2 Deterministic Trajectory

If we were to setup Monte Carlo simulations for a target tracking scenario with multiple-switching dynamics, we would need to generate at each run the target trajectory $\mathbf{Y}_k = \{\mathbf{y}_0, \mathbf{y}_1, \ldots, \mathbf{y}_k\}$ that is random both due to process noise in (4.52) and due to the random Markov transitions of the regime variable.

However, this is rarely done in practice. Instead, the books on target tracking [16, 22], the well-publicized benchmark (tracking) problem [23], and even most tracking system specifications consider purely *deterministic* trajectories. A typical

---

[7] In a typical TPM for tracking applications, the diagonal elements are much greater than the nondiagonal elements. This means that the regime sequences with a large number of switches are very unlikely and hence can be ignored in the computation of the weighted sum (4.66).

trajectory consists of a few fixed-length nonmaneuvering and maneuvering motion segments. For these types of problems, we can compute a kind of a CRLB which assumes that the regime sequence R_k^* is *known* a priori. By setting the probability $P(R_k^*)$ to 1 and the probabilities of other regime sequences to zero, it follows from (4.66) that $\text{CRLB}(\mathbf{x}_k) = [\mathbf{J}_k^*]^{-1}$, where $\mathbf{J}_k^*$ is computed via (4.57),

$$\mathbf{J}_k^* = \mathbf{D}_{k-1}^{*,22} - \left[\mathbf{D}_{k-1}^{*,12}\right]^T \left(\mathbf{J}_{k-1}^* + \mathbf{D}_{k-1}^{*,11}\right)^{-1} \mathbf{D}_{k-1}^{*,12}, \qquad (4.67)$$

with terms $\mathbf{D}_{k-1}^{*,22}$, $\mathbf{D}_{k-1}^{*,12}$, $\mathbf{D}_{k-1}^{*,11}$ and $\mathbf{D}_{k-1}^{*,12}$ defined accordingly. This bound is conservative (overly optimistic, not achievable by any filter) because it assumes the knowledge of the regime sequence, which the filtering algorithm actually needs to estimate. Nevertheless, the bound has been found very useful in assessing the error performance of bearings-only algorithms for tracking maneuvering targets (see, for example, Chapter 6 and [24]).

4.5 SUMMARY AND FURTHER READING

The chapter presented a derivation of the posterior Cramér-Rao lower bound for nonlinear filtering. The key result was (4.13), which provided a recursive method for the computation of the information matrix. The bound will be used in Part II of this book, whenever possible, being very useful for system design, performance prediction, and algorithm assessment.

The development of posterior CRLBs is still an active area of research and one of the favorite topics in the tracking research community. More recent developments are addressing the (possibly nonlinear) filtering problem where the origin of measurements is unknown. As it was pointed out in Section 1.5, this can happen in the presence of false measurements ($P_{FA} > 0$), missed detections ($P_D < 1$), and multiple targets. Some recent references on this subject are [25, 26, 27, 28]. The posterior CRLB for nonlinear filtering with hard constraints is another topic of interest – a recent attempt to solve this problem was reported in [29]. Finally, a CRLB for target tracking using unthresholded (raw) sensor data is presented in Chapter 11.

References

[1] N. Nehorai and M. Hawkes, "Performance bounds for estimating vector systems," *IEEE Trans. Signal Processing*, vol. 46, no. 6, pp. 1737–1749, 2000.

[2] M. L. Hernandez, T. Kirubarajan, and Y. Bar-Shalom, "Efficient multi-sensor resource management using Cramér-Rao lower bounds," in *Proc. SPIE (Signal and Data Processing of Small Targets)*, vol. 4728, 2002.

[3] I. Leibowicz, P. Nicolas, and L. Ratton, "Radar/ESM tracking of constant velocity target: Comparison of batch (MLE) and EKF performance," in *Proc. 3rd Int. Conf. Information Fusion*, (Paris, France), pp. TuC2-3 – TuC2-8, 2000.

[4] J. P. Helferty and D. R. Mugett, "Optimal observer trajectories for bearings only tracking by minimizing the trace of the Cramer-Rao lower bound," in *Proc. 32nd Conf. Decision and Control*, pp. 936–939, 1993.

[5] J. M. Passerieux and D. V. Cappel, "Optimal observer maneuver for bearings-only tracking," *IEEE Trans. Aerospace Electron. Syst.*, vol. 34, no. 3, pp. 777–788, 1998.

[6] S. Paris and J.-P. L. Cadre, "Planification for terrain aided navigation," in *Proc. 5th Int. Conf. Information Fusion*, (Annapois, MD), July 2002.

[7] T. H. Kerr, "Status of CR-like lower bounds for nonlinear filtering," *IEEE Trans. Aerosp. Electr. Syst.*, vol. 25, pp. 590–601, September 1989.

[8] B. Z. Bobrovsky and M. Zakai, "A lower bound on the estimation error for Markov processes," *IEEE Trans. Automatic Control*, vol. 20, pp. 765–788, 1975.

[9] J. H. Taylor, "The Cramer-Rao estimation error lower bound computation for deterministic nonlinear systems," *IEEE Trans. Automatic Control*, vol. 24, pp. 343–344, April 1979.

[10] J. I. Galdos, "A Cramer-Rao bound for multi-dimensional discrete-time dynamical systems," *IEEE Trans. Automatic Control*, vol. 25, pp. 117–119, February 1980.

[11] B. Z. Bobrovsky, E. Mayer-Wolf, and M. Zakai, "Some classes of global Cramer-Rao bounds," *Ann. Statist.*, vol. 15, no. 4, pp. 1421–1438, 1987.

[12] P. Tichavsky, C. H. Muravchik, and A. Nehorai, "Posterior Cramér-Rao bounds for discrete-time nonlinear filtering," *IEEE Trans. Signal Processing*, vol. 46, pp. 1386–1396, May 1998.

[13] N. Bergman, "Posterior Cramér-Rao bounds for sequential estimation," in *Sequential Monte Carlo Methods in Practice* (A. Doucet, N. de Freitas, and N. J. Gordon, eds.), New York: Springer, 2001.

[14] M. Šimandl, J. Královec, and P. Tichavský, "Filtering, predictive and smoothing Cramér-Rao bounds for discrete-time nonlinear dynamic systems," *Automatica*, vol. 37, pp. 1703–1716, 2001.

[15] H. L. VanTrees, *Detection, Estimation and Modulation Theory*. New York: John Wiley & Sons, 1968.

[16] Y. Bar-Shalom, X. R. Li, and T. Kirubarajan, *Estimation with Applications to Tracking and Navigation*. New York: John Wiley & Sons, 2001.

[17] R. D. Gill and B. Y. Levit, "Applications of the Van Trees inequality: a Bayesian Cramér-Rao bound," *Bernoulli*, vol. 1, pp. 59–79, 1995.

[18] Y. Bar-Shalom and X. R. Li, *Estimation and Tracking*. Norwood, MA: Artech House, 1993.

[19] B. Ristic, S. Arulampalam, and C. Musso, "The influence of communication bandwidth on target tracking with angle only measurements from two platforms," *Signal Processing*, vol. 81, pp. 1801–1811, 2001.

[20] B. D. O. Anderson and J. B. Moore, *Optimal Filtering*. Englewood Cliffs, NJ: Prentice-Hall, 1979.

[21] A. Bessell, B. Ristic, A. Farina, X. Wang, and M. S. Arulampalam, "Error performance bounds for tracking a manoeuvering target," in *Proc. Sixth Int. Conf. on Information Fusion (Fusion 2003)*, (Cairns, Australia), July 2003.

[22] A. Farina and F. A. Studer, *Radar Data Processing*. New York: John Wiley, 1985.

[23] W. D. Blair, G. A. Watson, T. Kirubarajan, and Y. Bar-Shalom, "Benchmark for radar allocation and tracking in ecm," *IEEE Trans. Aerospace and Electronic Systems*, vol. 34, pp. 1097–1113, October 1998.

[24] B. Ristic and S. Arulampalam, "Tracking a maneuvering target using angle-only measurements: Algorithms and performance," *Signal Processing*, vol. 83, pp. 1223–1238, 2003.

[25] M. L. Hernandez, A. D. Marrs, N. J. Gordon, S. R. Maskell, and C. M. Reed, "Cramér-Rao bounds for non-linear filtering with measurement origin uncertainty," in *Proc. 5th Int. Conf. Information Fusion*, vol. 1, (Annapolis, MD), pp. 18–25, 2002.

[26] C. Hue, J.-P. L. Cadre, and P. Pérez, "Performance Analysis of Two Sequential Monte Carlo Methods and Posterior Cramér-Rao Bounds for Multi-Target Tracking," Tech. Rep. 1457, IRISA, Campus de Beaulieu, Rennes, France, April 2002. Also in *Proc. 5th Int. Conf. Inform. Fusion*, Annapolis, MD, 2002.

[27] A. Farina, B. Ristic, and L. Timmoneri, "Cramér-Rao bound for nonlinear filtering with $P_d < 1$ and its application to target tracking," *IEEE Trans. Signal Processing*, vol. 50, pp. 854–867, August 2002.

[28] M. Hernandez, B. Ristic, A. Farina, and L. Timmoneri, "A comparison of two Cramer-Rao bounds for nonlinear filtering with $P_d < 1$," *IEEE Trans. Signal Processing*, (to be published in 2004).

[29] M. L. Hernandez and K. Hermiston, "Performance bounds for GMTI tracking," in *Proc. 6th Int. Conf. on Information Fusion*, (Cairns, Australia), July 2003.

Part II

Tracking Applications

Chapter 5

Tracking a Ballistic Object on Reentry

5.1 INTRODUCTION

Tracking ballistic objects is one of the most extensively studied applications considered by the aerospace engineering community. The ultimate goal of this research is to track, intercept, and destroy the ballistic objects before they hit the ground and cause some damage. A recent survey of ballistic target models [1], for example, lists some 57 unclassified references over the period of more than 30 years. No doubt the interest in the subject is driven by the proliferation of ballistic missiles. According to the Center for Defence and International Security Studies [2], ballistic missiles are currently manufactured by more than 20 and in possession of some 36 countries worldwide.

In addition to surveillance for missile defence, the on-line estimation of kinematic parameters of a ballistic object, reentering the atmosphere, is important for safety against aging satellites. It is known that the number of objects orbiting the Earth has been continuously increasing since the launch of the first satellite in 1957. The old satellites and their pieces, if big enough, can reenter the atmosphere and should be accurately tracked to anticipate their landing points on Earth [3].

The flight of a ballistic object, from launch to impact, is commonly modelled by three different dynamic regimes (phases): the boost phase, the coast (ballistic) phase, and the reentry phase. The boost phase is the rocket-powered endo-atmospheric flight, where the object is subjected to large thrust acceleration, as well as atmospheric drag (air resistance) and Earth gravity. After the rocket burnout, the object is in the ballistic phase. This regime is characterized by unpowered exo-atmospheric flight under gravity alone (free-flight motion). The last and the shortest phase is the reentry, where the object reenters the atmosphere. During this phase two significant forces act on the object, Earth gravity and atmospheric drag. Dynamic models commonly used for each of the three phases are *nonlinear* and hence various forms of nonlinear tracking filters have been proposed in the literature (see for example [1, 4, 5, 6] and references within). In this chapter we limit our analysis to a simplified model of ballistic object motion on reentry. The model is based on [7, 8]. For simplicity and without loss of generality we consider one-dimensional vertical motion, where a ballistic object (characterized by an unknown value of the ballistic coefficient) is falling on a straight line path directly towards a surface based tracking radar. The problem of two-dimensional motion when the value of the ballistic coefficient is known a priori has been studied in [6]. Technically speaking, the target dynamics is described by a nonlinear stochastic differential equation while the observations are in discrete-time [9]. The solution, in the Bayesian framework, is based on a stochastic nonlinear filter. The optimal nonlinear filter for this application would hence require an infinite dimensional system for its realization and consequently a number of approximations have been proposed in the past. The approximations based on the extended Kalman filter (EKF) [8, 10], and the second-order Taylor series based EKF [11] have been traditionally used. More recently, it was claimed that the unscented Kalman filter (UKF) [12] outperforms the EKF in this application.

The chapter adopts a relatively simple but convenient dynamic model of a ballistic object on reentry. Within this framework, it presents the theoretical Cramér-Rao bounds for sequential ballistic target state estimation. Furthermore, the performances of three nonlinear tracking filters, designed for this problem, are compared against each other and against the theoretically derived CRLB. The considered filters include the standard EKF, the UKF, and a particle filter.

5.2 TARGET DYNAMICS AND MEASUREMENTS

When a ballistic target reenters the atmosphere, after having traveled a long distance, its speed is high and the remaining time to ground impact is relatively short. In the example we adopt for analysis, a ballistic target is falling vertically as shown in Figure 5.1. Under the assumption that drag and gravity are the only forces acting on the object, the following differential equations govern its motion [8]:

$$\dot{h} = -v \tag{5.1}$$

$$\dot{v} = -\frac{\rho(h) \cdot g \cdot v^2}{2\beta} + g \tag{5.2}$$

$$\dot{\beta} = 0 \tag{5.3}$$

where h is altitude, v is velocity, $\rho(h)$ is air density, $g = 9.81$ m/s^2 is acceleration due to gravity and β is the ballistic coefficient. Air density, measured in kg/m^3, is modeled as an exponentially decaying function of altitude h: $\rho = \gamma e^{-\eta h}$ with

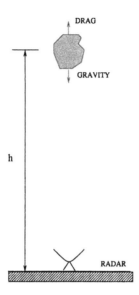

Figure 5.1 Experimental setup for ballistic target tracking.

$\gamma = 1.754$, and $\eta = 1.49 \cdot 10^{-4}$. Ballistic coefficient β depends on the object mass, shape, and cross-sectional area.

Target kinematic state is determined by height and velocity, but since β is unknown it needs to be included in the state vector as well [13]. By selecting the state vector as $\mathbf{x} = [h \ v \ \beta]^T$, the state dynamics in the continuous time t can be expressed by:

$$\dot{\mathbf{x}}_t = \mathbf{g}(\mathbf{x}_t) \tag{5.4}$$

where the expression for the matrix function $\mathbf{g}$ follows from (5.1)–(5.3). The discrete-time version of target dynamics is obtained using the Euler approximation [14] of (5.4) with a very small integration step τ; that is,

$$\begin{aligned}\mathbf{x}_{k+1} &= [\mathbf{x}_k + \tau \, \mathbf{g}(\mathbf{x}_k)] \\ &= \mathbf{f}(\mathbf{x}_k).\end{aligned} \tag{5.5}$$

By incorporating process noise in target dynamics, we finally obtain the state equation in discrete time as:

$$\mathbf{x}_{k+1} = \mathbf{f}(\mathbf{x}_k) + \mathbf{v}_k, \tag{5.6}$$

where

$$\mathbf{f}(\mathbf{x}_k) = \Phi \mathbf{x}_k - \mathbf{G}[D(\mathbf{x}_k) - g] \tag{5.7}$$

with matrices

$$\Phi = \begin{bmatrix} 1 & -\tau & 0 \\ 0 & 1 & 0 \\ 0 & 0 & 1 \end{bmatrix}, \qquad G = [0 \ \tau \ 0]^T$$

and drag

$$D(\mathbf{x}_k) = \frac{g \cdot \rho(\mathbf{x}_k[1]) \cdot \mathbf{x}_k^2[2]}{2\,\mathbf{x}_k[3]}. \tag{5.8}$$

As before, $\mathbf{x}_k[i]$ denotes the ith element of state vector $\mathbf{x}_k$. Note that drag is the only nonlinear term in the state dynamic equation (5.6). Process noise $\mathbf{v}_k$ in (5.6) accounts for imperfections in the kinematic model (e.g., unaccounted forces that act on the ballistic object, such as the lift force, small variations in the ballistic coefficient, and spinning motion). We assume that process noise is zero-mean Gaussian with a covariance matrix approximated by [15]:

$$\mathbf{Q} \approx \begin{bmatrix} q_1 \frac{\tau^3}{3} & q_1 \frac{\tau^2}{2} & 0 \\ q_1 \frac{\tau^2}{2} & q_1 \tau & 0 \\ 0 & 0 & q_2 \tau \end{bmatrix}. \tag{5.9}$$

Parameters q_1 and q_2 control the amount of process noise in target dynamics and the ballistic coefficient, respectively.

A radar, positioned on the surface of Earth below the object, is assumed to measure the target range (in this case object height) at regular intervals of T seconds. The measurement equation is then:

$$z_k = \mathbf{H}\mathbf{x}_k + w_k \tag{5.10}$$

where $\mathbf{H} = [1 \ 0 \ 0]$, and w_k is zero-mean white Gaussian measurement noise with variance $R = \sigma_r^2$, independent from process noise $\mathbf{v}_k$.

An example of kinematic parameters for a typical target trajectory is shown in Figure 5.2. The following parameters were used: initial height of 61 km; initial velocity of 3048 m/s; ballistic coefficient of 19161 kg/ms^2, $\tau = 0.1$ s, and $q_1 = q_2 = 0$. Observe how velocity is almost constant in the first 10 seconds of the ballistic flight on reentry, followed by a sharp drop due to deceleration caused by the air resistance at lower altitudes.

5.3 CRAMÉR-RAO BOUND

The recursive computation of information matrix $\mathbf{J}_k$ in the general case can be done using (4.13). The described problem of tracking a ballistic object on reentry,

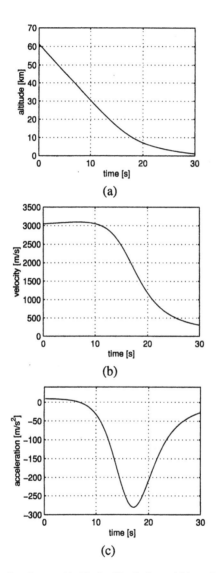

Figure 5.2 An example of a trajectory: (a) altitude, (b) velocity, and (c) acceleration of a ballistic object versus time.

involves additive Gaussian noise. Hence, matrices $\mathbf{D}_k^{11}$, $\mathbf{D}_k^{12}$, and $\mathbf{D}_k^{22}$ can be computed using (4.34), (4.36), and (4.37), respectively. Since the measurement equation is linear, we have:

$$\begin{aligned}\mathbf{D}_k^{11} &= \mathbb{E}\{\tilde{\mathbf{F}}_k^T \mathbf{Q}^{-1} \tilde{\mathbf{F}}_k\} \\ \mathbf{D}_k^{12} &= -\mathbb{E}\{\tilde{\mathbf{F}}_k^T\} \mathbf{Q}^{-1} = [\mathbf{D}_k^{21}]^T \\ \mathbf{D}_k^{22} &= \mathbf{Q}^{-1} + \mathbf{H}^T R^{-1} \mathbf{H}.\end{aligned} \quad (5.11)$$

The substitution of (5.11) into (4.13) then yields:

$$\mathbf{J}_{k+1} = \mathbf{Q}^{-1} + \mathbf{H}^T R^{-1}\mathbf{H} - \mathbf{Q}^{-1}\mathbb{E}\{\tilde{\mathbf{F}}_k\}\left[\mathbf{J}_k + \mathbb{E}\{\tilde{\mathbf{F}}_k^T \mathbf{Q}^{-1}\tilde{\mathbf{F}}_k\}\right]^{-1}\mathbb{E}\{\tilde{\mathbf{F}}_k^T\}\mathbf{Q}^{-1}. \quad (5.12)$$

In the absence of process noise this simplifies to (4.50); that is

$$\mathbf{J}_{k+1} = [\tilde{\mathbf{F}}_k^{-1}]^T \mathbf{J}_k \tilde{\mathbf{F}}_k^{-1} + \mathbf{H}^T R^{-1}\mathbf{H}. \quad (5.13)$$

The initial information matrix $\mathbf{J}_0$ is calculated from the initial density $p(\mathbf{x}_0)$. Assume the Gaussian distribution

$$p(\mathbf{x}_0) = \mathcal{N}\left(\mathbf{x}_0; \hat{\mathbf{x}}_{0|0}, \mathbf{P}_{0|0}\right)$$

where $\hat{\mathbf{x}}_{0|0}$ and $\mathbf{P}_{0|0}$ are the mean and the covariance typically used to initialize the Bayesian recursive estimators. For this initial density we have seen that $\mathbf{J}_0$ equals the inverse of the initial covariance matrix (i.e., $\mathbf{J}_0 = [\mathbf{P}_{0|0}]^{-1}$). In this chapter we will assume two-point differencing for initialization [15, 16] of the first two components of the state vector (height and velocity). The ballistic coefficient will assume a positive value that corresponds to a certain expected value of the body mass. The initial covariance matrix in this case can be written as:

$$\mathbf{P}_{0|0} = \begin{bmatrix} R & R/T & 0 \\ R/T & 2R/T^2 & 0 \\ 0 & 0 & \sigma_\beta^2 \end{bmatrix} \quad (5.14)$$

where σ_β is chosen to cover a range of possible body masses of a ballistic object. We will further discuss the choice of σ_β is Section 5.5.

By differentiation one can obtain the following expression for the Jacobian of $\mathbf{f}(\mathbf{x}_k)$:

$$\tilde{\mathbf{F}}_k = \left[\nabla_{\mathbf{x}_k} \mathbf{f}^T(\mathbf{x}_k)\right]^T \bigg|_{\mathbf{x}_k} = \begin{bmatrix} 1 & -\tau & 0 \\ f_{21}\tau & 1 - f_{22}\tau & f_{23}\tau \\ 0 & 0 & 1 \end{bmatrix} \quad (5.15)$$

where

$$f_{21} = \frac{\eta g \rho_k \mathbf{x}_k^2[2]}{2\mathbf{x}_k[3]} \qquad (5.16)$$

$$f_{22} = \frac{g \rho_k \mathbf{x}_k[2]}{\mathbf{x}_k[3]} \qquad (5.17)$$

$$f_{23} = \frac{g \rho_k \mathbf{x}_k^2[2]}{2\mathbf{x}_k^2[3]} \qquad (5.18)$$

and $\rho_k = \gamma \exp\{-\eta \mathbf{x}_k[1]\}$. Here, as usual, $\mathbf{x}_k[i]$ denotes the ith component of the state vector $\mathbf{x}$ at a discrete-time k.

Figure 5.3 shows the square-root of theoretical CRLBs corresponding to the components of the state vector: (a) altitude, (b) velocity, and (c) the ballistic coefficient. These bounds are obtained as:

$$\text{CRLB}(\mathbf{x}_k[i]) = \mathbf{J}_k^{-1}[i,i], \qquad (i = 1, 2, 3).$$

The CRLBs are calculated for the trajectory shown in Figure 5.2 (using the same initial values) with the following additional parameters: $R = (200\text{m})^2$; $T = \tau = 0.1$ s; $\sigma_\beta = 7184$ kg/ms^2. The bounds in the absence of process noise ($\mathbf{Q} = 0$) are computed via (4.50) and shown by solid lines. In the presence of process noise, the computation of the bounds requires the evaluation of the expectation operator in (5.12); this was implemented by averaging the appropriate terms over 100 Monte Carlo realizations of the state vector. These bounds (at process noise level $q_1 = q_2 = 5$) are plotted with dashed lines in Figure 5.3. Observe how in the interval between 0 and 10 s, $\sqrt{\text{CRLB}}$ is almost constant for the ballistic coefficient and decrease for altitude and velocity. This is due to the absence of drag in a high altitude regime (the state dynamics is practically linear and no additional information available to estimate β). During the interval between 10 and 20 seconds, due to the higher levels of air density, the value of drag (and therefore the degree of nonlinearity) is increasing while the velocity decreases rapidly. The CRLBs for altitude and velocity estimation grow, but the variance bounds for the ballistic coefficient decrease sharply (estimation of β becomes possible). In the last segment, after 20 seconds, all variance bounds decrease. Note that in the presence of process noise, the CRLBs are higher. This effect, however, is very small and becomes noticeable only at the end of the observation interval (the log scale in Figure 5.3 is applied to emphasize this difference).

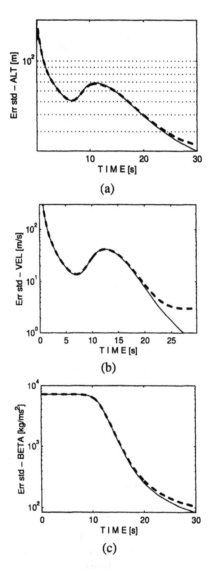

Figure 5.3 $\sqrt{\text{CRLB}}$ for (a) altitude, (b) velocity, and (c) ballistic coefficient (solid line corresponds to $q_1 = q_2 = 0$, dashed line to $q_1 = q_2 = 5$).

5.4 TRACKING FILTERS

Three nonlinear tracking filters are applied to the described problem: EKF, UKF and the particle filter.

Extended Kalman Filter

The EKF equations were given in Section 2.1 and rewritten here for convenience. The prediction is done using:

$$\hat{\mathbf{x}}_{k+1|k} = \mathbf{f}(\hat{\mathbf{x}}_{k|k}) \qquad (5.19)$$
$$\mathbf{P}_{k+1|k} = \hat{\mathbf{F}}_k \mathbf{P}_{k|k} \hat{\mathbf{F}}^T + \mathbf{Q} \qquad (5.20)$$

where $\hat{\mathbf{F}}_k$ is the Jacobian of $\mathbf{f}(\mathbf{x}_k)$ evaluated at the filter estimate $\hat{\mathbf{x}}_{k|k}$. The expression for the Jacobian is given in (5.15), with $\mathbf{x}_k$ replaced by $\hat{\mathbf{x}}_{k|k}$. Since the measurement equation is linear, the filter update is performed using the standard Kalman filter equations given by (1.16)–(1.19), rewritten here in a slightly different form:

$$\mathbf{K}_{k+1} = \mathbf{P}_{k+1|k}\mathbf{H}^T \left[\mathbf{H}\mathbf{P}_{k+1|k}\mathbf{H}^T + R\right]^{-1} \qquad (5.21)$$
$$\hat{\mathbf{x}}_{k+1|k+1} = \hat{\mathbf{x}}_{k+1|k} + \left[z_{k+1} - \mathbf{H}\hat{\mathbf{x}}_{k+1|k}\right] \qquad (5.22)$$
$$\mathbf{P}_{k+1|k+1} = (\mathbf{I} - \mathbf{K}_{k+1}\mathbf{H})\mathbf{P}_{k+1|k}. \qquad (5.23)$$

Unscented Kalman Filter

The standard UKF was described in Section 2.4. For the filtering problem that we consider in this chapter, however, a version of the UKF is developed that applies the unscented transform (Section 2.4.2) only to the state prediction step. The state update is done using the KF equations (5.21)–(5.23), because the measurement equation is linear. Since the state vector dimension is $n_x = 3$, the UKF uses $N = 7$ sample points. The value of the scaling parameter is chosen as $\kappa = 0$.

Particle Filter

The particle filter for the described problem was originally implemented using the SIR algorithm described in Table 3.4, with a "jittering" of resampled particles using the Epanechnikov kernel (Section 3.5.3) in order to prevent the degeneracy of particles [17]. This implementation of the PF was based on $N = 8000$ samples.

Since the measurement equation is linear, we developed another more efficient particle filter for this application. This version of the PF is implemented using the optimal importance density function (see Section 3.4.1):

$$q(\mathbf{x}_k|\mathbf{x}_{k-1},\mathbf{z}_k) = \mathcal{N}(\mathbf{x}_k; \mathbf{a}_k, \mathbf{\Sigma}_k) \qquad (5.24)$$

where the expressions for mean $\mathbf{a}_k$ and covariance $\mathbf{\Sigma}_k$ were given by (3.27) and (3.28), respectively. Recall that with the optimal importance density, the particle weights are updated using (3.21), that is, $w_k^i \propto w_{k-1}^i p(\mathbf{z}_k|\mathbf{x}_{k-1}^i)$, as a convenient special case of (3.16). The expression for density $p(\mathbf{z}_k|\mathbf{x}_{k-1}^i)$ is given in (3.26). An important feature of this type of particle filter is that the resampling step can (and should) be performed before sampling from (5.24). Due to a relatively small amount of process noise, it was necessary to improve the particle diversity after resampling. This was done by regularization, described in Section 3.5.3. A single cycle of the developed PF is described by the pseudocode in Table 5.1.

The particle filter based on the optimal importance density achieves the same error performance (to be described in the next section) as the original SIR particle filter developed for this application. However, the former is using only half the number of particles required by the SIR (i.e., $N = 4000$). This reduction in the number of particles may not appear as dramatic as one would expect, but again this is due to the small amount of process noise involved. Note that the optimal density (3.27) and the transitional prior (3.31) tend to be very similar when process noise is small; in a special case when process noise is zero, they become identical.

5.5 NUMERICAL RESULTS

This section investigates the error performance of the three nonlinear filters. However, before we engage in Monte Carlo simulations, let us identify the regions in the state space where the described tracking problem exhibits a high degree of nonlinearity. In those regions the EKF may introduce large estimation errors while the PF and possibly the UKF may still produce good performance. The only cause of nonlinearity in the dynamic equation (5.6) is the drag, given by (5.8). As a simple heuristic measure of the "degree" of nonlinearity (with respect to the ballistic coefficient) we adopt the modulus of the second partial derivative of the drag[1]:

$$\left|\frac{\partial^2 D}{\partial \beta^2}\right| = \frac{g\,\rho(h)\,v^2}{\beta^3}.$$

This function is plotted in Figure 5.4 for $v = 2000$ m/s, computed along the true trajectory. It appears that there is a higher degree of nonlinearity for smaller values of the ballistic coefficient, that is, for lighter tactical ballistic missiles.

---

1   The main drawback of this heuristic measure is that it does not take into account the support of the pdf. In general, when the support is tight, the nonlinearity appears less severe.

Table 5.1

PF for Tracking a Ballistic Object on Reentry

$[\{\mathbf{x}_k^{i\,*}\}_{i=1}^N] = $ PF-TBOR $[\{\mathbf{x}_{k-1}^i,\}_{i=1}^N, z_k]$

1. Compute normalized importance weights
   - FOR $i = 1 : N$
     - Weights up to a normalizing constant: $\tilde{w}_k^i = \exp\left\{-\frac{1}{2R}[z_k - \mathbf{H}\mathbf{f}(\mathbf{x}_{k-1}^i)]^2\right\}$
   - END FOR
   - FOR $i = 1 : N$
     - Normalize: $w_k^i = \tilde{w}_k^i / $ SUM $[\{\tilde{w}_k^i\}_{i=1}^N]$
   - END FOR

2. Resample using the algorithm in Table 3.2:
   $[\{\mathbf{x}_{k-1}^i, -, -\}_{i=1}^N] = $ RESAMPLE $[\{\mathbf{x}_{k-1}^i, w_k^i\}_{i=1}^N]$

3. Sample from the optimal importance density
   - FOR $i = 1 : N$
     - Draw $\mathbf{x}_k^i \sim \mathcal{N}(\mathbf{x}_k; \mathbf{a}_k^i, \mathbf{\Sigma})$ where
     $$\mathbf{\Sigma} = \mathbf{Q} - \mathbf{Q}\mathbf{H}^T(\mathbf{H}\mathbf{Q}\mathbf{H}^T + R)^{-1}\mathbf{H}\mathbf{Q}$$
     $$\mathbf{a}_k^i = \mathbf{f}(\mathbf{x}_{k-1}^i) + \mathbf{\Sigma}\mathbf{H}^T R^{-1}[z_k - \mathbf{H}\mathbf{f}(\mathbf{x}_{k-1}^i)]$$
   - END FOR

4. Regularization step
   - Calculate the empirical covariance matrix $\mathbf{S}_k$ of $\{\mathbf{x}_k^i\}_{i=1}^N$
   - Compute $\mathbf{D}_k$ such that $\mathbf{D}_k \mathbf{D}_k^T = \mathbf{S}_k$.
   - FOR $i = 1 : N$
     - Draw $\epsilon^i \sim K$ from the Epanechnikov (or Gaussian) kernel
     - $\mathbf{x}_k^{i\,*} = \mathbf{x}_k^i + h_{opt} \mathbf{D}_k \epsilon^i$
   - END FOR

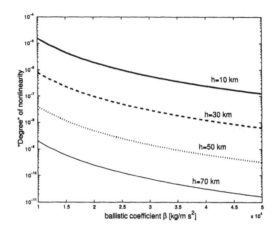

Figure 5.4 A degree of nonlinearity in the drag as a function of the ballistic coefficient and the object height.

The initialization of filters is described next. The initial values of the first two state components (altitude and velocity) are computed using the two-point differencing and assuming the Gaussian initial density. This approach is matched to the computation of the CRLBs as explained below (5.13). The initial value of the ballistic coefficient β_0, however, is not appropriate to model with a Gaussian initial pdf, because it has a finite support. Based on prior knowledge of possible object masses and cross sections (see, for example, [2]), one can identify an interval such that $\beta_0 \in [\beta_L, \beta_U]$. Having the sample space limited in a finite range, the most suitable distribution for modeling β_0 is the beta distribution [18]. Its pdf is given by:

$$p(\beta_0) = \frac{\Gamma(\lambda_1 + \lambda_2)}{(\beta_U - \beta_L)\Gamma(\lambda_1)\Gamma(\lambda_2)} \left(\frac{\beta_0 - \beta_L}{\beta_U - \beta_L}\right)^{\lambda_1 - 1} \left(1 - \frac{\beta_0 - \beta_L}{\beta_U - \beta_L}\right)^{\lambda_2 - 1} \quad (5.25)$$

where λ_1, λ_2 are the shape parameters and symbol Γ represents the Gamma function. In simulations we generate β_0 according to (5.25). Note, however, that both the EKF and the UKF require the initial density to be modeled as Gaussian. In addition, (5.13) for the initial information matrix is valid only for a Gaussian initial pdf. Therefore we need to approximate the density of β_0 by a Gaussian density. The

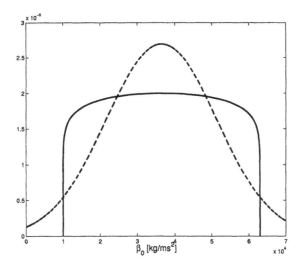

Figure 5.5 The probability density of β_0: Beta (solid line) and approximating Gaussian (dashed line).

mean and standard deviation of this approximating Gaussian is given by [18]:

$$\mathbb{E}[\beta_0] = \beta_L + (\beta_U - \beta_L)\frac{\lambda_1}{\lambda_1 + \lambda_2} \qquad (5.26)$$

$$\text{std}[\beta_0] = \sqrt{\frac{(\beta_U - \beta_L)^2 \lambda_1 \lambda_2}{(\lambda_1 + \lambda_2)^2 (\lambda_1 + \lambda_2 + 1)}} \qquad (5.27)$$

The following parameters are used in simulations: $\lambda_1 = \lambda_2 = 1.1$, $\beta_L = 10000$ kg/ms^2, $\beta_U = 63000$ kg/ms^2. Then $\mathbb{E}[\beta_0] = 36500$ kg/ms^2 and $\text{std}[\beta_0] = 14814$ (kg/ms^2). Figure 5.5 shows the true distribution of β_0 (with the specified parameters) and the approximating Gaussian. The initial "cloud" of particles for the third component of the state vector in the PF is generated according to the true density for β_0. The initialization of the third component of the state vector in the EKF and the UKF is performed as follows:

$$\mathbf{x}_{0|0}[3] = \mathbb{E}[\beta_0]$$
$$\mathbf{P}_{0|0}[3,3] = \sigma_\beta^2 = (\text{std}[\beta_0])^2.$$

The initial information matrix $\mathbf{J}_0$ for the computation of the CRLB is as in (5.13) and (5.14) with $\mathbf{P}_{0|0}[3,3] = \sigma_\beta^2 = (\text{std}[\beta_0])^2$. The error performance comparison is carried out using the ballistic target trajectory with the initial height of 60.96 km, initial velocity of 3048 m/s and with a random initial value of the ballistic coefficient (beta distributed, with the same parameters as in Figure 5.5). The amount of process noise, used both for the trajectory generation and in all three nonlinear filters, is $q_1 = q_2 = 5$.

All error curves corresponding to three filters (EKF, UKF, and PF) were obtained by Monte Carlo simulations with 200 independent runs, assuming perfect detection (i.e., the probability of detection $P_D = 1$ and no false detections, $P_{FA} = 0$). The results for the mean and the standard deviation of the estimation error are shown in Figures 5.6 and 5.7, respectively. Observe from Figure 5.6 that the EKF introduces large bias in the estimation of altitude, velocity, and the ballistic coefficient. The UKF and PF are unbiased in altitude and velocity throughout the observation period; their estimates of the ballistic coefficient are approaching zero bias in the second half of the observation period. The standard deviation of error in altitude and velocity using the EKF is much worse than the theoretical bound in the second half of the observation period, while the UKF and particularly the PF are very close to the theoretical CRLB bound. Although the UKF and the PF demonstrate similar accuracy, the errors are smaller using the PF; in fact the PF appears to be an efficient estimator for this application (unbiased with standard deviation practically equal to $\sqrt{\text{CRLB}}$).

In the considered example the initial density of the ballistic coefficient was selected deliberately to be very broad. This case corresponds to the situation where prior knowledge of the type of the ballistic object one can expect is very poor. We have shown that the EKF is not a suitable algorithm in these situations (biased with a large variance). If we somehow know a priori the type of the ballistic object to expect, we can make the initial pdf narrower. If this narrow density is concentrated in the region of higher values of the ballistic coefficient, then all three filters would demonstrate a similar error performance. This case was investigated in [6].

In relation to the computational load of the three algorithms the following results were obtained (based on MATLAB implementation). The UKF requires 5 times more CPU time than the EKF. The PF requires 100 times more CPU time than the EKF.

5.6 CONCLUDING REMARKS

The problem of tracking a ballistic object on reentry with an unknown ballistic coefficient has been studied. The Cramér-Rao lower bound has been derived and the performance of three nonlinear filters has been investigated. The fact that the ballistic coefficient of the target is unknown makes this problem very challenging. The simulation results indicate that the EKF is fast but unreliable; it is characterized by a significant bias in altitude and ballistic coefficient. Both the UKF and PF show similar error performance in this application; they are unbiased and very close to

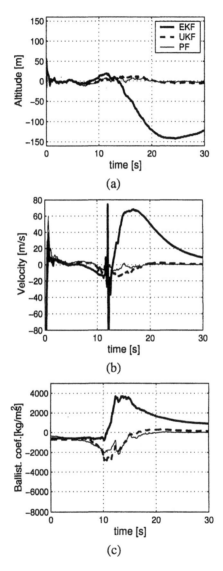

Figure 5.6 Mean error in estimating: (a) altitude, (b) velocity, and (c) ballistic coefficient (EKF - solid thick line; UKF - dashed line; PF - solid thin line).

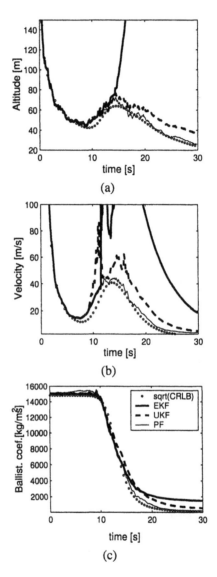

Figure 5.7 Standard deviation of estimation error (a) altitude, (b) velocity, and (c) ballistic coefficient (EKF - solid thick line; UKF - dashed line; PF - solid thin line; $\sqrt{\text{CRLB}}$ - circles).

the theoretical CRLB. The PF, however, is slightly more accurate than the UKF (although the UKF is faster) and could be a realistic option in practice, given its excellent statistical performance.

In general, when prior knowledge of one of the components in the state vector is very imprecise, the divergence problems in the EKF can be avoided using the static multiple model estimator described in Section 2.3.1. This filter would use several EKFs in parallel, each with a different initial value of the ballistic coefficient. This principle will be further explored in Chapters 6 and 7, in the context of other applications.

In this chapter the posterior CRLB has been applied to establish the best achievable error performance. Recently, the CRLB has been explored as a tool to predict the accuracy in estimation of the impact point, based on a segment of the ballistic trajectory measured by a radar [19]. This accuracy depends on target/sensor geometry, measurement accuracy, and the quality of the initial state estimate (usually supplied by another cueing sensor).

References

[1] X. R. Li and V. P. Jilkov, "A survey of maneuvering target tracking, Part II: Ballistic target models," in *Proc. SPIE*, vol. 4473, pp. 559–581, 2001.

[2] Centre for Defence and International Security Studies, http://www.cdiss.org/bmthreat.htm, *Ballistic Missiles by Producer Country*.

[3] S. Chamberlain and T. Slauenwhite, "United States space command space surveillance network overview," in *Proc. 1st European Conf. on Space Debris*, (Darmstad), pp. 37–42, April 1993.

[4] M. E. Hough, "Nonlinear recursive filter for boost trajectories," *Journal of Guidance, Control and Dynamics*, vol. 24, pp. 991–997, September-October 2001.

[5] J. R. V. Zandt, "Boost phase tracking with an unscented filter," in *Proc. SPIE*, vol. 4728, 2002.

[6] A. Farina, B. Ristic, and D. Benvenuti, "Tracking a ballistic target: comparison of several non linear filters," *IEEE Trans. Aerospace and Electronic Systems*, vol. 38, no. 3, pp. 854–867, 2002.

[7] A. Gelb, *Applied Optimal Estimation*. Cambridge, MA: MIT Press, 1974.

[8] P. Zarchan, *Tactical and Strategic Missile Guidance*, vol. 157 of *Progress in Astronautics and Aeronautics*. Washington, D.C.: AIAA, 2nd ed., 1994.

[9] A. H. Jazwinski, *Stochastic Processes and Filtering Theory*. New York: Academic Press, 1970.

[10] R. K. Mehra, "A comparison of several nonlinear filters for reentry vehicle tracking," *IEEE Trans. Automatic Control*, vol. AC-16, no. 4, pp. 307–319, 1971.

[11] M. Athans, R. P. Wishner, and A. Bertolini, "Suboptimal state estimation for continuous-time nonlinear systems from discrete noise measurements," *IEEE Trans. Automatic Control*, vol. AC-13, pp. 504–515, October 1968.

[12] S. Julier, J. Uhlmann, and H. F. Durrant-White, "A new method for nonlinear transformation of means and covariances in filters and estimators," *IEEE Trans. Automatic Control*, vol. 45, pp. 477–482, March 2000.

[13] G. P. Cardillo, A. V. Mrstik, and T. Plambeck, "A track filter for reentry objects with uncertain drag," *IEEE Trans. Aerospace and Electronic Systems*, vol. 35, pp. 394–409, April 1999.

[14] P. E. Kloeden and E. Platen, *Numerical Solution of Stochastic Differential Equations*. New York: Springer, 1995.

[15] Y. Bar-Shalom, X. R. Li, and T. Kirubarajan, *Estimation with Applications to Tracking and Navigation*. New York: John Wiley & Sons, 2001.

[16] A. Farina and F. A. Studer, *Radar Data Processing*. New York: John Wiley, 1985.

[17] B. Ristic, A. Farina, D. Benvenuti, and S. Arulampalam, "Performance bounds and comparison of nonlinear filters for tracking a ballistic object on reentry," *IEE Proc.-Radar Sonar Navig.*, vol. 150, pp. 65–70, April 2003.

[18] K. Bury, *Statistical Distributions in Engineering*. Cambridge, U.K.: Cambridge University Press, 1999.

[19] A. Farina, D. Benvenuti, and B. Ristic, "Estimation accuracy of a landing point of a ballistic target," in *Proc. 5th Int. Conf. Information Fusion*, vol. 1, (Annapolis, MD), pp. 1–8, 2002.

Chapter 6

Bearings-Only Tracking

6.1 INTRODUCTION

The problem of bearings-only tracking arises in a variety of important practical applications. Typical examples are submarine tracking (using a passive sonar) or aircraft surveillance (using a radar in a passive mode or an electronic warfare device) [1, 2, 3]. The problem is sometimes referred to as target motion analysis (TMA), and its objective is to track the kinematics (typically position and velocity) of a moving target using noise-corrupted bearing measurements. In the case of autonomous TMA (single observer only), which is the focus of a major part of this chapter, the observation platform needs to maneuver in order to estimate the target range [1, 3]. This requirement for the ownship maneuver and its impact on target state observability have been explored extensively in [4, 5, 6, 7].

Most researchers in the field of bearings-only tracking have concentrated on tracking a nonmaneuvering target. Due to inherent nonlinearity and observability issues, it is difficult to construct a finite-dimensional optimal Bayesian filter even for this relatively simple problem. Various forms of algorithms have been proposed [2, 8, 9, 10, 11], and they mainly fall into two categories: batch processing type and recursive type. The batch processing type involves delayed processing of all measurements – these algorithms tend to be computationally rather demanding and will not be explored in this chapter. The recursive type algorithms have traditionally been based on the Kalman filter and generally show instability and filter divergence, particularly in highly nonlinear scenarios (e.g., close target-observer encounters).

As for the bearings-only tracking of a maneuvering target, the problem is much more difficult and so far very limited research has been published in the open literature. For example, an interacting multiple model (IMM) based tracker was proposed in [12, 13] for this problem. These algorithms employ a constant velocity (CV) model along with maneuver models to capture the dynamic behavior of a maneuvering target scenario. Le Cadre and Tremois [14] modeled the maneuvering target using the CV model with process noise and developed a tracking filter in the hidden Markov model framework.

This chapter presents the application of both particle filters (PFs) and conventional (KF-based) filters for autonomous bearings-only tracking of nonmaneuvering and maneuvering targets. The error performance of the developed filters is analyzed by Monte Carlo simulations and compared to the theoretical Cramér-Rao lower bounds (CRLBs).

In addition to the autonomous bearings-only tracking problem, two further cases are investigated in the chapter: (1) multiple sensor bearings-only tracking, and (2) tracking with hard constraints. The multiple sensor bearings-only problem involves a slight modification to the original problem where a second static sensor provides bearing measurements to the original platform. The problem of tracking with hard constraints involves the use of prior knowledge, such as speed constraints, to improve tracker performance.

The organization of the chapter is as follows. Section 6.2 presents the mathematical formulation for the bearings-only tracking problem for each of the four different cases investigated: (1) nonmaneuvering case, (2) maneuvering case, (3) multiple sensor case, and (4) tracking with hard constraints. In Section 6.3 the relevant CRLBs are derived for all but case (4) for which the analytic derivation is difficult (due to the non-Gaussian prior and process noise vectors). The tracking algorithms for each case are then presented in Section 6.4 followed by simulation results in Section 6.5.

6.2 PROBLEM FORMULATION

6.2.1 Nonmaneuvering Case

Conceptually, the basic problem in angle-only tracking is to estimate the trajectory of a target (i.e., position and velocity) from noise-corrupted sensor angle data. For the case of a single-sensor problem, this angle data is obtained from a single moving observer (ownship). We define the problem mathematically in the Cartesian coordinates by considering a typical target-observer encounter depicted in Figure 6.1. The target, located at coordinates (x^t, y^t) moves with a nearly constant velocity vector $(\dot{x}^t, \dot{y}^t)$ and is defined to have the state vector

$$\mathbf{x}^t = [x^t \ y^t \ \dot{x}^t \ \dot{y}^t]^T.$$

The ownship state is similarly defined as

$$\mathbf{x}^o = [x^o \ y^o \ \dot{x}^o \ \dot{y}^o]^T,$$

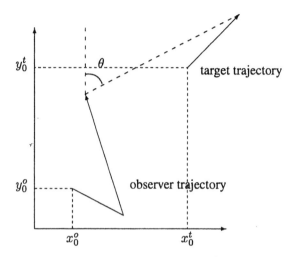

Figure 6.1 A typical two-dimensional target-observer geometry (θ is bearing).

where the velocity vector is typically constant for periods termed "legs." By introducing the relative state vector defined by

$$\mathbf{x} \triangleq \mathbf{x}^t - \mathbf{x}^o = [x \ y \ \dot{x} \ \dot{y}]^T,$$

the discrete-time state equation for this problem can be written as

$$\mathbf{x}_{k+1} = \mathbf{F}_k \mathbf{x}_k + \mathbf{\Gamma}_k \mathbf{v}_k - \mathbf{U}_{k,k+1}, \tag{6.1}$$

where

$$\mathbf{F}_k = \begin{bmatrix} 1 & 0 & T & 0 \\ 0 & 1 & 0 & T \\ 0 & 0 & 1 & 0 \\ 0 & 0 & 0 & 1 \end{bmatrix}, \quad \mathbf{\Gamma}_k = \begin{bmatrix} T^2/2 & 0 \\ 0 & T^2/2 \\ T & 0 \\ 0 & T \end{bmatrix}, \tag{6.2}$$

T is the sampling interval, $\mathbf{v}_k$ is a 2×1 i.i.d. process noise vector with $\mathbf{v}_k \sim \mathcal{N}(\mathbf{0}, \mathbf{Q})$, and

$$\mathbf{U}_{k,k+1} = \begin{bmatrix} u_{k_1} \\ u_{k_2} \\ u_{k_3} \\ u_{k_4} \end{bmatrix} = \begin{bmatrix} x^o_{k+1} - x^o_k - T\dot{x}^o_k \\ y^o_{k+1} - y^o_k - T\dot{y}^o_k \\ \dot{x}^o_{k+1} - \dot{x}^o_k \\ \dot{y}^o_{k+1} - \dot{y}^o_k \end{bmatrix} \tag{6.3}$$

is a vector of deterministic inputs that account for the effects of observer accelerations. Here, $\mathbf{Q} = \sigma_a I_2$, where σ_a is a scalar and I_2 is the 2×2 identity matrix.

Vector $\mathbf{U}_{k,k+1}$ is assumed to be known at every instant of time since the observer state vector $\mathbf{x}_k^o$ is usually provided by an on-board inertial navigation system (INS) aided by a global positioning system (GPS).

The available measurement at time k is the angle from the observer's platform to the target, referenced (clockwise positive) to the y-axis (i.e., angle between y-axis and line of sight; see Figure 6.1), and is given by

$$z_k = h(\mathbf{x}_k) + w_k \qquad (6.4)$$

where w_k is a zero mean independent Gaussian noise with variance σ_θ^2 and

$$h(\mathbf{x}_k) = \arctan\left(\frac{x_k}{y_k}\right) \qquad (6.5)$$

is the true bearing angle.[1]

Given a sequence of measurements z_k, $k = 1, 2, \ldots$, defined by (6.4) and (6.5), and target motion model described in (6.1), (6.2), and (6.3), the angle-only tracking problem is to obtain estimates of the state vector $\mathbf{x}_k$.

There are two features to note about the described tracking problem. First, the problem is nonlinear as the measurements in (6.4) are nonlinearly related to the state vector. Second, the state is unobservable for certain target-observer geometries. To track the target with angle-only measurements, it has been found [1, 6] that the observer must "outmaneuver" the target. Although certain observer maneuvers will optimize tracking performance in terms of reduced estimation errors, this subject (optimization of observer trajectory) will not be explored in this chapter.

6.2.2 Maneuvering Case

The dynamics of a maneuvering target is modeled by multiple switching regimes (also known as a jump Markov system), as described in Section 1.4. We make the assumption that at any time in the observation period, the target motion obeys one of $s = 3$ dynamic behavior models: (1) CV motion model, (2) clockwise coordinated turn (CT) model, and (3) anticlockwise CT model. Let $S \triangleq \{1, 2, 3\}$ denote the set of three models for the dynamic motion, and let r_k be the regime (mode) variable in effect in the interval $(k-1, k]$. Then, the target dynamics can be mathematically written as

$$\mathbf{x}_{k+1} = \mathbf{f}(\mathbf{x}_k, \mathbf{x}_k^o, \mathbf{x}_{k+1}^o, r_{k+1}) + \mathbf{\Gamma}_k \mathbf{v}_k \qquad (6.6)$$

---

[1] Strictly speaking, bearing angle is measured from the north (N) or south (S) ends of the reference meridian, while azimuth is measured clockwise positive from north. Thus, azimuths of 135° and 330° correspond to bearings of S45°E and N30°W, respectively (E-east, W-west).

where $r_{k+1} \in S$, $\mathbf{\Gamma}_k$ and $\mathbf{v}_k$ were defined earlier, and $\mathbf{f}(\cdot, \cdot, \cdot, r_{k+1})$ is the mode-conditioned transition function given by:

$$\mathbf{f}(\mathbf{x}_k, \mathbf{x}_k^o, \mathbf{x}_{k+1}^o, r_{k+1}) = \mathbf{F}^{(r_{k+1})}(\mathbf{x}_k) \cdot (\mathbf{x}_k + \mathbf{x}_k^o) - \mathbf{x}_{k+1}^o. \quad (6.7)$$

Here $\mathbf{F}^{(r_{k+1})}(\cdot)$ is the transition matrix corresponding to mode r_{k+1}, which, for the particular problem of interest can be specified as follows. $\mathbf{F}^{(1)}(\cdot)$ corresponds to CV motion and is thus given by (6.2), that is

$$\mathbf{F}^{(1)}(\mathbf{x}_k) = \begin{bmatrix} 1 & 0 & T & 0 \\ 0 & 1 & 0 & T \\ 0 & 0 & 1 & 0 \\ 0 & 0 & 0 & 1 \end{bmatrix}. \quad (6.8)$$

The next two transition matrices correspond to coordinated turn transitions (clockwise and anticlockwise, respectively). These are given by

$$\mathbf{F}^{(j)}(\mathbf{x}_k) = \begin{bmatrix} 1 & 0 & \frac{\sin(\Omega_k^{(j)}T)}{\Omega_k^{(j)}} & -\frac{(1-\cos(\Omega_k^{(j)}T))}{\Omega_k^{(j)}} \\ 0 & 1 & \frac{(1-\cos(\Omega_k^{(j)}T))}{\Omega_k^{(j)}} & \frac{\sin(\Omega_k^{(j)}T)}{\Omega_k^{(j)}} \\ 0 & 0 & \cos(\Omega_k^{(j)}T) & -\sin(\Omega_k^{(j)}T) \\ 0 & 0 & \sin(\Omega_k^{(j)}T) & \cos(\Omega_k^{(j)}T) \end{bmatrix}, \quad j = 2, 3 \quad (6.9)$$

where the mode-conditioned turning rates are

$$\Omega_k^{(2)} = \frac{a_m}{\sqrt{(\dot{x}_k + \dot{x}_k^o)^2 + (\dot{y}_k + \dot{y}_k^o)^2}}, \quad \Omega_k^{(3)} = \frac{-a_m}{\sqrt{(\dot{x}_k + \dot{x}_k^o)^2 + (\dot{y}_k + \dot{y}_k^o)^2}}. \quad (6.10)$$

Here $a_m > 0$ is a typical maneuver acceleration. Note that the turning rate is expressed as a function of target speed (a nonlinear function of the state vector $\mathbf{x}_k$) and thus models 2 and 3 are clearly nonlinear transitions.

The regime variable r_k is modeled by a time homogeneous three-state first-order Markov chain with transition and initial probabilities defined by (1.36) and (1.38), respectively.

The available measurement at time k is the same as for the case of nonmaneuvering target and so is given by (6.4). The state variable of interest for estimation is the hybrid state vector $\mathbf{y}_k = (\mathbf{x}_k^T, r_k)^T$. Thus, given a sequence of measurements $\mathbf{Z}_k$, and the jump Markov system model (6.6), the problem is to obtain estimates of the hybrid state vector $\mathbf{y}_k$. In particular, we are interested in computing the kinematic state estimate $\hat{\mathbf{x}}_{k|k} = \mathbb{E}[\mathbf{x}_k|\mathbf{Z}_k]$, and mode probabilities $P(r_k = j|\mathbf{Z}_k)$, for every $j \in S$.

6.2.3 Multiple Sensor Case

Suppose there is a possibility of ownship receiving additional (secondary) bearing measurements from a sensor located at (x_k^s, y_k^s). For simplicity, we assume that (1) additional measurements are synchronous[2] to the primary sensor measurements; and (2) there is a zero transmission delay from the sensor at (x_k^s, y_k^s) to the ownship at (x_k^o, y_k^o). The secondary measurement can be modeled as

$$z_k' = h'(\mathbf{x}_k) + w_k' \tag{6.11}$$

where

$$h'(\mathbf{x}_k) = \arctan\left(\frac{x_k + x_k^o - x_k^s}{y_k + y_k^o - y_k^s}\right) \tag{6.12}$$

and w_k' is a zero-mean white Gaussian noise sequence with variance $\sigma_{\theta'}^2$. If the additional bearing measurement is not received at time k, we set $z_k' = \emptyset$. The bearings-only tracking problem for this multisensor case is then to estimate the state vector $\mathbf{x}_k$ given a sequence of measurements $\mathbf{Z}_k = \{z_1, z_1', \ldots, z_k, z_k'\}$.

6.2.4 Tracking with Constraints

In many tracking problems, there are some hard constraints in the state vector that can be a valuable source of information in the estimation process. For example, we may know the minimum and maximum speeds of the target, given by the constraint

$$s_{\min} \leq \sqrt{(\dot{x}_k + \dot{x}_k^o)^2 + (\dot{y}_k + \dot{y}_k^o)^2} \leq s_{\max}. \tag{6.13}$$

Suppose some constraint (such as the speed constraint) is imposed on the state vector, and denote the set of constrained state vectors by Ψ. Let the initial distribution of the state vector in the absence of constraints be $\mathbf{x}_0 \sim p(\mathbf{x}_0)$. With constraints, this initial distribution becomes a truncated density $\tilde{p}(\mathbf{x}_0)$, that is,

$$\tilde{p}(\mathbf{x}_0) = \begin{cases} \dfrac{p(\mathbf{x}_0)}{\int_{\mathbf{x}_0 \in \Psi} p(\mathbf{x}_0) d\mathbf{x}_0}, & \mathbf{x}_0 \in \Psi \\ 0, & \text{otherwise.} \end{cases} \tag{6.14}$$

Likewise, the dynamics model should be modified in such a way that $\mathbf{x}_k$ is always constrained to Ψ. In the absence of hard constraints, suppose that process noise

---

2 Tracking with asynchronous multiplatform bearings measurements is a trivial extension, considered in [15].

$\mathbf{v}_k \sim g(\mathbf{v})$ is used in the filter. With constraints, the pdf of $\mathbf{v}_k$ becomes a state-dependent truncated density given by

$$\tilde{g}(\mathbf{v};\mathbf{x}_k) = \begin{cases} \dfrac{g(\mathbf{v})}{\int_{\mathbf{v}\in G(\mathbf{x}_k)} g(\mathbf{v})d\mathbf{v}}, & \mathbf{v} \in G(\mathbf{x}_k) \\ 0, & \text{otherwise} \end{cases} \quad (6.15)$$

where $G(\mathbf{x}_k) = \{\mathbf{v} : \mathbf{x}_k \in \Psi\}$.

For the bearings-only tracking problem we will consider hard constraints in target dynamics only. The measurement model remains the same as that for the unconstrained problem. Given a sequence of measurements $\mathbf{Z}_k$ and some constraint Ψ on the state vector, the aim is to obtain estimates of the state vector $\mathbf{x}_k$; that is,

$$\begin{aligned}\hat{\mathbf{x}}_{k|k} &= \mathbb{E}\left[\mathbf{x}_k | \mathbf{Z}_k, \Psi\right] \\ &= \int \mathbf{x}_k p(\mathbf{x}_k | \mathbf{Z}_k, \Psi) d\mathbf{x}_k,\end{aligned} \quad (6.16)$$

where $p(\mathbf{x}_k|\mathbf{Z}_k, \Psi)$ is the posterior density of the state, given the measurements and hard constraints.

6.3 CRAMÉR-RAO LOWER BOUNDS

Realistic amounts of process noise in the target trajectory make a very small impact on the Cramér-Rao lower bound for bearings-only tracking [16]. Hence the CRLBs developed in this chapter will assume zero process noise; that is, $\mathbf{v}_k = 0$ in (6.1). This will eliminate the need for the expectation $\mathbb{E}$ with respect to the state vector, as described in Section 4.3.

6.3.1 Nonmaneuvering Case

The CRLBs applicable to the described bearings-only tracking problem are computed using the recursive calculation of the information matrix $\mathbf{J}_k$ given in Section 4.3.3. The key equation is (4.50), repeated here for convenience:

$$\mathbf{J}_{k+1} = \left[\mathbf{F}_k^{-1}\right]^T \mathbf{J}_k \mathbf{F}_k^{-1} + \tilde{\mathbf{H}}_{k+1}^T \mathbf{R}_{k+1}^{-1} \tilde{\mathbf{H}}_{k+1}, \quad (6.17)$$

where $\mathbf{F}_k$ is given by (6.2), $\mathbf{R}_{k+1} = \sigma_\theta^2$ is the variance of the bearing measurements, and

$$\tilde{\mathbf{H}}_{k+1} = \left[\nabla_{\mathbf{x}_{k+1}} h_{k+1}^T(\mathbf{x}_{k+1})\right]^T \tag{6.18}$$

$$= \left[\begin{array}{cccc} \dfrac{y_{k+1}}{x_{k+1}^2 + y_{k+1}^2} & \dfrac{-x_{k+1}}{x_{k+1}^2 + y_{k+1}^2} & 0 & 0 \end{array}\right]. \tag{6.19}$$

The recursion (6.17) is initialized by

$$\mathbf{J}_1 = \mathbf{P}_1^{-1} \tag{6.20}$$

where $\mathbf{P}_1$ is the initial covariance matrix of the state estimate. This can be computed using the expression (6.69), where we replace the first measurement θ_1 by the true value of the initial bearing.

6.3.2 Maneuvering Case

For a deterministic maneuvering target trajectory simulation, a conservative CRLB was developed in Section 4.4. Let

$$R_k^* = \{r_1^*, r_2^*, \ldots, r_k^*\} \tag{6.21}$$

denote the true regime history of the target trajectory. Then, the required CRLB can be computed by using the appropriate transition $\tilde{\mathbf{F}}_{k-1}^{(r_k^*)}$ corresponding to r_k^* at each time index k. In particular, we let

$$\tilde{\mathbf{F}}_k^{(r_{k+1}^*)} \triangleq \left[\nabla_{\mathbf{x}_k} \mathbf{f}^T(\mathbf{x}_k, \mathbf{x}_k^o, \mathbf{x}_{k+1}^o, r_{k+1}^*)\right]^T \tag{6.22}$$

denote the Jacobian of the transition at k corresponding to model r_{k+1}^*. Then, the only change required in the CRLB computations for the maneuvering target case is to replace matrix $\mathbf{F}_k$ with $\tilde{\mathbf{F}}_k^{(r_{k+1}^*)}$ in recursion (6.17):

$$\mathbf{J}_{k+1}^* = \left[\left[\tilde{\mathbf{F}}_k^{(r_{k+1}^*)}\right]^{-1}\right]^T \mathbf{J}_k^* \left[\tilde{\mathbf{F}}_k^{(r_{k+1}^*)}\right]^{-1} + \tilde{\mathbf{H}}_{k+1}^T \mathbf{R}_{k+1}^{-1} \tilde{\mathbf{H}}_{k+1}. \tag{6.23}$$

For $r_{k+1}^* = 1$, the matrix $\tilde{\mathbf{F}}_k^{(r_{k+1}^*)}$ is identical to (6.8). The Jacobians $\tilde{\mathbf{F}}_k^{(r_{k+1}^*)}$, corresponding to maneuver modes $r_{k+1}^* = 2$ and $r_{k+3}^* = 3$, can be computed as follows. Let $f_i^{(j)}$ denote the ith element of the dynamics model function $\mathbf{f}(\cdot, \cdot, \cdot, j)$ and let

$$\Omega_k^{(2)} = \frac{a_m}{\sqrt{(\dot{x}_k^t)^2 + (\dot{y}_k^t)^2}}, \quad \Omega_k^{(3)} = \frac{-a_m}{\sqrt{(\dot{x}_k^t)^2 + (\dot{y}_k^t)^2}} \tag{6.24}$$

where $\dot{x}_k^t$ and $\dot{y}_k^t$ are the target velocity components. Then it can be shown that:

$$\tilde{\mathbf{F}}_k^{(j)} = \begin{bmatrix} 1 & 0 & \frac{\partial f_1^{(j)}}{\partial \dot{x}_k} & \frac{\partial f_1^{(j)}}{\partial \dot{y}_k} \\ 0 & 1 & \frac{\partial f_2^{(j)}}{\partial \dot{x}_k} & \frac{\partial f_2^{(j)}}{\partial \dot{y}_k} \\ 0 & 0 & \frac{\partial f_3^{(j)}}{\partial \dot{x}_k} & \frac{\partial f_3^{(j)}}{\partial \dot{y}_k} \\ 0 & 0 & \frac{\partial f_4^{(j)}}{\partial \dot{x}_k} & \frac{\partial f_4^{(j)}}{\partial \dot{y}_k} \end{bmatrix}, \quad j = 2, 3 \quad (6.25)$$

where

$$\frac{\partial f_1^{(j)}}{\partial \dot{x}_k} = \frac{\sin(\Omega_k^{(j)} T)}{\Omega_k^{(j)}} + g_1^{(j)}(k) \frac{\partial \Omega_k^{(j)}}{\partial \dot{x}_k} \quad (6.26)$$

$$\frac{\partial f_1^{(j)}}{\partial \dot{y}_k} = \frac{-(1 - \cos(\Omega_k^{(j)} T))}{\Omega_k^{(j)}} + g_1^{(j)}(k) \frac{\partial \Omega_k^{(j)}}{\partial \dot{y}_k} \quad (6.27)$$

$$\frac{\partial f_2^{(j)}}{\partial \dot{x}_k} = \frac{(1 - \cos(\Omega_k^{(j)} T))}{\Omega_k^{(j)}} + g_2^{(j)}(k) \frac{\partial \Omega_k^{(j)}}{\partial \dot{x}_k} \quad (6.28)$$

$$\frac{\partial f_2^{(j)}}{\partial \dot{y}_k} = \frac{\sin(\Omega_k^{(j)} T)}{\Omega_k^{(j)}} + g_2^{(j)}(k) \frac{\partial \Omega_k^{(j)}}{\partial \dot{y}_k} \quad (6.29)$$

$$\frac{\partial f_3^{(j)}}{\partial \dot{x}_k} = \cos(\Omega_k^{(j)} T) + g_3^{(j)}(k) \frac{\partial \Omega_k^{(j)}}{\partial \dot{x}_k} \quad (6.30)$$

$$\frac{\partial f_3^{(j)}}{\partial \dot{y}_k} = -\sin(\Omega_k^{(j)} T) + g_3^{(j)}(k) \frac{\partial \Omega_k^{(j)}}{\partial \dot{y}_k} \quad (6.31)$$

$$\frac{\partial f_4^{(j)}}{\partial \dot{x}_k} = \sin(\Omega_k^{(j)} T) + g_4^{(j)}(k) \frac{\partial \Omega_k^{(j)}}{\partial \dot{x}_k} \quad (6.32)$$

$$\frac{\partial f_4^{(j)}}{\partial \dot{y}_k} = \cos(\Omega_k^{(j)} T) + g_4^{(j)}(k) \frac{\partial \Omega_k^{(j)}}{\partial \dot{y}_k} \quad (6.33)$$

with

$$g_1^{(j)}(k) = \frac{T\cos(\Omega_k^{(j)}T)\dot{x}_k^t}{\Omega_k^{(j)}} - \frac{\sin(\Omega_k^{(j)}T)\dot{x}_k^t}{(\Omega_k^{(j)})^2} - \frac{T\sin(\Omega_k^{(j)}T)\dot{y}_k^t}{\Omega_k^{(j)}}$$
$$+ \frac{(-1+\cos(\Omega_k^{(j)}T))\dot{y}_k^t}{(\Omega_k^{(j)})^2} \qquad (6.34)$$

$$g_2^{(j)}(k) = \frac{T\sin(\Omega_k^{(j)}T)\dot{x}_k^t}{\Omega_k^{(j)}} - \frac{(1-\cos(\Omega_k^{(j)}T))\dot{x}_k^t}{(\Omega_k^{(j)})^2} + \frac{T\cos(\Omega_k^{(j)}T)\dot{y}_k^t}{\Omega_k^{(j)}}$$
$$- \frac{\sin(\Omega_k^{(j)}T)\dot{y}_k^t}{(\Omega_k^{(j)})^2} \qquad (6.35)$$

$$g_3^{(j)}(k) = -\sin(\Omega_k^{(j)}T)T\dot{x}_k^t - \cos(\Omega_k^{(j)}T)T\dot{y}_k^t \qquad (6.36)$$
$$g_4^{(j)}(k) = \cos(\Omega_k^{(j)}T)T\dot{x}_k^t - \sin(\Omega_k^{(j)}T)T\dot{y}_k^t \qquad (6.37)$$

$$\frac{\partial \Omega_k^{(j)}}{\partial \dot{x}_k} = \frac{(-1)^{j+1}a_m \dot{x}_k^t}{[(\dot{x}_k^t)^2 + (\dot{y}_k^t)^2]^{3/2}}, \qquad (6.38)$$

$$\frac{\partial \Omega_k^{(j)}}{\partial \dot{y}_k} = \frac{(-1)^{j+1}a_m \dot{y}_k^t}{[(\dot{x}_k^t)^2 + (\dot{y}_k^t)^2]^{3/2}} \qquad (6.39)$$

for $j = 2, 3$.

6.3.3 Multiple Sensor Case

In this section we discuss the computation of the CRLB for the multiple sensor case problem defined in Section 6.2.3. If $z'_{k+1} = \emptyset$, this reduces the problem to a single sensor case. If however, $z'_{k+1} \neq \emptyset$, we define an augmented measurement vector $[z_{k+1}, z'_{k+1}]^T$ and again apply recursion (6.17) to compute the information matrix J_k. This time, however, the Jacobian $\tilde{H}'_{k+1}$ and the measurement error covariance R'_{k+1} are as follows:

$$\tilde{H}'_{k+1} = \left[\nabla_{\mathbf{x}_{k+1}}[h(\mathbf{x}_{k+1})\ h'(\mathbf{x}_{k+1})]\right]^T \qquad (6.40)$$

$$R'_{k+1} = \begin{bmatrix} \sigma_\theta^2 & 0 \\ 0 & \sigma_{\theta'}^2 \end{bmatrix} \qquad (6.41)$$

where $h'(\cdot)$ given by (6.12) is the measurement function applicable to the second sensor, and $\sigma_{\theta'}^2$ is its noise variance. By evaluating the gradient operation in (6.40), we get

$$\tilde{\mathbf{H}}'_{k+1} = \begin{bmatrix} \dfrac{\partial h}{\partial x_{k+1}} & \dfrac{\partial h}{\partial y_{k+1}} & \dfrac{\partial h}{\partial \dot{x}_{k+1}} & \dfrac{\partial h}{\partial \dot{y}_{k+1}} \\ \dfrac{\partial h'}{\partial x_{k+1}} & \dfrac{\partial h'}{\partial y_{k+1}} & \dfrac{\partial h'}{\partial \dot{x}_{k+1}} & \dfrac{\partial h'}{\partial \dot{y}_{k+1}} \end{bmatrix} \quad (6.42)$$

where

$$\frac{\partial h}{\partial x_{k+1}} = \frac{y_{k+1}}{x_{k+1}^2 + y_{k+1}^2}, \quad \frac{\partial h}{\partial y_{k+1}} = \frac{-x_{k+1}}{x_{k+1}^2 + y_{k+1}^2} \quad (6.43)$$

$$\frac{\partial h'}{\partial x_{k+1}} = \frac{y_{k+1} + y_{k+1}^o - y_{k+1}^s}{(x_{k+1} + x_{k+1}^o - x_{k+1}^s)^2 + (y_{k+1} + y_{k+1}^o - y_{k+1}^s)^2} \quad (6.44)$$

$$\frac{\partial h'}{\partial y_{k+1}} = \frac{-(x_{k+1} + x_{k+1}^o - x_{k+1}^s)}{(x_{k+1} + x_{k+1}^o - x_{k+1}^s)^2 + (y_{k+1} + y_{k+1}^o - y_{k+1}^s)^2} \quad (6.45)$$

$$\frac{\partial h}{\partial \dot{x}_{k+1}} = \frac{\partial h}{\partial \dot{y}_{k+1}} = \frac{\partial h'}{\partial \dot{x}_{k+1}} = \frac{\partial h'}{\partial \dot{y}_{k+1}} = 0. \quad (6.46)$$

A few examples of CRLBs are presented in Section 6.5.

6.4 TRACKING ALGORITHMS

This section presents suboptimal algorithms for recursive state estimation of both nonmaneuvering and maneuvering targets, using bearings-only measurements. The modifications necessary to handle the multisensor problem with the incorporation of hard constraints will be also described.

6.4.1 Nonmaneuvering Case

Bearings-only tracking of a nonmaneuvering target is a well-studied problem, with its research originating in the sonar tracking community [2, 3]. Many nonlinear filters such as the EKF in the Cartesian and modified polar (MP) coordinates [9], the pseudolinear estimator [8], maximum likelihood estimator [3], modified gain EKF [10], and, recently, particle filters [15, 17] have been proposed as solutions. In this section, we focus our attention on four specific algorithms: (1) modified polar coordinates EKF, (2) range-parameterized EKF, (3) unscented Kalman filter, and (4) particle filter. Most of these algorithms have been described in general terms in Part I of this book. Hence only the design details necessary for their application to the bearings-only tracking problem are described next.

6.4.1.1 Modified Polar Coordinates EKF

One of the first algorithms employed to solve the angle-only tracking problem was the EKF in Cartesian coordinates [3, 18]. Simulation studies have shown, however, that the EKF in Cartesian coordinates exhibits unstable behavior. To overcome these difficulties, a new EKF was proposed [9], which is formulated in modified polar coordinates. This coordinate system was shown to be well suited for angle-only target motion analysis because it automatically decouples the observable and unobservable components of the estimated state vector. Such decoupling prevents covariance matrix ill-conditioning, which is the primary cause of filter instability and occasional divergence.

The MP state vector is comprised of the following four components: bearing rate, range rate divided by range, bearing, and reciprocal of range. In theory, the first three can be estimated from the single-sensor angular data without a need for an ownship maneuver; the fourth component, however, remains unobservable until a maneuver occurs.

The modified polar coordinates EKF (MP-EKF) is developed by considering the state dynamics and measurement equations in these coordinates. Thus, it is necessary to write down the MP analogs of the Cartesian model equations (6.1) and (6.4). Consider the MP state vector $\mathbf{y}_k$ defined by

$$\mathbf{y}_k = [y_{k_1}\ y_{k_2}\ y_{k_3}\ y_{k_4}]^T$$
$$= \left[\dot{\theta}_k\ \frac{\dot{r}_k}{r_k}\ \theta_k\ \frac{1}{r_k}\right]^T. \quad (6.47)$$

It has been shown [9] that in these coordinates, the equivalent state dynamics equation is given by

$$\mathbf{y}_{k+1} = \mathbf{f}^{mp}(\mathbf{y}_k)$$
$$= \begin{bmatrix} (\alpha_2\alpha_3 - \alpha_1\alpha_4)/(\alpha_1^2 + \alpha_2^2) \\ (\alpha_1\alpha_3 + \alpha_2\alpha_4)/(\alpha_1^2 + \alpha_2^2) \\ y_{k_3} + \tan^{-1}(\alpha_1/\alpha_2) \\ y_{k_4}/(\alpha_1^2 + \alpha_2^2)^{1/2} \end{bmatrix}, \quad (6.48)$$

where α_i, $i = 1, \ldots, 4$ are functions of $\mathbf{y}_k$ and $\mathbf{U}_{k,k+1}$, given by

$$\alpha_1 = Ty_{k_1} - y_{k_4}[u_{k_1}\cos\theta_k - u_{k_2}\sin\theta_k] \quad (6.49)$$
$$\alpha_2 = 1 + Ty_{k_2} - y_{k_4}[u_{k_1}\sin\theta_k + u_{k_2}\cos\theta_k] \quad (6.50)$$
$$\alpha_3 = y_{k_1} - y_{k_4}[u_{k_3}\cos\theta_k - u_{k_4}\sin\theta_k] \quad (6.51)$$
$$\alpha_4 = y_{k_2} - y_{k_4}[u_{k_3}\sin\theta_k - u_{k_4}\cos\theta_k] \quad (6.52)$$

Similarly, the measurement equation can be expressed in the MP coordinates as

$$z_k = \mathbf{H}_y \mathbf{y}_k + w_k \tag{6.53}$$

where $\mathbf{H}_y = [0\ 0\ 1\ 0]$. Equations (6.48) and (6.53) are exact analogs of (6.1) and (6.4), respectively. Observe that in the MP coordinates, the state dynamics equation is nonlinear while the measurement equation is linear, whereas the reverse is true in the Cartesian coordinates. Let us denote the Jacobian (linearized transition matrix) as:

$$\hat{\mathbf{F}}_k^{mp} = \left[\nabla_{\mathbf{y}_k}\left[\mathbf{f}^{mp}(\mathbf{y}_k)\right]^T\right]^T \bigg|_{\mathbf{y}=\hat{\mathbf{y}}_{k|k}} \tag{6.54}$$

whose derivation is given in Section 6.7. Straightforward application of EKF equations (see Section 2.1) to (6.48) and (6.53) will now yield the MP-EKF:

$$\begin{aligned}
\hat{\mathbf{y}}_{k+1|k} &= \mathbf{f}^{mp}\left(\hat{\mathbf{y}}_{k|k}\right) \\
\mathbf{P}_{k+1|k} &= \hat{\mathbf{F}}_k^{mp} \mathbf{P}_{k|k} (\hat{\mathbf{F}}_k^{mp})^T \\
\mathbf{K}_{k+1} &= \mathbf{P}_{k+1} \mathbf{H}_y^T \left[\mathbf{H}_y \mathbf{P}_{k+1|k} \mathbf{H}_y^T + \sigma_\theta^2\right]^{-1} \\
\hat{\mathbf{y}}_{k+1|k+1} &= \hat{\mathbf{y}}_{k+1|k} + \mathbf{K}_{k+1}\left[z_{k+1} - \mathbf{H}_y \hat{\mathbf{y}}_{k+1|k}\right] \\
\mathbf{P}_{k+1|k+1} &= \left[\mathbf{I} - \mathbf{K}_{k+1} \mathbf{H}_y\right] \mathbf{P}_{k+1|k}.
\end{aligned}$$

Once the state estimates have been computed in the MP coordinates, it is necessary to transform them back into the Cartesian coordinates for comparison with other filters. The MP to Cartesian transformation is given by [9]

$$\begin{aligned}
\mathbf{x}_k &= \mathbf{g}^{-1}(\mathbf{y}_k) \\
&= \frac{1}{y_{k_4}} \begin{bmatrix} \sin y_{k_3} \\ \cos y_{k_3} \\ y_{k_2} \sin y_{k_3} + y_{k_1} \cos y_{k_3} \\ y_{k_2} \cos y_{k_3} - y_{k_1} \sin y_{k_3} \end{bmatrix}.
\end{aligned} \tag{6.55}$$

Filter Initialization

For a fair simulation performance comparison (presented in Section 6.5), it is important that all filters have the same (or equivalent) initialization. A Cartesian EKF initialization is adopted as the baseline for all filters. Hence we first present initialization in the Cartesian coordinates, and then derive the MP-EKF initialization from this baseline.

The vector of position components of the state vector, $\mathbf{x}_{pos} = (x,y)^T$ is initialized based on the first bearing measurement θ_1 and the prior knowledge of the initial target range. Suppose the initial range prior is $r \sim \mathcal{N}(\bar{r}, \sigma_r^2)$; that is, $(r - \bar{r}) \sim \mathcal{N}(0, \sigma_r^2)$. Also, since the bearing measurement noise is zero mean Gaussian, we have $(\theta - \theta_1) \sim \mathcal{N}(0, \sigma_\theta^2)$, where θ and θ_1 are the true and measured bearing at time $k = 1$. Now, the x and y components of the Cartesian state vector are related to r and θ by

$$x = r \sin \theta, \quad y = r \cos \theta. \tag{6.56}$$

From the distributions of $(r - \bar{r})$ and $(\theta - \theta_1)$ and the above transformations, it can be shown that an approximate mean and covariance of $\mathbf{x}_{pos}$ are [1, Section 3.7.2]

$$\bar{\mathbf{x}}_{pos} = \begin{bmatrix} \bar{x} \\ \bar{y} \end{bmatrix} = \begin{bmatrix} \bar{r} \sin \theta_1 \\ \bar{r} \cos \theta_1 \end{bmatrix} \tag{6.57}$$

$$\begin{aligned} \mathbf{P}_{pos} &= \mathbb{E}\left[(\mathbf{x}_{pos} - \bar{\mathbf{x}}_{pos})(\mathbf{x}_{pos} - \bar{\mathbf{x}}_{pos})^T\right] \\ &= \begin{bmatrix} P_{xx} & P_{xy} \\ P_{yx} & P_{yy} \end{bmatrix} \end{aligned} \tag{6.58}$$

where

$$\begin{aligned} P_{xx} &= \bar{r}^2 \sigma_\theta^2 \cos^2 \theta_1 + \sigma_r^2 \sin^2 \theta_1 \tag{6.59} \\ P_{yy} &= \bar{r}^2 \sigma_\theta^2 \sin^2 \theta_1 + \sigma_r^2 \cos^2 \theta_1 \tag{6.60} \\ P_{xy} &= P_{yx} = (\sigma_r^2 - \bar{r}^2 \sigma_\theta^2) \sin \theta_1 \cos \theta_1. \tag{6.61} \end{aligned}$$

Similarly, suppose we have some prior knowledge of target speed s and course c given by $s \sim \mathcal{N}(\bar{s}, \sigma_s^2)$ and $c \sim \mathcal{N}(\bar{c}, \sigma_c^2)$, respectively. Now, the velocity component of the state vector, $\mathbf{x}_{vel} = (\dot{x}, \dot{y})^T$, is given by

$$\mathbf{x}_{vel} = \begin{bmatrix} \dot{x} \\ \dot{y} \end{bmatrix} = \begin{bmatrix} \dot{x}^t - \dot{x}^o \\ \dot{y}^t - \dot{y}^o \end{bmatrix} = \begin{bmatrix} s \sin(c) - \dot{x}^o \\ s \cos(c) - \dot{y}^o \end{bmatrix} \tag{6.62}$$

Therefore, using a similar argument to the computations for $\bar{\mathbf{x}}_{pos}$ and $\mathbf{P}_{pos}$, the corresponding velocity component computations yield

$$\bar{\mathbf{x}}_{vel} = \begin{bmatrix} \bar{\dot{x}} \\ \bar{\dot{y}} \end{bmatrix} = \begin{bmatrix} \bar{s} \sin(\bar{c}) - \dot{x}_0^o \\ \bar{s} \cos(\bar{c}) - \dot{y}_0^o \end{bmatrix} \tag{6.63}$$

$$\begin{aligned} \mathbf{P}_{vel} &= \mathbb{E}\left[(\mathbf{x}_{vel} - \bar{\mathbf{x}}_{vel})(\mathbf{x}_{vel} - \bar{\mathbf{x}}_{vel})^T\right] \\ &= \begin{bmatrix} P_{\dot{x}\dot{x}} & P_{\dot{x}\dot{y}} \\ P_{\dot{y}\dot{x}} & P_{\dot{y}\dot{y}} \end{bmatrix} \end{aligned} \tag{6.64}$$

where

$$P_{\dot{x}\dot{x}} = \bar{s}^2 \sigma_c^2 \cos^2(\bar{c}) + \sigma_s^2 \sin^2(\bar{c}) \tag{6.65}$$
$$P_{\dot{y}\dot{y}} = \bar{s}^2 \sigma_c^2 \sin^2(\bar{c}) + \sigma_s^2 \cos^2(\bar{c}) \tag{6.66}$$
$$P_{\dot{x}\dot{y}} = P_{\dot{y}\dot{x}} = (\sigma_s^2 - \bar{s}^2 \sigma_c^2) \sin(\bar{c}) \cos(\bar{c}) \tag{6.67}$$

Now, combining the results of $\mathbf{x}_{pos}$ and $\mathbf{x}_{vel}$, the initialization of the state vector and its covariance in Cartesian coordinates is

$$\hat{\mathbf{x}}_1 = \begin{bmatrix} \bar{x} \\ \bar{y} \\ \bar{\dot{x}} \\ \bar{\dot{y}} \end{bmatrix} = \begin{bmatrix} \bar{r} \sin \theta_1 \\ \bar{r} \cos \theta_1 \\ \bar{s} \sin(\bar{c}) - \dot{x}_0^o \\ \bar{s} \cos(\bar{c}) - \dot{y}_0^o \end{bmatrix} \tag{6.68}$$

$$\mathbf{P}_1 = \begin{bmatrix} P_{xx} & P_{xy} & 0 & 0 \\ P_{yx} & P_{yy} & 0 & 0 \\ 0 & 0 & P_{\dot{x}\dot{x}} & P_{\dot{x}\dot{y}} \\ 0 & 0 & P_{\dot{y}\dot{x}} & P_{\dot{y}\dot{y}} \end{bmatrix} \tag{6.69}$$

where the elements of $\mathbf{P}_1$ are given by (6.59)–(6.61) and (6.65)–(6.67).

The above initialization in the Cartesian coordinates can now be transformed into the MP coordinates to give an equivalent initialization as follows. First, we write down the nonlinear transformation from Cartesian to modified polar coordinates $\mathbf{y} = [\dot{\theta}, \dot{r}/r, \theta, 1/r]^T$ as

$$\mathbf{y} = \mathbf{g}(\mathbf{x})$$
$$= \left[\frac{y\dot{x} - x\dot{y}}{x^2 + y^2}, \frac{x\dot{x} + y\dot{y}}{x^2 + y^2}, \tan^{-1}\left(\frac{x}{y}\right), \frac{1}{\sqrt{x^2 + y^2}} \right]^T \tag{6.70}$$

Then, the mean and covariance of this transformed vector can be approximated as

$$\bar{\mathbf{y}} \approx \mathbf{g}(\mathbb{E}(\mathbf{x})) = \mathbf{g}(\hat{\mathbf{x}}_1) \tag{6.71}$$
$$\mathbf{P}_1^\mathbf{y} = \mathbb{E}\left[(\mathbf{y} - \bar{\mathbf{y}})(\mathbf{y} - \bar{\mathbf{y}})^T\right] \approx \mathbf{G}\mathbf{P}_1\mathbf{G}^T \tag{6.72}$$

where

$$\mathbf{G} = \nabla_\mathbf{x} \left[\mathbf{g}^T(\mathbf{x})\right]^T \bigg|_{\mathbf{x}=\hat{\mathbf{x}}_1} = \begin{bmatrix} \frac{\partial g_1}{\partial x} & \frac{\partial g_1}{\partial y} & \frac{\partial g_1}{\partial \dot{x}} & \frac{\partial g_1}{\partial \dot{y}} \\ \frac{\partial g_2}{\partial x} & \frac{\partial g_2}{\partial y} & \frac{\partial g_2}{\partial \dot{x}} & \frac{\partial g_2}{\partial \dot{y}} \\ \frac{\partial g_3}{\partial x} & \frac{\partial g_3}{\partial y} & \frac{\partial g_3}{\partial \dot{x}} & \frac{\partial g_3}{\partial \dot{y}} \\ \frac{\partial g_4}{\partial x} & \frac{\partial g_4}{\partial y} & \frac{\partial g_4}{\partial \dot{x}} & \frac{\partial g_4}{\partial \dot{y}} \end{bmatrix} \tag{6.73}$$

and $\mathbf{P}_1$ is the initial covariance in Cartesian coordinates. By defining $\bar{r} = \sqrt{\bar{x}^2 + \bar{y}^2}$, the individual elements of $\mathbf{G}$ can be computed as

$$\frac{\partial g_1}{\partial x} = \frac{-\bar{y} - 2\bar{x}\bar{\theta}}{(\bar{r})^2}, \quad \frac{\partial g_1}{\partial y} = \frac{\bar{x} - 2\bar{y}\bar{\theta}}{(\bar{r})^2} \tag{6.74}$$

$$\frac{\partial g_2}{\partial x} = \frac{\bar{x} - 2\bar{x}\left(\frac{\dot{r}}{r}\right)}{(\bar{r})^2}, \quad \frac{\partial g_2}{\partial y} = \frac{\bar{y} - 2\bar{y}\left(\frac{\dot{r}}{r}\right)}{(\bar{r})^2} \tag{6.75}$$

$$\frac{\partial g_1}{\partial \dot{x}} = \frac{\partial g_2}{\partial \dot{y}} = \frac{\partial g_3}{\partial x} = -\frac{\partial g_4}{\partial y} = \frac{\bar{y}}{(\bar{r})^2} \tag{6.76}$$

$$\frac{\partial g_1}{\partial \dot{y}} = \frac{\partial g_3}{\partial y} = \frac{\partial g_4}{\partial x} = -\frac{\partial g_2}{\partial \dot{x}} = \frac{-\bar{x}}{(\bar{r})^2} \tag{6.77}$$

$$\frac{\partial g_3}{\partial \dot{x}} = \frac{\partial g_3}{\partial \dot{y}} = \frac{\partial g_4}{\partial \dot{x}} = \frac{\partial g_4}{\partial \dot{y}} = 0. \tag{6.78}$$

6.4.1.2 Range-Parameterized EKF

The range-parameterized EKF (RP-EKF) is based on the static MM filter described in Section 2.3.1. The basic idea is to use a number of independent EKF trackers in parallel, each with a different initial range estimate [19, 20]. To do so, the range interval of interest is divided into a number of subintervals, and each subinterval is dealt with an independent EKF. Suppose the range interval of interest is $(r_{\min}, r_{\max})$, and we wish to track using N_F EKF filters. For a particular EKF, we note that the tracking performance is highly dependent on the *coefficient of variation* of the range estimate [19], C_r, given by $\sigma_{\hat{r}}/\hat{r}$, where $\hat{r}$ and $\sigma_{\hat{r}}$ are the range estimate and its standard deviation, respectively. In order to maintain a comparable performance for all N_F filters, it is desirable to subdivide the interval $(r_{\min}, r_{\max})$ such that C_r is the same for each subinterval. Note that C_r for each subinterval may be computed approximately as σ_{r_i}/r_i, where r_i is the mean of subinterval i and σ_{r_i} is the range standard deviation for that subinterval. Assuming the range errors to be uniformly distributed in each subinterval, the desirable subdivision can be obtained if the subinterval boundaries are chosen as a geometrical progression. If ρ is the common ratio, we have the relation

$$r_{\max} = r_{\min}\rho^{N_F},$$

which gives ρ as

$$\rho = \left(\frac{r_{\max}}{r_{\min}}\right)^{1/N_F}.$$

For the above division of range, it is easily established [19] that the coefficient of variation is given by

$$C_r = \frac{\sigma_{r_i}}{r_i} = \frac{2(\rho - 1)}{\sqrt{12}(\rho + 1)}. \tag{6.79}$$

To determine how the state estimates of parallel filters are combined, we need to compute the weights associated with each EKF. According to (2.23), the weight of filter i at time k is given by:

$$w_k^i = \frac{p(z_k|i)\, w_{k-1}^i}{\sum_{j=1}^{N_F} p(z_k|j)\, w_{k-1}^j}, \tag{6.80}$$

where $p(z_k|i)$ is the likelihood of measurement z_k, given that the target originated in subinterval i. If one uses N_F parallel filters, in the absence of prior information about the true target range, all initial weights w_1^i are set to $1/N_F$. Typically, in the RP-EKF one uses EKFs in the Cartesian coordinates. Then, assuming Gaussian statistics, the likelihood $p(z_k|i)$ that features in (6.80) can be computed as [19]

$$p(z_k|i) = \frac{1}{\sqrt{2\pi\sigma_i^2}} \exp\left[-\frac{1}{2}\left(\frac{z_k - \hat{z}_{k|k-1}^i}{\sigma_i}\right)^2\right] \tag{6.81}$$

where $\hat{z}_{k|k-1}^i$ is the predicted angle at k for filter i, and σ_i^2 is the innovation variance for filter i given by

$$\sigma_i^2 = \hat{\mathbf{H}}_k^i \mathbf{P}_{k|k-1}^i \hat{\mathbf{H}}_k^{i\,T} + \sigma_\theta^2. \tag{6.82}$$

Here $\hat{\mathbf{H}}_k^i$ is the Jacobian of nonlinear measurement function $h(\cdot)$, evaluated at the predicted state vector of filter i (i.e., at $\hat{\mathbf{x}}_{k|k-1}^i$):

$$\hat{\mathbf{H}}_k^i = \left[\begin{array}{cccc} \dfrac{\hat{y}_{k|k-1}^i}{(\hat{x}_{k|k-1}^i)^2 + (\hat{y}_{k|k-1}^i)^2} & \dfrac{-\hat{x}_{k|k-1}^i}{(\hat{x}_{k|k-1}^i)^2 + (\hat{y}_{k|k-1}^i)^2} & 0 & 0 \end{array}\right]. \tag{6.83}$$

Similarly, $\mathbf{P}_{k|k-1}^i$ in (6.82) is the predicted covariance for filter i.

If we denote the state estimate coming from filter i by $\hat{\mathbf{x}}_{k|k}$ and its associated covariance by $\mathbf{P}_{k|k}^i$, the combined output of the RP-EKF is computed using the Gaussian mixture formulas (2.26) and (2.27).

The initialization for RP-EKF is carried out according to the initialization techniques discussed in the previous section. In particular, filter i of the RP-EKF is initialized according to (6.68) and (6.69), where $\bar{r}$ and σ_r are replaced by $\bar{r}_i$ and σ_{r_i} that correspond to interval i. For the chosen subdivision of $[r_{\min}, r_{\max}]$, assuming a uniform distribution of range errors in interval i, the mean and variance of range prior applicable to interval $i = 1, \ldots, N_F$ can be evaluated as

$$\bar{r}_i = r_{\min} \frac{(\rho^{i-1} + \rho^i)}{2} \tag{6.84}$$

$$\sigma_{r_i} = r_{\min} \frac{(\rho^i - \rho^{i-1})}{\sqrt{12}}. \tag{6.85}$$

The RP-EKF error performance is reported to be superior to that of a single EKF, for the case of very vague prior information on target range [19]. This is mainly due to a much smaller coefficient of variation used by parallel filters. The performance improvement, however, comes at the expense of an N_F-fold increase in computations, if all the range subintervals are processed throughout. In a majority of target-observer scenarios, however, the weighting of mismatched filters rapidly reduces to zero after the ownship maneuver and hence the corresponding filters can be removed from the tracking process without loss of accuracy.

Note that the described concept of *range parametrization* can be extended to *speed parametrization* if prior knowledge of target speed is vague. Specifically, if the only knowledge on target speed is that it belongs to some interval $(s_{\min}, s_{\max})$, then one can subdivide this into N_F^* subintervals and associate a filter for each interval. Thus, the resultant filter would consist of $N = N_F \times N_F^*$ EKFs. For this formulation, the velocity components of the initial state and covariance will be tailored to correspond to the appropriate speed interval in a similar manner described above for range intervals.

6.4.1.3 The Unscented Kalman Filter

This algorithm, which was described in Section 2.4, is implemented for the bearings-only tracking problem in Cartesian coordinates. For the nonmaneuvering target tracking problem, it uses the unscented transform (Section 2.4.2) only for the measurement prediction that is a nonlinear function. The state vector dimension is $n_x = 4$ and hence $N = 9$ sample points are used. The value of the scaling parameter is chosen as $\kappa = -1$. Finally, for the purpose of fair comparison, the initialization is carried out according to the Cartesian state vector initialization specified by (6.68) and (6.69).

6.4.1.4 Particle Filter

The particle filter (PF) applicable to this problem is based on the SIR particle filter described in Table 3.4 of Section 3.5.1. The filter is initialized by drawing a sample $\mathbf{x}^i$, $i = 1, \ldots, N$ such that $\mathbf{x}_1^i \sim \mathcal{N}(\hat{\mathbf{x}}_1, \mathbf{P}_1)$ where $\hat{\mathbf{x}}_1$ and $\mathbf{P}_1$ are the Cartesian state vector initialization parameters given by (6.68) and (6.69), respectively. Resampling is carried out if the effective sample size $\hat{N}_{eff}$ is below $N_{thr} = N/3$.

Furthermore, if resampling is required, we also apply the regularization step to the resampled particles to provide particle diversity (see Section 3.5.3). Since $n_x = 4$ in this application, constant c_{n_x} that features in (3.51) is equal to $\pi^2/2$.

6.4.2 Maneuvering Target Case

This section describes five recursive algorithms designed for tracking a maneuvering target using bearings-only measurements. Two of the algorithms are IMM-based algorithms and the other three are particle-filter-based schemes. The algorithms considered are: (1) IMM-EKF, (2) IMM-UKF, (3) MMPF, (4) AUX-MMPF, and (5) JMS-PF. Essentially, these algorithms attempt to solve the jump Markov system filtering problem formulated in Section 6.2.2. The details of the algorithms are given below. All filters are initialized assuming $\mathbf{x}_1 \sim \mathcal{N}(\mathbf{x}_1; \hat{\mathbf{x}}_1, \mathbf{P}_1)$ where $\hat{\mathbf{x}}_1$ and $\mathbf{P}_1$ are given by (6.68) and (6.69), respectively.

6.4.2.1 IMM-EKF Algorithm

Recall from Section 2.3.2 that when the target dynamics is described by multiple switching models, the posterior density of the state vector is a mixture density [21]. The goal of the IMM algorithm is to merge all mixture components in such a way that the first and second moments are matched. The main point is that for each dynamic model, a separate filter is used; in the case of IMM-EKF for the maneuvering target problem, we have three parallel EKFs, one for CV motion and the other two for the clockwise and anticlockwise coordinated motion dynamics.

The Jacobian of the measurement function was given in (6.83), where in this case index $i = 1, 2, 3$ corresponds to the index of a model-matched EKF. In addition, dynamic models for regimes 2 and 3 are nonlinear functions of the state $\mathbf{x}_k$ and the Jacobians corresponding to these transformations are given by (6.25). Note however that these two Jacobians are evaluated at the filtered state $\hat{\mathbf{x}}_{k|k}$; they are required for the computation of the prediction covariance $\mathbf{P}^i_{k+1|k}$.

The sources of approximation in the IMM-EKF algorithm are twofold. First, the EKF approximates nonlinear transformations by linear transformations about some operating point. If the nonlinearity is severe or if the operating point is not chosen properly, the resultant approximation can be poor, leading to filter divergence. Second, the IMM approximates the exponentially growing Gaussian mixture with a finite Gaussian mixture. The above two approximations can cause filter instability in certain scenarios.

6.4.2.2 IMM-UKF Algorithm

This algorithm is similar to the IMM-EKF with the main difference being that the model matched EKFs are replaced by model-matched unscented Kalman filters (see Section 2.4). The UKF for model 1 uses the unscented transform (UT) only for the filter update (because only the measurement equation is nonlinear). The UKFs for models 2 and 3 use the UT for both the prediction and the update stage of the filter.

6.4.2.3 Multiple-Model Particle Filter

The details of this algorithm are given in Section 3.5.5. Essentially, the state vector $\mathbf{x}_k$ is augmented to include the regime variable $r_k \in \{1,2,3\}$ that characterizes the dynamic behavior of the target. Thus, at time index k, a sample $\{(\mathbf{x}_k^i, r_k^i)\}_{i=1}^N$ represents the posterior density of the augmented state $\mathbf{y}_k = \begin{bmatrix} \mathbf{x}_k^T, r_k \end{bmatrix}^T$ from which the relevant quantities such as $\hat{\mathbf{x}}_{k|k}$ and mode probabilities $\mathrm{P}\{r_k = j | \mathbf{Z}_k\}$, $j = 1, 2, 3$ can be computed.

6.4.2.4 Auxiliary Multiple-Model Particle Filter

The auxiliary SIR particle filter and the multiple-model particle filter were described in Sections 3.5.2 and 3.5.5, respectively. The auxiliary multiple-model particle filter (AUX-MMPF) [22] is a hybrid of the two and is presented here for completeness. The AUX-MMPF focuses on the characterization of pdf $p(\mathbf{x}_k, i, r_k | \mathbf{Z}_k)$, where i refers to the ith particle at $k - 1$. Then, this density is marginalized to obtain a representation of $p(\mathbf{x}_k | \mathbf{Z}_k)$.

A proportionality for the joint probability density $p(\mathbf{x}_k, i, r_k | \mathbf{Z}_k)$ can be written using Bayes' rule as

$$\begin{aligned}
p(\mathbf{x}_k, i, r_k | \mathbf{Z}_k) &\propto p(z_k | \mathbf{x}_k) p(\mathbf{x}_k, i, r_k | \mathbf{Z}_{k-1}) \\
&= p(z_k | \mathbf{x}_k) p(\mathbf{x}_k | r_k, i, \mathbf{Z}_{k-1}) p(r_k | i, \mathbf{Z}_{k-1}) p(i | \mathbf{Z}_{k-1}) \\
&= p(z_k | \mathbf{x}_k) p(\mathbf{x}_k | \mathbf{x}_{k-1}^i, r_k) p(r_k | r_{k-1}^i) w_{k-1}^i \quad (6.86)
\end{aligned}$$

where $p(r_k | r_{k-1})$ is an element of the transitional probability matrix Π defined by (1.36). To sample directly from $p(\mathbf{x}_k, i, r_k | \mathbf{Z}_k)$ as given by (6.86) is not practical. Hence, we use *importance sampling* (see Section 3.1) to first obtain a sample from a density that closely resembles (6.86), and then weigh the samples appropriately to produce an MC representation of $p(\mathbf{x}_k, i, r_k | \mathbf{Z}_k)$. This can be done by introducing the function $q(\mathbf{x}_k, i, r_k | \mathbf{Z}_k)$ with proportionality

$$q(\mathbf{x}_k, i, r_k | \mathbf{Z}_k) \propto p(z_k | \mu_k^i(r_k)) p(\mathbf{x}_k | \mathbf{x}_{k-1}^i, r_k) p(r_k | r_{k-1}^i) w_{k-1}^i \quad (6.87)$$

where

$$\mu_k^i(r_k) = \mathbb{E}\{\mathbf{x}_k|\mathbf{x}_{k-1}^i, r_k\} \quad (6.88)$$
$$= \mathbf{f}(\mathbf{x}_{k-1}^i, \mathbf{x}_{k-1}^o, \mathbf{x}_k^o, r_k). \quad (6.89)$$

Importance density $q(\mathbf{x}_k, i, r_k|\mathbf{Z}_k)$ differs from (6.86) only in the first factor. Now, we can write $q(\mathbf{x}_k, i, r_k|\mathbf{Z}_k)$ as

$$q(\mathbf{x}_k, i, r_k|\mathbf{Z}_k) = q(i, r_k|\mathbf{Z}_k)\, q(\mathbf{x}_k|i, r_k, \mathbf{Z}_k) \quad (6.90)$$

and define

$$q(\mathbf{x}_k|i, r_k, \mathbf{Z}_k) \triangleq p(\mathbf{x}_k|\mathbf{x}_{k-1}^i, r_k). \quad (6.91)$$

In order to obtain a sample from the density $q(\mathbf{x}_k, i, r_k|\mathbf{Z}_k)$, we first integrate (6.87) with respect to $\mathbf{x}_k$ to get an expression for $q(i, r_k|\mathbf{Z}_k)$,

$$q(i, r_k|\mathbf{Z}_k) \propto p(z_k|\mu_k^i(r_k))\, p(r_k|r_{k-1}^i)\, w_{k-1}^i. \quad (6.92)$$

A random sample can now be obtained from the density $q(\mathbf{x}_k, i, r_k|\mathbf{Z}_k)$ as follows. First, a sample $\{i^j, r_k^j\}_{j=1}^N$ is drawn from the discrete distribution $q(i, r_k|\mathbf{Z}_k)$ given by (6.92). This can be done by splitting each of the N particles at $k-1$ into s groups (s is the number of possible modes), each corresponding to a particular mode. Each of the sN particles is assigned a weight proportional to (6.92), and N points $\{i^j, r_k^j\}_{j=1}^N$ are then sampled from this discrete distribution. From (6.90) and (6.91) it is seen that samples $\{\mathbf{x}_k^j\}_{j=1}^N$ from the joint density $q(\mathbf{x}_k, i, r_k|\mathbf{Z}_k)$ can now be generated from $p(\mathbf{x}_k|\mathbf{x}_{k-1}^{i^j}, r_k^j)$. The resultant triplet sample $\{\mathbf{x}_k^j, i^j, r_k^j\}_{j=1}^N$ is a random sample from the density $q(\mathbf{x}_k, i, r_k|\mathbf{Z}_k)$. To use these samples to characterize the density $p(\mathbf{x}_k, i, r_k|\mathbf{Z}_k)$, we attach the weights w_k^j to each particle, where w_k^j is a ratio of (6.87) and (6.86), evaluated at $\{\mathbf{x}_k^j, i^j, r_k^j\}$; that is,

$$w_k^j = \frac{p(z_k|\mathbf{x}_k^j)\, p(\mathbf{x}_k^j|\mathbf{x}_{k-1}^{i^j}, r_k^j)\, p(r_k^j|r_{k-1}^{i^j})\, w_{k-1}^{i^j}}{p(z_k|\mu_k^{i^j}(r_k))\, p(\mathbf{x}_k^j|\mathbf{x}_{k-1}^{i^j}, r_k^j)\, p(r_k^j|r_{k-1}^{i^j})\, w_{k-1}^{i^j}}$$
$$= \frac{p(z_k|\mathbf{x}_k^j)}{p(z_k|\mu_k^{i^j}(r_k))} \quad (6.93)$$

By defining the augmented vector $\mathbf{y}_k \triangleq (\mathbf{x}_k^T, i, r_k)^T$, we can write down an MC representation of the pdf $p(\mathbf{x}_k, i, r_k|\mathbf{Z}_k)$ as

$$p(\mathbf{x}_k, i, r_k|\mathbf{Z}_k) = p(\mathbf{y}_k) \approx \sum_{j=1}^N w_k^j \delta(\mathbf{y}_k - \mathbf{y}_k^j).$$

Observe that by omitting the $\{i^j, r_k^j\}$ components in the triplet sample, we have a representation of the marginalized density $p(\mathbf{x}_k|\mathbf{Z}_k)$; that is,

$$p(\mathbf{x}_k|\mathbf{Z}_k) \approx \sum_{j=1}^{N} w_k^j \delta(\mathbf{x}_k - \mathbf{x}_k^j).$$

A single cycle of the AUX-MMPF is described in Table 6.1. The AUX-MMPF is initialized by generating N samples $\{\mathbf{x}_k^i\}_{i=1}^{N}$ from the initial density $p(\mathbf{x}_1)$ with uniform initial weights $w_k^i = 1/N$, $i = 1, \ldots, N$.

6.4.2.5 Jump Markov System Particle Filter

The jump Markov system particle filter (JMS-PF) is based on the JMLS particle filter proposed in [23, 24] for a jump Markov linear system (a more efficient version of this PF was reported in [25]). As usual, let

$$\begin{aligned}\mathbf{X}_k &= \{\mathbf{x}_0, \mathbf{x}_1, \ldots, \mathbf{x}_k\} \\ R_k &= \{r_1, \ldots, r_k\}\end{aligned}$$

denote the sequences of states and modes up to time index k. Standard particle filtering techniques focused on the estimation of the pdf of the state vector $\mathbf{x}_k$. However, in the JMS particle filter, we place emphasis on the estimation of the pdf of the mode sequence R_k, given measurements $\mathbf{Z}_k = \{z_1, \ldots, z_k\}$. The density $p(\mathbf{X}_k, R_k|\mathbf{Z}_k)$ can be factorized into

$$p(\mathbf{X}_k, R_k|\mathbf{Z}_k) = p(\mathbf{X}_k|R_k, \mathbf{Z}_k)p(R_k|\mathbf{Z}_k). \quad (6.94)$$

Given a specific mode sequence R_k and measurements $\mathbf{Z}_k$, the first term on the right hand side of (6.94), $p(\mathbf{X}_k|R_k, \mathbf{Z}_k)$, can easily be estimated using an EKF or some other nonlinear filter. Therefore we focus our attention on $p(R_k|\mathbf{Z}_k)$ and for estimation of this density we propose to use a particle filter. This approach where we apply the PF only to a part of the state space is an instance of the Rao-Blackwellization concept, described in Section 3.6.

Using Bayes' rule, we note that

$$p(R_k|\mathbf{Z}_k) = \frac{p(z_k|\mathbf{Z}_{k-1}, R_k)p(r_k|r_{k-1})}{p(z_k|\mathbf{Z}_{k-1})} p(R_{k-1}|\mathbf{Z}_{k-1}). \quad (6.95)$$

Equation (6.95) is equivalent to (1.46) and provides a useful recursion for the estimation of $p(R_k|\mathbf{Z}_k)$ using a particle filter. We describe a general recursive algorithm that generates N particles $\{R_k^i\}_{i=1}^{N}$ at time k, which characterizes the

Table 6.1

Auxiliary Multiple-Model Particle Filter for Maneuvering Target Tracking

---

$[\{\mathbf{x}_k^i, w_k^i\}_{i=1}^N]$ = AUX-MMPF $[\{\mathbf{x}_{k-1}^i, w_{k-1}^i\}_{i=1}^N, \mathbf{z}_k]$

1. Create a discrete importance density
   - FOR $i = 1 : N$
     - FOR $r_k = 1 : s$
       * Compute support points as $\mu_k^i(r_k) = \mathbf{f}(\mathbf{x}_{k-1}^i, \mathbf{x}_{k-1}^o, \mathbf{x}_k^o, r_k)$
       * Compute weights $q(i, r_k | \mathbf{Z}_k) = p(z_k | \mu_k^i(r_k)) \, p(r_k | r_{k-1}^i) \, w_{k-1}^i$
     - END FOR
   - END FOR

2. Draw N samples $\{i^j, r_k^j\}_{j=1}^N$ from the discrete distribution created in step 1 (this can be done by suitably modifying the resampling algorithm given in Table 3.2)

3. Predict the particles selected in Step 2 and compute unnormalized weights
   - FOR $j = 1 : N$
     - $\mathbf{x}_k^j = \mathbf{f}(\mathbf{x}_{k-1}^{i^j}, \mathbf{x}_{k-1}^o, \mathbf{x}_k^o, r_k^j) + \Gamma_k \mathbf{v}_k^j$
     - $\tilde{w}_k^j \propto \dfrac{p(z_k | \mathbf{x}_k^j)}{p(z_k | \mu_k^{i^j}(r_k^j))}$
   - END FOR

4. Normalize weights
   - FOR $j = 1 : N$
     - Normalize: $w_k^j = \tilde{w}_k^j / \text{SUM}[\{\tilde{w}_k^i\}_{i=1}^N]$
   - END FOR

5. Calculate $\hat{N}_{eff}$ using (3.18)

6. IF $\hat{N}_{eff} < N_{thr}$
   - Resample using the algorithm in Table 3.2:
     $[\{\mathbf{x}_k^i, w_k^i, -\}_{i=1}^N]$ = RESAMPLE $[\{\mathbf{x}_k^i, w_k^i\}_{i=1}^N]$

pdf $p(R_k|\mathbf{Z}_k)$. The algorithm requires the introduction of an importance function $q(r_k|\mathbf{Z}_k, R_{k-1})$. Suppose at time $k-1$, we have a set of particles $\{R_{k-1}^i\}_{i=1}^N$ that characterizes the pdf $p(R_{k-1}|\mathbf{Z}_{k-1})$. That is,

$$p(R_{k-1}|\mathbf{Z}_{k-1}) \approx \frac{1}{N} \sum_{i=1}^N \delta(R_{k-1} - R_{k-1}^i). \tag{6.96}$$

Now draw N samples $r_k^i \sim q(r_k|\mathbf{Z}_k, R_{k-1}^i)$. Then, from (6.95) and the principle of importance sampling (see Section 3.1), we can write

$$p(R_k|\mathbf{Z}_k) \approx \sum_{i=1}^N w_k^i \delta(R_k - R_k^i), \tag{6.97}$$

where $R_k^i \equiv \{R_{k-1}^i, r_k^i\}$, and weight

$$w_k^i \propto \frac{p(z_k|\mathbf{Z}_{k-1}, R_k^i) p(r_k^i|r_{k-1}^i)}{q(r_k^i|\mathbf{Z}_k, R_{k-1}^i)}. \tag{6.98}$$

From (6.97) we note that one can perform resampling (if required) to obtain an approximate i.i.d. sample from $p(R_k|\mathbf{Z}_k)$. The recursion can be initialized according to the specified initial state distribution of the Markov chain for r_k; see (1.38).

How do we choose the importance density $q(r_k|\mathbf{Z}_k, R_{k-1})$? A sensible selection criterion is to choose a proposal that minimizes the variance of the importance weights at time k, given R_{k-1} and $\mathbf{Z}_k$. According to this strategy, it was shown in [23] that the optimal importance density is $p(r_k|\mathbf{Z}_k, R_{k-1}^i)$. Now, it is easy to see that this density satisfies

$$p(r_k|\mathbf{Z}_k, R_{k-1}^i) = \frac{p(z_k|\mathbf{Z}_{k-1}, R_{k-1}^i, r_k) p(r_k|r_{k-1}^i)}{p(z_k|\mathbf{Z}_{k-1}, R_{k-1}^i)}. \tag{6.99}$$

Note that $p(r_k|\mathbf{Z}_k, R_{k-1}^i)$ is proportional to the numerator of (6.99) as the denominator is independent of r_k. Also, the term $p(r_k|r_{k-1})$ is simply the Markov chain transitional probability (specified by the transitional probability matrix, $\mathbf{\Pi}$). The term $p(z_k|\mathbf{Z}_{k-1}, R_k)$, which features in the numerator of (6.99), can be approximated by one-step-ahead EKF outputs, that is we can write:

$$p(z_k|\mathbf{Z}_{k-1}, R_k) \approx \mathcal{N}(\nu_k(R_k, \mathbf{Z}_{k-1}); 0, \mathbf{S}_k(R_k, \mathbf{Z}_{k-1})) \tag{6.100}$$

where $\nu_k(\cdot, \cdot)$ and $\mathbf{S}_k(\cdot, \cdot)$ are the mode history conditioned innovation and its covariance, respectively. Thus, $p(r_k|r_{k-1})$ and (6.100) allow the computation of the optimal importance density.

Using (6.99) as the importance density $q(\cdot|\cdot,\cdot)$ in (6.98), we find that the weight

$$w_k^i \propto p(z_k|\mathbf{Z}_{k-1}, R_{k-1}^i). \qquad (6.101)$$

As before, $r_k \in \{1,\ldots,s\}$. Then, the importance weights given above can be computed as

$$w_k^i \propto p(z_k|\mathbf{Z}_{k-1}, R_{k-1}^i) = \sum_{j=1}^{s} p(z_k|\mathbf{Z}_{k-1}, R_{k-1}^i, r_k = j) p(r_k = j|r_{k-1}^i). \qquad (6.102)$$

Note that the computation of the importance weights in (6.102) requires s one-step-ahead EKF innovations and their covariances.

This completes the description of the PF for estimation of the Markov chain distribution $p(R_k|\mathbf{Z}_k)$. As mentioned earlier, given a particular mode sequence, the state estimates are easily obtained using a standard EKF. One cycle of the JMS-PF is given in Table 6.2.

6.4.3 Multiple Sensor Case

This problem was described in Section 6.2.3. The extensions to EKF, UKF, and particle-filter-based trackers for the case of multiple sensor measurements are fairly straightforward given the functional relationship (6.12) of the additional measurement to the target state. Note that tracking will be carried out in the relative coordinate system (relative to ownship) and function (6.12) provides the relationship between this relative state vector and the secondary measurement.

6.4.4 Tracking with Hard Constraints

The problem of bearings-only tracking with hard constraints was described in Section 6.2.4. Recall that for constraint $\mathbf{x}_k \in \Psi$, the state estimate is given by the mean of the posterior density $p(\mathbf{x}_k|\mathbf{Z}_k, \Psi)$. This density cannot be easily constructed by the standard Kalman-filter-based techniques. However, since particle filters make no restrictions on the initial density or the distributions of process and measurement noise, it turns out that $p(\mathbf{x}_k|\mathbf{Z}_k, \Psi)$ can be constructed easily using the particle filters. The only modifications required in the particle filters for the case of constraint $\mathbf{x}_k \in \Psi$ are:

- The initial distribution needs to be $\tilde{p}(\mathbf{x})$ defined in (6.14) and the filter needs to be able to sample from this density.

- In the prediction step, samples are drawn from the constrained process noise density $\tilde{g}(\mathbf{v}; \mathbf{x}_k)$ instead of the standard process noise pdf.

Table 6.2
JMS Particle Filter for Maneuvering Target Tracking

$[\{\hat{x}^i_{k|k}, P^i_{k|k}, R^i_k\}_{i=1}^N]$ = JMS-PF $[\{\hat{x}^i_{k-1|k-1}, P^i_{k-1|k-1}, R^i_{k-1}\}_{i=1}^N, z_k]$

- FOR $i = 1 : N$
    - Construct the sampling distribution for mode r_k given by (6.99)
    - Draw $\tilde{r}^i_k \sim p(r_k|Z_k, R^i_{k-1})$
    - Set $\tilde{R}^i_k \triangleq \{R^i_{k-1}, \tilde{r}^i_k\}$
    - Compute unnormalized importance weights according to (6.102), that is,
    $$\tilde{w}^i_k = p(z_k|Z_{k-1}, R^i_{k-1})$$
- END FOR
- Calculate total weight: t = SUM $[\{\tilde{w}^i_k\}_{i=1}^N]$
- FOR $i = 1 : N$
    - Normalize: $w^i_k = t^{-1}\tilde{w}^i_k$
    - Apply one-step-ahead EKF
    $$[\tilde{\hat{x}}^i_{k|k}, \tilde{P}^i_{k|k}] = \text{EKF-STEP}[\hat{x}^i_{k-1|k-1}, P^i_{k-1|k-1}, \tilde{r}^i_k, z_k]$$
- END FOR
- Resample using the algorithm in Table 3.2:
$[\{\hat{x}^i_{k|k}, -, i^j\}_{j=1}^N]$ = RESAMPLE $[\{\tilde{\hat{x}}^i_{k|k}, w^i_k\}_{i=1}^N]$
- FOR $j = 1 : N$
    - Assign: $P^j_k = \tilde{P}^{i^j}_k$ and $R^j_k = \tilde{R}^{i^j}_k$.
- END FOR

Both changes require the ability to sample from a truncated density. A simple method to sample from a generic truncated density $\tilde{t}(\mathbf{x})$ defined by

$$\tilde{t}(\mathbf{x}) = \begin{cases} \dfrac{t(\mathbf{x})}{\int_{\mathbf{x} \in \Psi} t(\mathbf{x}) d\mathbf{x}}, & \mathbf{x} \in \Psi \\ 0, & \text{otherwise} \end{cases} \qquad (6.103)$$

is as follows. Suppose we can easily sample from $t(\mathbf{x})$. Then, to draw $\mathbf{x} \sim \tilde{t}(\mathbf{x})$, we can use the rejection sampling from $t(\mathbf{x})$, until the condition $\mathbf{x} \in \Psi$ is satisfied. The resulting sample is then distributed according to $\tilde{t}(\mathbf{x})$. This simple technique will be adopted in the modifications required in the particle filter for the constrained problem. With the above modifications, the particle filter leads to a cloud of particles that characterize the posterior density $p(\mathbf{x}_k | \mathbf{Z}_k, \Psi)$, from which the state estimate $\hat{\mathbf{x}}_{k|k}$ and its covariance $\mathbf{P}_{k|k}$ can be obtained.

6.5 SIMULATION RESULTS

In this section, we present a performance comparison of the various tracking algorithms described in the previous section. The comparison will be based on a set of Monte-Carlo (MC) simulations and where possible, the CRLB will be used to indicate the best possible performance that one can expect for a given scenario and a set of parameters. Before proceeding, we give a description of the four performance metrics that will be used in this analysis: (1) the root-mean square (RMS) position error, (2) efficiency η, (3) root time averaged mean square (RTAMS) error, and (4) number of divergent tracks.

To define each of the above performance metrics, let (x_k^i, y_k^i) and $(\hat{x}_k^i, \hat{y}_k^i)$ denote the true and estimated target positions at time k at the ith MC run. Let M be the total number of independent MC runs. Then, the RMS position error at k can be computed as

$$\text{RMS}_k = \sqrt{\dfrac{1}{M} \sum_{i=1}^{M} (\hat{x}_k^i - x_k^i)^2 + (\hat{y}_k^i - y_k^i)^2}. \qquad (6.104)$$

Now, if $\mathbf{J}_k^{-1}[i,j]$ denotes the ijth element of the inverse information matrix for the problem at hand, then the corresponding CRLB for the metric (6.104) can be written as

$$\text{CRLB}(\text{RMS}_k) = \sqrt{\mathbf{J}_k^{-1}[1,1] + \mathbf{J}_k^{-1}[2,2]} \qquad (6.105)$$

The second metric stated above is the efficiency parameter η defined as

$$\eta_k \triangleq \dfrac{\text{CRLB}(\text{RMS}_k)}{\text{RMS}_k} \times 100\% \qquad (6.106)$$

which indicates "closeness" to the CRLB. Thus, $\eta_k = 100\%$ implies an efficient estimator that achieves the CRLB exactly.

For a particular scenario and parameters, the overall performance of a filter is evaluated using the third metric that is the RTAMS error. This is defined as

$$\text{RTAMS} = \sqrt{\frac{1}{(t_{\max} - \ell)M} \sum_{k=\ell+1}^{t_{\max}} \sum_{i=1}^{M} (\hat{x}_k^i - x_k^i)^2 + (\hat{y}_k^i - y_k^i)^2} \qquad (6.107)$$

where $t_{\max}$ is the total number of observations (or time epochs) and ℓ is a time index after which the averaging is carried out. Typically ℓ is chosen to coincide with the end of the first ownship maneuver.

The final metric stated above is the number of divergent tracks. A track is declared divergent if its estimated position error at any time index exceeds a threshold that is set to be 20 km in our simulations. It must be noted that the first three metrics described above are computed only on nondivergent tracks.

6.5.1 Nonmaneuvering Case

The scenario to be investigated, shown in Figure 6.2, involves bearings-only tracking of a submarine from an ownship that is also a submarine. The target, which is initially 5 km away from the ownship, maintains a constant speed of 4 knots (1 knot ≈ 0.514 m/s) and a steady course of $-140°$. Target motion, described by (6.1), in simulations is subjected to a very small amount of process noise with $\sigma_a = 10^{-6}$ km/s^2 (Figure 6.2 illustrates the case with zero process noise). The observer traveling at a fixed speed of 5 knots and an initial course of $140°$ executes a maneuver in the interval from 13 to 17 minutes to attain a new course of $20°$. It maintains this new course for the rest of the observation period, which lasts 30 minutes. Bearing measurements are received every $T = 1$ minute with a nominal accuracy $\sigma_\theta = 1.5°$.

Unless otherwise mentioned, the following nominal filter parameters were used in the simulations. The initial range and speed priors were set to $\sigma_r = 2$ km, $\sigma_s = 2$ knots. The initial course and its standard deviation were set to $\bar{c} = \theta_1 + \pi$ and $\sigma_c = \pi/\sqrt{12}$, where θ_1 is the initial bearing measurement. All Cartesian-based filters were implemented with a small amount of process noise with $\sigma_a = 1.6 \times 10^{-6}$ km/s^2. The particle filter used $N = 5000$ particles with resampling only when $\hat{N}_{eff} < N_{thr}$ where the threshold was set to $N_{thr} = N/3$. Furthermore, if a resampling step is carried out, then the particles were regularized according to the technique discussed in Section 3.5.3. For the range-parameterized EKF we divide the range interval ($r_{\min} = 1, r_{\max} = 25$) km into $N_F = 5$ subintervals and associate each with an EKF. Also, the speed interval ($s_{\min} = 2, s_{\max} = 15$) knots is divided into $N_F^* = N_F = 5$ intervals and we associate the ith speed interval with

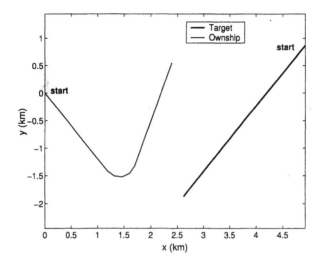

Figure 6.2 A typical bearings-only tracking scenario.

the ith range interval. That is, as in a typical sonar context, we associate targets detected at long range with higher speeds and those detected at short range with lower speeds. This association implies that the total number of filters for the RP-EKF is just N_F rather than $N_F \times N_F^*$. The coefficient of variation for the RP-EKF was set to $C_R = 0.18$.

The first trial examines the RMS position error versus time for the MP-EKF, PF, UKF, and RP-EKF, and compares it with the computed CRLB. The performance results are shown in Figure 6.3 where we see that the PF achieved the best performance with the RMS error very close to the CRLB. A summary of these results are also tabulated in Table 6.3. The first three columns of the table list the metrics corresponding to the performance at the end of the observation period (i.e., error at time epoch $k = 30$ minutes). Note that the second column expresses the RMS error as a percentage of the final range, which for this scenario is 2.43 km. The fourth column gives the RTAM error computed using (6.107). The fifth columns lists the improvement in performance relative to a baseline filter that was chosen to be the MP-EKF. Thus, the improvement factor, based on the RTAMS error, is defined as

$$\text{Improvement} = \frac{\text{RTAMS(filter)} - \text{RTAMS(MP-EKF)}}{\text{RTAMS(MP-EKF)}} \times 100\%. \qquad (6.108)$$

The final column reports the number of divergent tracks in 100 MC runs carried out in the study. From this table we see that except for the MP-EKF that resulted in 11 divergent tracks, the other three algorithms produced no divergences. For this scenario and parameters, the overall performance of the PF and RP-EKF are comparable, achieving about 56% improvement over MP-EKF. Note that although the UKF RMS error performance appears slightly worse than that of MP-EKF, it produced no divergent tracks. Thus its overall performance may be considered better than that of the MP-EKF.

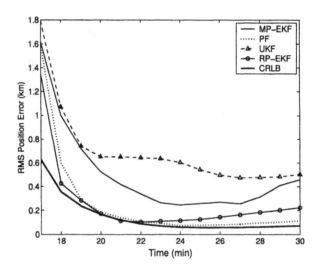

Figure 6.3 RMS position error versus time for a nonmaneuvering scenario.

Table 6.3
Performance Comparison for Nonmaneuvering Case

| Algorithm/ | RMS Error (final) | | | RTAMS | Improvement | Divergent |
CRLB	(km)	(%)	η	(km)	(%)	Tracks
MP-EKF	0.46	19	16	0.47	0	11
PF	0.11	5	64	0.21	56	0
UKF	0.50	21	14	0.63	-34	0
RP-EKF	0.22	9	32	0.20	57	0
CRLB	0.07	3	100	0.14	70	-

The second trial examines the error performance of the particle filter for varying values of the number of particles. Figure 6.4 shows the efficiency at final time, η_{30}, as a function of the number of particles for $100 \leq N \leq 10000$. As expected, we see an improvement in performance as the number of particles is increased. However, note that as N is increased beyond 5000, there is insignificant improvement in performance. Thus we used $N = 5000$ in our simulations, although be aware that the optimal value of N is scenario and parameter dependent.

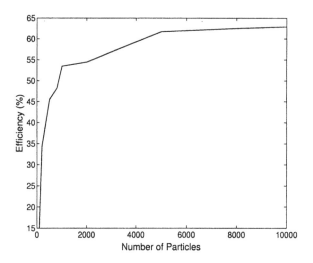

Figure 6.4 RMS Efficiency at the final time versus the number of particles.

The third trial examines the performance of the considered filters as a function of measurement accuracy. Figure 6.5 shows the RMS error curves at the final observation time versus the standard deviation of measurements, for $0.5° \leq \sigma_\theta \leq 6°$. As expected, we note a degradation in performance as σ_θ is increased. The particle filter achieved the best performance in this study, achieving a 5% to 13% RMS error (as the percentage of the final range) as σ_θ is increased from 0.5° to 6°. A detailed comparison of the filters is given in Tables 6.4 and 6.5, which show results for $\sigma_\theta = 0.5°$ and $\sigma_\theta = 6°$, respectively. We see that the MP-EKF resulted in divergent tracks and this increased from 5 to 15 as the measurement accuracy is decreased. For a small value of σ_θ we note that both the PF and the RP-EKF exhibit similar performance followed by UKF and then MP-EKF. Note that as in the earlier case, although the RMS error of the MP-EKF is smaller than that of other filters, since it resulted in divergent tracks while the others did not, the performance of the MP-EKF is the worst in this set of trials. For a large value of σ_θ, we note that except for the PF and UKF, the other filters show a degradation in performance. In particular, the RMS error at the end of the observation period for the PF and UKF was within 13% and 15%, respectively, of the final target range. For the RP-EKF and MP-EKF, the same metric was 33% and 92%, respectively.

The behavior of the EKF-based trackers in the above study can be explained as follows. As σ_θ is increased, the observability of the target decreases, and in particular, the cross-range observability decreases. This decrease in observability means that the chosen operating points for local linearization in these filters are likely to be far from ideal, resulting in erroneous Jacobian $\mathbf{H}$ computations. This in turn affects the estimate of the innovation and state covariances that also affects the gain computations and consequently the state estimates. Furthermore, for the bearings-only tracking problem, as observability decreases, the posterior density becomes more non-Gaussian and so the characterization of this by a Gaussian (or finite Gaussian mixture, in the case of the RP-EKF) becomes less effective. These two reasons explain the rapid degradation in performance of the EKF-based trackers as σ_θ is increased. As for the PF, due to its ability to handle non-Gaussian densities and since it does not incorporate linearization, it does well even as σ_θ is increased.

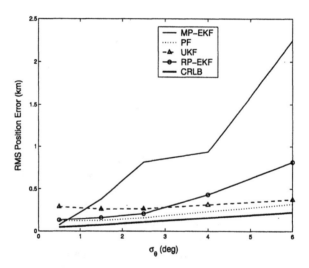

Figure 6.5 RMS error at the final time versus the bearing measurement accuracy.

The next trial examines the performance of the filters as a function of the initial target range. Figure 6.6 shows a comparison of the RMS error at the end of the observation time versus the initial range r_0, where $4\,\text{km} \leq r_0 \leq 12\,\text{km}$. We note that again the PF achieved the best accuracy with no divergent tracks. A detailed comparison of the filters for $r_0 = 4$ km and $r_0 = 12$ km is given in Tables 6.6 and 6.7, respectively. Observe that due to the high degree of nonlinearity at short range, the MP-EKF performance is very poor, resulting in 14 divergent tracks. The particle filter, due to its ability to cope well in nonlinear scenarios, achieves a final

Table 6.4
Performance Comparison for $\sigma_\theta = 0.5°$

| Algorithm/ | RMS Error (final) | | | RTAMS | Improvement | Divergent |
CRLB	(km)	(%)	η	(km)	(%)	Tracks
MP-EKF	0.07	3	64	0.08	0	5
PF	0.12	5	38	0.10	-16	0
UKF	0.29	12	16	0.26	-206	0
RP-EKF	0.13	5	35	0.10	-17	0
CRLB	0.05	2	100	0.06	29	-

Table 6.5
Performance Comparison for $\sigma_\theta = 6°$

| Algorithm/ | RMS Error (final) | | | RTAMS | Improvement | Divergent |
CRLB	(km)	(%)	η	(km)	(%)	Tracks
MP-EKF	2.25	92	10	1.82	0	15
PF	0.32	13	69	0.98	46	0
UKF	0.37	15	60	0.98	46	0
RP-EKF	0.81	33	28	1.72	5	0
CRLB	0.22	9	100	0.45	75	-

RMS error that is within 5% of the final range. As the initial range is increased, the degree of nonlinearity is decreased and hence at $r_0 = 12$ km all filters exhibit similar performance. For this range no divergent tracks were reported for any filter.

The behavior of the EKF-based trackers in the above study can be explained more or less as before. For long-range scenarios, the degree of nonlinearity is less severe and so the EKF-based trackers perform as well as the PF. However, as the target range is decreased, the scenario becomes more nonlinear and so the linear approximations of the EKF-based trackers become less effective. This affects the estimate of the innovation and state covariances that in turn affects gain computations and consequently state estimates. Furthermore, as the degree of nonlinearity increases, the posterior density becomes more non-Gaussian and so the characterization of this by a Gaussian (or finite Gaussian mixture) becomes less effective. For these reasons the performance of the EKF-based trackers at short range is poor. The particle filter, again due to its ability to handle non-Gaussian densities and its ability to handle nonlinear transformations without resorting to linearization, is able to cope well even for the short-range scenarios.

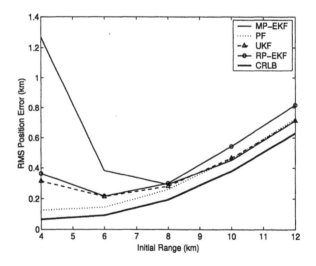

Figure 6.6 RMS efficiency at the final time versus the initial target range.

The final set of trials examines the performance of the filters for various initial densities. Figure 6.7 and Table 6.8 show the performance results for filters initialized with a good (well-concentrated) initial pdf. In particular the range and speed accuracies were set to be $\sigma_r = 0.5$ km and $\sigma_s = 0.5$ knots, respectively. For this case we note that all filters achieved a comparable performance with no divergent tracks.

Table 6.6
Performance Comparison for $r_0 = 4$ km

Algorithm/	RMS Error (final)			RTAMS	Improvement	Divergent
CRLB	(km)	(%)	η	(km)	(%)	Tracks
MP-EKF	1.27	47	5	0.80	0	14
PF	0.13	5	52	0.09	88	0
UKF	0.31	12	21	0.25	68	0
RP-EKF	0.36	13	18	0.22	72	0
CRLB	0.07	2	100	0.06	92	-

Table 6.7
Performance Comparison for $r_0 = 12$ km

Algorithm/	RMS Error (final)			RTAMS	Improvement	Divergent
CRLB	(km)	(%)	η	(km)	(%)	Tracks
MP-EKF	0.71	10	88	1.63	0	0
PF	0.73	10	87	1.26	23	0
UKF	0.71	10	88	1.26	23	0
RP-EKF	0.82	11	77	1.42	13	0
CRLB	0.63	9	100	1.21	26	-

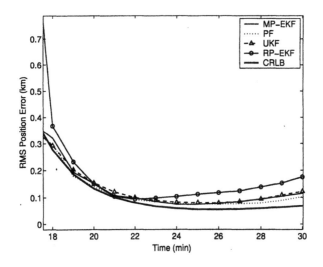

Figure 6.7 RMS position error versus time for the case of a good initial density.

Table 6.8
Performance Comparison for the Case of a Good Initial Density

| Algorithm/ | RMS Error (final) | | | RTAMS | Improvement | Divergent |
CRLB	(km)	(%)	η	(km)	(%)	Tracks
MP-EKF	0.12	5	59	0.14	0	0
PF	0.10	4	67	0.13	7	0
UKF	0.12	5	56	0.14	1	0
RP-EKF	0.18	7	39	0.17	-21	0
CRLB	0.07	3	100	0.12	16	-

Next, we consider the case of a bad initial pdf, and in particular we set $\sigma_r = 5$ km and $\sigma_s = 3$ knots. For this case, from Figure 6.8 and Table 6.9 we see that the MP-EKF shows a rapid degradation in performance with 38 divergent tracks while the particle filter and the RP-EKF achieved excellent error performance. The MP-EKF fails in the case of bad prior because of the poor choice of the operating point about which the linearization is carried out. Second, bad initial pdf implies low observability that in turn results in more non-Gaussian pdfs for this problem, which the MP-EKF approximates with a Gaussian. Since the PF does not use any linearization procedure and is able to deal with non-Gaussian posterior pdfs, it does not suffer from this problem. Observe that the RP-EKF does well in this case despite the low observability. The explanation is as follows. Note that for this case, since σ_r is large, there is a large uncertainty in the line of bearing direction. But the RP-EKF uses multiple filters, and for each filter the effective uncertainty in the line of bearing direction is reduced by N_F times. Hence the RP-EKF is able to cope well in this scenario. Furthermore, note that the initial density parameters σ_r and σ_s are not inputs to the RP-EKF as it only relies on the specification of the minimum and maximum value of target range and speed.

6.5.2 Maneuvering Case

The target-observer geometry for this case is shown in Figure 6.9. The target that is initially 5 km away from ownship maintains an initial course of $-140°$. It executes a maneuver in the interval 20 to 25 minutes to attain a new course of $100°$. It maintains this new course for the rest of the observation period, which lasts 40 minutes. The ownship motion for the first 30 minutes is exactly the same as that for the nonmaneuvering case. It continues with constant speed and course until it starts a second maneuver at $k = 32$ minutes which ends at $k = 35$ minutes and attains

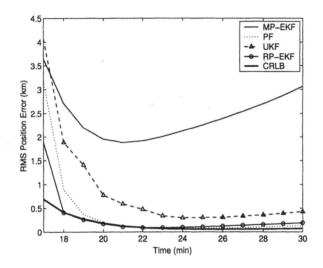

Figure 6.8 RMS position error versus time for the case of a bad initial density.

Table 6.9
Performance Comparison for the Case of a Bad Initial Density

Algorithm/	RMS Error (final)			RTAMS	Improvement	Divergent
CRLB	(km)	(%)	η	(km)	(%)	Tracks
MP-EKF	3.07	126	2	2.39	0	38
PF	0.13	5	55	0.29	88	0
UKF	0.43	18	17	0.77	68	1
RP-EKF	0.19	8	38	0.18	92	0
CRLB	0.07	3	100	0.16	93	-

a new course of 155°. The final target-observer range for this case is 2.91 km. The sampling time $T = 1$ minute and measurement accuracy $\sigma_\theta = 1.5°$.

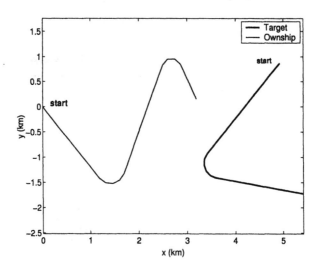

Figure 6.9 A typical bearings-only tracking scenario with a maneuvering target.

The transition probability matrix required for the jump Markov process was chosen to be

$$\Pi = \begin{bmatrix} 0.9 & 0.05 & 0.05 \\ 0.4 & 0.5 & 0.1 \\ 0.4 & 0.1 & 0.5 \end{bmatrix} \tag{6.109}$$

and the typical maneuver acceleration parameter for the filters was set to $a_m = 1.08 \times 10^{-5}$ km/s^2. Unless otherwise mentioned, all other filter parameters were identical to the nonmaneuvering case.

Figure 6.10 shows the RMS error curves corresponding to the five filters designed for the maneuvering case: IMM-EKF, IMM-UKF, MMPF, AUX-MMPF, and JMS-PF. A detailed comparison is also given in Table 6.10. From the graph and the table, it is clear that the performance of the IMM-EKF and IMM-UKF is poor compared to the other three filters. Though the final RMS error performance of the IMM-UKF is comparable to the JMS-PF, since it had one divergent track, its overall performance is considered worse than that of the JMS-PF. It is clear that the best filters for this case were the MMPF and AUX-MMPF, which achieved a 59% and 56% improvement, respectively, over the IMM-EKF. Also note that the JMS-PF performance is between that of IMM-EKF/IMM-UKF and MMPF/AUX-MMPF. The reason for this can be explained as follows. There are two sources of approximations in both IMM-EKF and IMM-UKF. First, the probability of mode history is approximated by the IMM routine that merges mode histories. Second, the mode-conditioned filter estimates are obtained using an EKF and UKF (for the IMM-EKF and IMM-UKF, respectively), both of which approximate the non-Gaussian posterior density by a Gaussian. In contrast, the MMPF and AUX-MMPF attempt to alleviate both sources of approximations: they estimate the mode probabilities with no merging of histories and they make no linearization (as in EKF) and characterize the non-Gaussian posterior density in a near optimal manner. Thus we observe the superior performance of the MMPF and AUX-MMPF. The JMS-PF on the other hand is worse than MMPF/AUX-MMPF but better than IMM-EKF/IMM-UKF as it attempts to alleviate only one of the sources of approximations discussed above. Specifically, while the JMS-PF attempts to compute the probability of mode history exactly, it uses an EKF (a local linearization approximation) to compute the mode-conditioned filtered estimates. It is interesting to note from the improvement figures for the JMS-PF and MMPF that the first source of approximation is more critical than the second one. In fact the contributions of the first and second sources of approximation appears to be in the ratio 2:1.

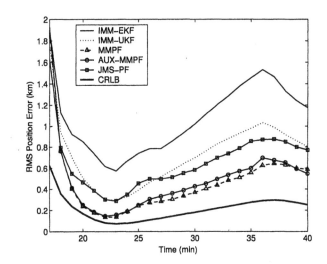

Figure 6.10 RMS position error versus time for a maneuvering target scenario.

Next, we illustrate a case where the IMM-EKF shows a tendency to diverge while the MMPF tracks the target well for the same set of measurements. Figure 6.11(a) shows the estimated track and 95% error ellipses (plotted every 8 minutes) for the IMM-EKF. Note that the IMM-EKF covariance estimate at 8 minutes is poor as it does not encapsulate the true target position. This has resulted in not only subsequent poor track estimates, but also inability to detect the target maneuver.

Table 6.10
Performance Comparison for the Maneuvering Target Case

| Algorithm/ | RMS Error (final) | | | RTAMS | Improvement | Divergent |
CRLB	(km)	(%)	η	(km)	(%)	Tracks
IMM-EKF	1.18	40	22	1.07	0	0
IMM-UKF	0.80	28	32	0.72	32	1
MMPF	0.59	20	43	0.44	59	0
AUX-MMPF	0.55	19	46	0.47	56	0
JMS-PF	0.77	27	33	0.64	40	0
CRLB	0.25	9	100	0.21	80	-

This is clear from the mode probability curves shown in Figure 6.11(b): although there is a slight bump in the mode probability for the correct maneuver model, the algorithm is unable to establish the occurrence of the maneuver. The overall result is a track that is showing tendency to diverge from the true track.

For the same set of measurements, the MMPF shows excellent performance as can be seen from Figure 6.12. Here we note that the 95% confidence ellipse of the PF encapsulates the true target position at all times. Notice that the size of the covariance matrix shortly after the target maneuver is small compared to other times. The reason for this is that the target observability is best at that instant compared to other times. For the given scenario, both the ownship maneuver and the target maneuver have resulted in a geometry that is very observable at that instant. After the target maneuver, the relative position of the target increases and this leads to a slight decrease in observability and hence slight enlargement of the covariance matrix. The mode probability curves for the MMPF shows that unlike the results of IMM-EKF, the MMPF mode probabilities indicate a higher probability of occurrence of a maneuver. The overall result is a much better tracker performance for the same set of measurements.

The next trial examines the robustness of the algorithm to a mismatch in the filter parameter a_m (typical maneuver acceleration) and the true parameter $\bar{a}_m$ corresponding to the scenario. For the particular scenario described for the maneuvering case, the target does a 24°/min maneuver at 4 knots, which corresponds to $\bar{a}_m = 1.44 \times 10^{-5}$km/s^2. Figure 6.13 shows the efficiency at final time for varying values of a_m, in particular for $\pm 50\%$ variations about the true value $\bar{a}_m$. We see that for $\pm 50\%$ variations in a_m, the efficiency varies by only $\pm 12\%$, indicating a fairly robust performance.

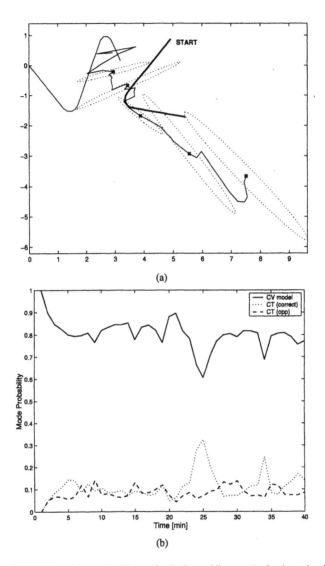

Figure 6.11 IMM-EKF tracker results: (a) x-y plot (units are kilometers) of estimated trajectories and the corresponding 95% confidence ellipses; and (b) mode probabilities.

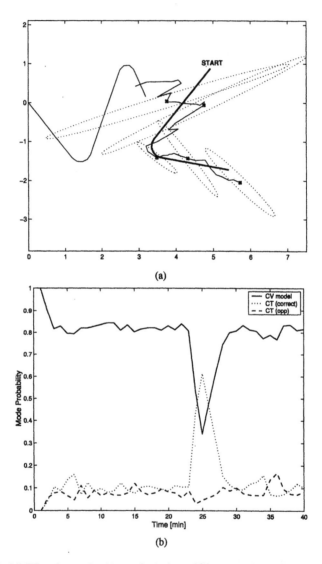

Figure 6.12 MMPF tracker results: (a) x-y plot (units are kilometers) of estimated trajectories and the corresponding 95% confidence ellipses; and (b) mode probabilities.

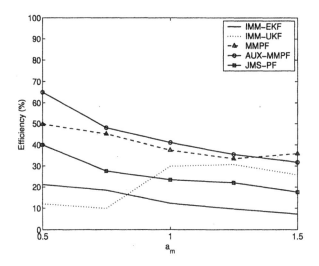

Figure 6.13 Efficiency at the final time versus parameter a_m.

6.5.3 Multiple Sensor Case

Here we consider the identical scenario to the one considered in Section 6.5.2, except that an additional static sensor, located at $(5 \text{ km}, -2 \text{ km})$ provides bearing measurements to the ownship at regular time intervals. These measurements with accuracy $\sigma_{\theta'} = 2°$ arrive at only 3 time epochs, namely at $k = 10, 20$, and 30. Figure 6.14 shows a comparison of IMM-EKF, IMM-UKF, and MMPF for this case. It is seen that the MMPF exhibits excellent performance, with RMS error results very close to the CRLB. The detailed comparison given in Table 6.11 shows that the MMPF achieves a final RMS error accuracy that is within 8% of the final range. Interestingly, the IMM-EKF and IMM-UKF performance is very poor, even worse than their corresponding performance when no additional measurement is received. Though this may seem counterintuitive, it is real and can be explained as follows. For the given geometry, at the time of first arrival of the bearing measurement from the second sensor, it is possible that due to nonlinearities and low observability in the time interval 0 to 10 minutes, the track estimates and filter calculated covariance of the IMM-based filters are in error, leading to a large innovation for the second measurement. The inaccurate covariance estimate results in an incorrect gain, computation for the second sensor measurement. In the update equations of these filters, the large innovation gets weighted by the computed gain, which does not properly reflect the contribution of the new measurement. The consequence of this is filter divergence. It turns out that for the ownship measurements-only case,

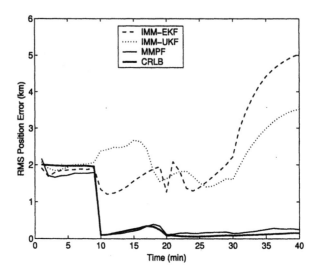

Figure 6.14 RMS position error versus time for a multisensor case.

even if the track and covariance estimates are in error, the errors introduced in the filter gain computation are not as severe as in the multisensor case. Furthermore, as the uncertainty is mainly along the line of bearing, the innovation for this case is not likely to be very large. Thus the severity of track and covariance error for this particular scenario is worse for the multisensor case than for the single-sensor case. Similar results have been observed in the context of an air surveillance scenario in [15].

Table 6.11
Performance Comparison for the Multisensor Case

Algorithm/ CRLB	RMS Error (final) (km)	(%)	η	RTAMS (km)	Improvement (%)	Divergent Tracks
IMM-EKF	5.03	173	3	3.16	0	17
IMM-UKF	3.51	121	4	2.32	27	7
MMPF	0.25	8	63	0.22	93	1
CRLB	0.15	5	100	0.13	96	-

6.5.4 Tracking with Hard Constraints

In this section we present the results for the case of bearings-only tracking with hard constraints. The scenario and parameters used for this case are identical to the ones considered in Section 6.5.2. This time, however, in addition to the available bearing measurements, we also impose some hard constraints on target speed. Specifically, assume that we have prior knowledge that target speed is in the range $3.5 \leq s \leq 4.5$ knots. This type of prior information is difficult to incorporate into the standard EKF-based algorithms (such as the IMM-EKF) and therefore, in the comparison below, the IMM-EKF will not utilize hard constraints information. The particle filter-based algorithms (in particular the MMPF and AUX-MMPF), can easily incorporate such nonstandard information according to the technique described in Section 6.4.4.

Figure 6.15 shows the RMS error in estimated position for the MMPF that incorporates prior knowledge of speed constraint (referred as MMPF-C). The figure also shows the performance curves of the IMM-EKF and the standard MMPF that do not utilize knowledge of hard constraints. A detailed numerical comparison is given in Table 6.12. It can be seen that the MMPF-C achieves 83% improvement in RTAMS over the IMM-EKF. Also, observe that by incorporating the hard constraints, the MMPF-C achieves a 50% reduction in RTAMS error over the standard MMPF that does not utilize hard constraints (emphasizing the significance of this nonstandard prior information). Incorporating such nonstandard information results in highly non-Gaussian posterior pdfs that the particle filter is effectively able to characterize.

Table 6.12
Performance Comparison for Tracking with Hard Constraints

Algorithm	RMS Error (final)		RTAMS	Improvement	Divergent
	(km)	(%)	(km)	(%)	Tracks
IMM-EKF	1.37	47	1.21	0	0
MMPF	0.53	18	0.44	64	0
MMPF-C	0.12	4	0.20	83	0

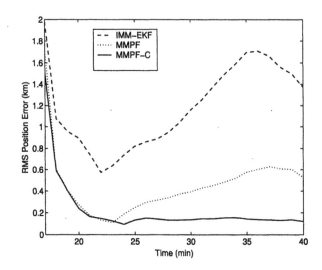

Figure 6.15 RMS position error versus time for the case of tracking with speed constraint $3.5 \leq s \leq 4.5$ knots.

6.6 SUMMARY

This chapter presented a comparative study of the bearings-only tracking problem. Four separate cases have been analyzed: nonmaneuvering target case; maneuvering target case, multiple-sensor case, and tracking with speed constraints. The results overwhelmingly confirm the superior performance of particle-filter-based algorithms against the conventional KF-based and IMM-based schemes. The key strength of the particle filter, demonstrated in this application, is its flexibility to handle nonstandard prior information along with the ability to deal with nonlinearities and non-Gaussianities.

6.7 APPENDIX: LINEARIZED TRANSITION MATRIX FOR MP-EKF

The Jacobian (linearized transition matrix) $\tilde{\mathbf{F}}_k^{mp}$, for the modified polar coordinate EKF, was defined in (6.54) as

$$\hat{\mathbf{F}}_k^{mp} = \left[\nabla_{\mathbf{y}_k} \left[\mathbf{f}^{mp}(\mathbf{y}_k) \right]^T \right]^T \bigg|_{\mathbf{y}=\hat{\mathbf{y}}_{k|k}} \quad (6.110)$$

The right side of (6.110) can be written as [2]

$$\left[\nabla_{\mathbf{y}_k} [\mathbf{f}^{mp}(\mathbf{y}_k)]^T\right]^T\bigg|_{\mathbf{y}=\hat{\mathbf{y}}_{k|k}} = \mathbf{C} + \mathbf{D} \cdot \mathbf{E} \tag{6.111}$$

where the elements of matrices $\mathbf{C}$, $\mathbf{D}$, and $\mathbf{E}$ are given by:

$$\mathbf{C}[i,j] = \frac{\partial f^{mp}[i]}{\partial y_k[j]},$$

$$\mathbf{D}[i,j] = \frac{\partial f^{mp}[i]}{\partial \alpha[j]},$$

$$\mathbf{E}[i,j] = \frac{\partial \alpha[i]}{\partial y_k[j]}.$$

where $\alpha = [\alpha_1 \ \alpha_2 \ \alpha_3 \ \alpha_4]^T$ was defined by (6.49)–(6.52). The elements of matrices $\mathbf{C}$, $\mathbf{D}$, and $\mathbf{E}$ can be evaluated with the result

$$\mathbf{C} = \begin{bmatrix} 0 & 0 & 0 & 0 \\ 0 & 0 & 0 & 0 \\ 0 & 0 & 1 & 0 \\ 0 & 0 & 0 & 1/(\alpha_1^2+\alpha_2^2)^{1/2} \end{bmatrix}, \tag{6.112}$$

$$\mathbf{D} = \begin{bmatrix} d_{11} & -d_{21} & d_{13} & d_{32} \\ d_{21} & d_{11} & -d_{32} & d_{13} \\ d_{31} & d_{32} & 0 & 0 \\ d_{41} & d_{42} & 0 & 0 \end{bmatrix} \tag{6.113}$$

where

$$\begin{aligned}
d_{11} &= [-\alpha_1(\alpha_2\alpha_3 - \alpha_1\alpha_4) - \alpha_2(\alpha_1\alpha_3 + \alpha_2\alpha_4)]/(\alpha_1^2+\alpha_2^2)^2 \\
d_{21} &= [-\alpha_1(\alpha_1\alpha_3 + \alpha_2\alpha_4) + \alpha_2(\alpha_2\alpha_3 - \alpha_1\alpha_4)]/(\alpha_1^2+\alpha_2^2)^2 \\
d_{31} &= \alpha_2/(\alpha_1^2+\alpha_2^2) \\
d_{41} &= -\alpha_1 \hat{y}_{k|k}[4]/(\alpha_1^2+\alpha_2^2)^{3/2} \\
d_{32} &= -\alpha_1/(\alpha_1^2+\alpha_2^2) \\
d_{42} &= -\alpha_2 \hat{y}_{k|k}[4]/(\alpha_1^2+\alpha_2^2)^{3/2} \\
d_{13} &= \alpha_2/(\alpha_1^2+\alpha_2^2)
\end{aligned} \tag{6.114}$$

and

$$\mathbf{E} = \begin{bmatrix} T & 0 & e_{13} & e_{14} \\ 0 & T & e_{23} & e_{24} \\ 1 & 0 & e_{33} & e_{34} \\ 0 & 1 & e_{43} & e_{44} \end{bmatrix} \tag{6.115}$$

where

$$\begin{aligned}
e_{14} &= -\left[u_{k_1}\cos\hat{y}_{k|k}[3] - u_{k_2}\sin\hat{y}_{k|k}[3]\right] \\
e_{24} &= -\left[u_{k_1}\sin\hat{y}_{k|k}[3] + u_{k_2}\cos\hat{y}_{k|k}[3]\right] \\
e_{34} &= -\left[u_{k_3}\cos\hat{y}_{k|k}[3] - u_{k_4}\sin\hat{y}_{k|k}[3]\right] \\
e_{44} &= -\left[u_{k_3}\sin\hat{y}_{k|k}[3] + u_{k_4}\cos\hat{y}_{k|k}[3]\right] \\
e_{13} &= -\hat{y}_{k|k}[4]\, e_{24} \\
e_{23} &= \hat{y}_{k|k}[4]\, e_{14} \\
e_{33} &= -\hat{y}_{k|k}[4]\, e_{44} \\
e_{43} &= \hat{y}_{k|k}[4]\, e_{34}
\end{aligned} \quad (6.116)$$

In the above equations $\hat{y}_{k|k}[j]$ denotes the jth element of the estimated state vector $\hat{\mathbf{y}}_{k|k}$ and u_{k_i} are the elements of vector $\mathbf{U}_{k,k+1}$ defined in (6.3).

References

[1] S. Blackman and R. Popoli, *Design and Analysis of Modern Tracking Systems*. Norwood, MA: Artech House, 1999.

[2] J. C. Hassab, *Underwater signal and data processing*. Boca Raton, FL: CRC Press, 1989.

[3] S. C. Nardone, A. G. Lindgren, and K. F. Gong, "Fundamental properties and performance of conventional bearings-only target motion analysis," *IEEE Trans. Automatic Control*, vol. 29, no. 9, pp. 775–787, 1984.

[4] E. Fogel and M. Gavish, "Nth-order dynamics target observability from angle measurements," *IEEE Trans. Aerospace and Electronic Systems*, vol. AES-24, pp. 305–308, May 1988.

[5] T. L. Song, "Observability of target tracking with bearings-only measurements," *IEEE Trans. Aerospace and Electronic Systems*, vol. 32, pp. 1468–1471, October 1996.

[6] C. Jauffret and D. Pillon, "Observability in passive target motion analysis," *IEEE Trans. Aerospace and Electronic Systems*, vol. 32, pp. 1290–1300, October 1996.

[7] P. Shar and X. R. Li, "A practical approach to observability of bearings-only target tracking," in *Proc. SPIE*, vol. 3809, pp. 514–520, July 1999.

[8] V. J. Aidala and S. C. Nardone, "Biased estimation properties of the pseudo-linear tracking filter," *IEEE Trans. Aerospace and Electronic Systems*, vol. AES-18, pp. 432–441, July 1982.

[9] V. J. Aidala and S. E. Hammel, "Utilization of modified polar coordinates for bearings-only tracking," *IEEE Trans. Automatic Control*, vol. AC-28, pp. 283–294, March 1983.

[10] T. A. Song and J. L. Speyer, "A stochastic analysis of the modified gain extended Kalman filter with application to estimation with bearings-only measurements," *IEEE Trans. Automatic Control*, vol. AC-30, pp. 940–949, October 1985.

[11] S. C. Nardone and M. L. Graham, "A closed form solution to bearings-only target motion analysis," *IEEE Journal of Oceanic Engineering*, vol. 22, pp. 168–178, January 1997.

[12] S. S. Blackman and S. H. Roszkowski, "Application of IMM filtering to passive ranging," in *Proc. SPIE*, vol. 3809, pp. 270–281, 1999.

[13] T. Kirubarajan, Y. Bar-Shalom, and D. Lerro, "Bearings-only tracking of maneuvering targets using a batch-recursive estimator," *IEEE Trans. Aerospace and Electronic Systems*, vol. 37, pp. 770–780, July 2001.

[14] J. P. L. Cadre and O. Tremois, "Bearings-only tracking for maneuvering sources," *IEEE Trans. Aerospace and Electronic Systems*, vol. 34, pp. 179–193, January 1998.

[15] B. Ristic and S. Arulampalam, "Tracking a maneuvering target using angle-only measurements: Algorithms and performance," *Signal Processing*, vol. 83, pp. 1223–1238, 2003.

[16] B. Ristic, S. Zollo, and S. Arulampalam, "Performance bounds for manoevering target tracking using asyncronous multi-platform angle-only measurements," in *Proc. 4th Int. Conf. Information Fusion (Fusion 2001)*, (Montreal, Canada), August 2001.

[17] S. Arulampalam and B. Ristic, "Comparison of the particle filter with range-parametrised and modified polar EKFs for angle-only tracking," in *Proc. of SPIE, Signal and Data Processing of Small Targets*, vol. 4048, pp. 288–299, 2000.

[18] V. J. Aidala, "Kalman filter behaviour in bearings-only tracking applications," *IEEE Trans. Aerospace and Electronic Systems*, vol. AES-15, pp. 29–39, January 1979.

[19] N. Peach, "Bearings-only tracking using a set of range-parametrised extended Kalman filters," *IEE Proc. Control Theory Appl.*, vol. 142, pp. 73–80, January 1995.

[20] T. R. Kronhamn, "Bearings-only target motion analysis based on a multi-hypothesis Kalman filter and adaptive ownship motion control," in *IEE Proc. on Radar, Sonar and Navigation*, vol. 145, pp. 247–252, August 1998.

[21] Y. Bar-Shalom, X. R. Li, and T. Kirubarajan, *Estimation with Applications to Tracking and Navigation*. New York: John Wiley & Sons, 2001.

[22] R. Karlsson and N. Bergman, "Auxiliary particle filters for tracking a manoeuvring target," in *Proc. IEEE Conf. Decision and Control*, (Sydney, Australia), December 2000.

[23] A. Doucet, N. Gordon, and V. Krishnamurthy, "Particle filters for state estimation of jump Markov linear systems," *IEEE Trans. Signal Processing*, vol. 49, pp. 613–624, March 2001.

[24] N. Bergman, A. Doucet, and N. Gordon, "Optimal estimation and Cramer-Rao bounds for partial non-Gaussian state space models," *Ann. Inst. Statist. Math.*, vol. 53, no. 1, pp. 97–112, 2001.

[25] C. Andrieu, M. Davy, and A. Doucet, "Efficient particle filtering for jump Markov systems," in *Proc. IEEE Int. Conf. Acoust. Speech Signal Proc. (ICASSP)*, vol. 2, pp. 1625–1628, May 2002.

Chapter 7

Range-Only Tracking

7.1 INTRODUCTION

One of the maritime surveillance modes of the DSTO Ingara multimode radar [1] is the inverse synthetic aperture radar (ISAR) mode [2]. In this mode the radar collects high-resolution range profiles of the target in order to generate its real-time ISAR images. The problem, however, is that during an extended collection of target range profiles (required to build up a database of ISAR imagery), the antenna spotlights the location on the sea surface where the target was initially detected, while in reality the target moves and at some stage disappears from the radar antenna beam (in azimuth or the sampled swath in range). For this reason there is a need to track the target in the ISAR mode using the information contained in the ISAR range profiles. This information can be presented in the form of target range and range-rate measurements for tracking. If tracking in this context proves to be possible, the output of the tracker can then be used to automatically control the range gate setting and antenna pointing.

Tracking targets with range-only measurements has attracted very little research interest in the past. To our best knowledge, the only publication devoted to this problem, [3], discusses the conditions for target observability from range-only measurements. It concludes that if the target moves with a constant acceleration, the observer dynamics must have a nonzero jerk in order to observe the target state. This condition appears to be the same as the observability criterion for the related problem of angle-only target motion analysis [4, 5], studied in the previous chapter.

This chapter starts by presenting a mathematical framework for the problem of range-only tracking. Subsequently the theoretical CRLBs are derived under the assumption of $P_d = 1$ and the absence of false alarms. The bounds are analyzed from the aspect of algorithm convergence, a critical factor in this application. This is followed by the development of three algorithms for range-only tracking and their error performance comparison against the theoretical bounds. The three algorithms are: (1) the maximum likelihood estimator (MLE) over a cumulative measurement set; (2) the extended Kalman filter (EKF) with an angle-parametrization, and (3) the particle filter. The MLE is selected as the statistically efficient estimator when target motion is deterministic. This is a batch algorithm, which in practical systems is applied to measurements collected over a sliding time interval ("window"). Since the range-only tracking problem can be formulated as a nonlinear filtering problem, the EKF and the particle filter are selected as approximations to the optimal recursive Bayesian solution of the nonlinear filtering problem. Finally the chapter presents an application of developed tools and algorithms to a set of real ISAR data collected in the recent trials with the Ingara radar. The material presented in this chapter is taken from [6].

7.2 PROBLEM DESCRIPTION

Two of the maritime surveillance modes of the DSTO Ingara radar are the scan mode [7] and the ISAR mode [2]. In a typical wide area maritime surveillance, an airborne observer uses the radar scan mode to detect and locate maritime surface vessels. Once the location of a target is known, the radar operator switches the radar to the ISAR mode to collect high-resolution range profiles of the target. The range profiles are processed to form ISAR images that are shown to the operator in realtime. The radar antenna is controlled in the ISAR mode in such a manner that it "spotlights" the ocean surface at the target coordinates that were obtained from the initial detection in the radar scan mode. As the target is moving, at some point in time it may disappear from the radar antenna beam either in the cross-range direction or the sampled swath in the range direction. Because the aim is to continuously collect the data in the ISAR mode, we would like to be able to track the target using the information contained in the ISAR range profiles, which can be presented in the form of target range and range-rate measurements. The idea is to use a range-only tracker output to steer the radar antenna towards the moving target. The rate of algorithm convergence is clearly of critical importance in this application.

A typical scenario for the ISAR data collection purposes is illustrated in Figure 7.1. An airborne observer is usually flying along a circular trajectory, while the target (surface vessel) is assumed to travel with a constant velocity along a straight line in the x-y plane. Although in reality the target-observer scenario is three-dimensional, it suffices to define and study the problem in two dimensions, because the slant range and the range-rate can be readily converted to the ground range and range-rate. For the typical ranges of interest, the curvature of the Earth can be neglected and hence we assume the local tangential plane (flat Earth) model.

Range-Only Tracking

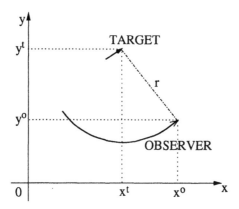

Figure 7.1 A typical range-only tracking scenario.

As always, the problem is to estimate the target kinematic state (position, heading, speed) from the noise-corrupted measurements. The target kinematic state can be fully described by the state vector defined in the discrete time as:

$$\mathbf{x}_k^t = [x_k^t \ \dot{x}_k^t \ y_k^t \ \dot{y}_k^t]^T \tag{7.1}$$

where x_k^t and y_k^t are the Cartesian target coordinates at time index k and $\dot{x}_k^t$ and $\dot{y}_k^t$ are their respective derivatives (velocities). For the assumed target motion model, $\dot{x}_k^t$ and $\dot{y}_k^t$ are constant in time. Knowing the target state vector it is then straightforward to calculate target heading or speed. The observer state, defined as

$$\mathbf{x}_k^o = [x_k^o \ \dot{x}_k^o \ y_k^o \ \dot{y}_k^o]^T \tag{7.2}$$

is assumed to be known (this information is provided, for example, by an on-board inertial navigation system, aided by the GPS receiver). The relative state vector is defined as:

$$\mathbf{x}_k \triangleq \mathbf{x}_k^t - \mathbf{x}_k^o = [x_k \ \dot{x}_k \ y_k \ \dot{y}_k]^T. \tag{7.3}$$

The measurements of target range and range-rate are available at time instants

$$t_k = t_0 + T_d + (k-1)T, \qquad k = 1, 2, \ldots \tag{7.4}$$

where t_0 is the time of initial detection, T is the (constant) sampling interval in the ISAR mode and $T_d \gg T$ is the period of time required for the radar to switch from the scan mode to the ISAR mode. The measurement vector at time instants t_k is defined by:

$$\mathbf{z}_k = \begin{bmatrix} r_k & \dot{r}_k \end{bmatrix}^T \qquad (k = 1, 2, \ldots) \tag{7.5}$$

and the measurement equation is given by:

$$\mathbf{z}_k = \mathbf{h}_I(\mathbf{x}_k) + \mathbf{w}_k \qquad (7.6)$$

where subscript I in $\mathbf{h}_I$ stands for the ISAR mode and

$$\mathbf{h}_I(\mathbf{x}_k) = \begin{bmatrix} h_r(\mathbf{x}_k) & h_{\dot r}(\mathbf{x}_k) \end{bmatrix}^T \qquad (7.7)$$

with

$$h_r(\mathbf{x}_k) = \sqrt{x_k^2 + y_k^2} \qquad (7.8)$$

$$h_{\dot r}(\mathbf{x}_k) = \frac{x_k \dot x_k + y_k \dot y_k}{\sqrt{x_k^2 + y_k^2}}. \qquad (7.9)$$

Measurement noise $\mathbf{w}_k$ in (7.6) is assumed to be white, zero-mean Gaussian, with covariance $\mathbf{R} = \text{diag}[\sigma_r^2, \sigma_{\dot r}^2]$.

The information about the target state at time t_0 is available in form of the measurement of target location (range and azimuth) supplied by the radar in the scan mode. This measurement is denoted by

$$\mathbf{z}_0 = \begin{bmatrix} r_0 & \theta_0 \end{bmatrix}^T \qquad (7.10)$$

and the measurement equation at t_0 is given by:

$$\mathbf{z}_0 = \mathbf{h}_s(\mathbf{x}_0) + \mathbf{w}_0 \qquad (7.11)$$

where

$$\mathbf{h}_s(\mathbf{x}_0) = \begin{bmatrix} h_r(\mathbf{x}_0) & h_\theta(\mathbf{x}_0) \end{bmatrix}^T \qquad (7.12)$$

with $h_r(\ )$ defined by (7.8) and

$$h_\theta(\mathbf{x}_0) = \arctan\left(\frac{x_0}{y_0}\right). \qquad (7.13)$$

The subscript S in $\mathbf{h}_s$ of (7.11) stands for the scan mode. In (7.13), x_0 and y_0 are the relative target Cartesian coordinates at $k = 0$ and can be worked out from the measurement $\mathbf{z}_0$. Note that azimuth θ_0 is defined as the angle from the observer to the target, referenced clockwise positive to the y-axis. Measurement noise $\mathbf{w}_0$ in (7.11) is white, independent from $\mathbf{w}_k$, $k = 1, 2, \ldots$, and distributed according to $p(\mathbf{w}_0) = \mathcal{N}(\mathbf{w}_0; 0, \mathbf{R}_0)$, where $\mathbf{R}_0 = \text{diag}[\sigma_r^2, \sigma_\theta^2]$.

The state equation for the described problem can be written as:

$$\mathbf{x}_{k+1} = \mathbf{F}_k \mathbf{x}_k - \mathbf{U}_{k+1,k} \tag{7.14}$$

where $\mathbf{F}_k$ is the transition matrix defined as

$$\mathbf{F}_k = \begin{bmatrix} 1 & T_k & 0 & 0 \\ 0 & 1 & 0 & 0 \\ 0 & 0 & 1 & T_k \\ 0 & 0 & 0 & 1 \end{bmatrix}, \tag{7.15}$$

T_k is the sampling interval, given by

$$T_k = \begin{cases} T_d & k = 0 \\ T & k = 1, 2, \dots \end{cases} \tag{7.16}$$

and

$$\mathbf{U}_{k+1,k} = \begin{bmatrix} x_{k+1}^o - x_k^o - T_k \dot{x}_k^o \\ \dot{x}_{k+1}^o - \dot{x}_k^o \\ y_{k+1}^o - y_k^o - T_k \dot{y}_k^o \\ \dot{y}_{k+1}^o - \dot{y}_k^o \end{bmatrix}$$

is a vector of deterministic inputs that accounts for the observer acceleration.

7.3 CRAMÉR-RAO BOUNDS

7.3.1 Derivations

The state equation, given by (7.14), is linear and deterministic. The measurement equation, given by (7.11) for $k = 0$ and (7.6) for $k = 1, 2, \dots$, is nonlinear. The CRLBs for this case can be computed recursively using a simplified form of (4.50) given by:

$$\mathbf{J}_{k+1} = [\mathbf{F}_k^{-1}]^T \mathbf{J}_k \mathbf{F}_k^{-1} + \tilde{\mathbf{H}}_{k+1}^T \mathbf{R}_{k+1}^{-1} \tilde{\mathbf{H}}_{k+1}, \tag{7.17}$$

for $k = 1, 2, \dots$, where

$$\mathbf{J}_1 = [\mathbf{F}_0^{-1}]^T \tilde{\mathbf{H}}_0^T \mathbf{R}_0^{-1} \tilde{\mathbf{H}}_0 \mathbf{F}_0^{-1} \tag{7.18}$$

and

$$\tilde{\mathbf{H}}_k = \begin{cases} \left[\nabla_{\mathbf{x}_0} \mathbf{h}_s^T(\mathbf{x}_0) \right]^T & \text{if } k = 0, \\ \left[\nabla_{\mathbf{x}_k} \mathbf{h}_1^T(\mathbf{x}_k) \right]^T & \text{if } k = 1, 2, \dots \end{cases} \tag{7.19}$$

By taking the first derivatives, it can be shown that the elements of Jacobian $\tilde{\mathbf{H}}_k$ are as follows:

$$\tilde{\mathbf{H}}_k[1,1] = \frac{x_k}{\sqrt{x_k^2 + y_k^2}}$$

$$\tilde{\mathbf{H}}_k[1,2] = 0$$

$$\tilde{\mathbf{H}}_k[1,3] = \frac{y_k}{\sqrt{x_k^2 + y_k^2}}$$

$$\tilde{\mathbf{H}}_k[1,4] = 0$$

for $k = 0, 1, 2, \ldots$ and

$$\tilde{\mathbf{H}}_k[2,1] = \begin{cases} \frac{y_0}{x_0^2 + y_0^2} & \text{if } k = 0, \\ \frac{\dot{x}_k\sqrt{x_k^2+y_k^2} - (x_k\dot{x}_k + y_k\dot{y}_k)\frac{x_k}{\sqrt{x_k^2+y_k^2}}}{x_k^2+y_k^2} & \text{if } k = 1, 2, \ldots \end{cases}$$

$$\tilde{\mathbf{H}}_k[2,2] = \begin{cases} 0 & \text{if } k = 0, \\ \frac{x_k}{\sqrt{x_k^2+y_k^2}} & \text{if } k = 1, 2, \ldots \end{cases}$$

$$\tilde{\mathbf{H}}_k[2,3] = \begin{cases} \frac{-x_0}{x_0^2 + y_0^2} & \text{if } k = 0, \\ \frac{\dot{y}_k\sqrt{x_k^2+y_k^2} - (x_k\dot{x}_k + y_k\dot{y}_k)\frac{y_k}{\sqrt{x_k^2+y_k^2}}}{x_k^2+y_k^2} & \text{if } k = 1, 2, \ldots \end{cases}$$

$$\tilde{\mathbf{H}}_k[2,4] = \begin{cases} 0 & \text{if } k = 0, \\ \frac{y_k}{\sqrt{x_k^2+y_k^2}} & \text{if } k = 1, 2, \ldots \end{cases}$$

The information matrix expressed by (7.17) depends on the geometry of the considered scenario, the measurements accuracy, the time delay T_d, and the sampling interval T. The CRLBs of the components of $\mathbf{x}_k$ are calculated as the diagonal elements of the inverse of the information matrix.

7.3.2 Analysis

As an illustration of the Cramér-Rao bounds, consider the scenario shown in Figure 7.2. The target is moving towards northeast with a speed of 8 m/s. The observer is flying at a speed of 150 m/s along a circular path with a radius of 15 km. At time t_0, the observer is at a location with coordinates (50 km, 35 km) in Figure 7.2. The covariances are assumed to be as follows: $\mathbf{R}_0 = \text{diag}\,[(20\text{m})^2,\ (3^\circ)^2]$ and $\mathbf{R} = \text{diag}\,[(20\,\text{m})^2,\ (2\,\text{m/s})^2]$, with a time delay of $T_d = 40\,s$ and the sampling interval $T = 1\,s$.

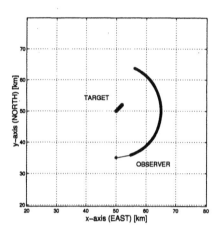

Figure 7.2 The range-only tracking scenario.

The square root of CRLB curves for $\hat{x}_k$, $\hat{y}_k$, $\hat{\dot{x}}_k$, and $\hat{\dot{y}}_k$ ($k = 1, 2, \ldots$) are shown in Figure 7.3 (starting from $t_1 = t_0 + T_d$). Since the initial range estimate is fairly accurate while the initial angular estimate is poor, for this particular observer-target scenario, the CRLB predicts a more accurate initial estimate of the target position in y than along the x-axis.

Next we analyze the effect of the sampling interval (T), the radius of the observer circular path (ρ) and the standard deviation of the range-rate measurements ($\sigma_{\dot{r}} = \sqrt{\mathbf{R}[2,2]}$) on the CRLB. The results are shown in Figure 7.4. The scenario and the parameters are the same as in Figures 7.2 and 7.3, except for those parameters whose effect on CRLBs is being investigated.

By reducing the sampling interval, the total number of measurements is increased, and the estimates of the target position in the x and y directions are expected to be better, as confirmed by the CRLBs in Figure 7.4(a,b). Note that this would increase the computational load of the algorithms.

The effect of the radius of the observer circular path, ρ, is shown in Figure 7.4(c,d). Since the target is moving with a constant velocity along a straight line, the observability criterion [3] requires the observer to have some nonzero acceleration.

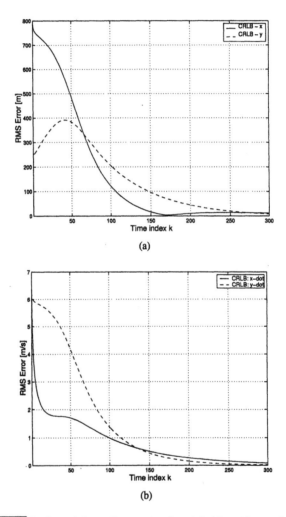

Figure 7.3 $\sqrt{\text{CRLB}}$ for $k = 1, 2, \ldots$ (i.e., starting from t_1): (a) position x_k (solid line) and y_k (dashed line); and (b) velocity $\dot{x}_k$ (solid line) and $\dot{y}_k$ (dashed line).

The smaller the radius, the larger the observer acceleration, and consequently the faster the convergence.[1]

The significance of the range-rate measurements, $\dot{r}_k$, is investigated in Figure 7.4(e,f). Two cases are considered, the first with totally imprecise range-rate measurements (i.e., very large $\mathbf{R}[2,2]$, mimicking the absence of $\dot{r}_k$ measurements), and the second with the precise range-rate measurements $\sigma_{\dot{r}} = 1$ m/s. The Cramér-Rao bounds indicate that there is some benefit in using the range-rate measurements, since the algorithm convergence can be improved.

The Significance of Prior Information

The CRLB that we presented and analyzed so far did not assume any prior knowledge about the target position or speed. The recursive Bayesian estimators, such as the EKF or the particle filter, make use of this prior knowledge (if available), as opposed to the MLE. Prior knowledge, in general, can speed up algorithm convergence. In this particular application we have some prior knowledge of the maximum target speed.

The recursive Bayesian estimators are typically initialized using the first measurement and prior knowledge. The initial measurement $\mathbf{z}_0 = [r_0, \ \theta_0]^T$ is in polar coordinates, hence the initial target state and its covariance are calculated by conversion from polar to the Cartesian coordinates.[2] The initial state vector is then

$$\hat{\mathbf{x}}_{0|0} = [r_0 \sin\theta_0 \ \ -\dot{x}_0^o \ \ r_0 \cos\theta_0 \ \ -\dot{y}_0^o]^T \qquad (7.20)$$

where $\dot{x}_0^o$ and $\dot{y}_0^o$ are the initial observer velocity components defined in (7.2). The covariance matrix of the initial estimate is given by

$$\mathbf{P}_{0|0} = \begin{bmatrix} P_{11} & 0 & P_{13} & 0 \\ 0 & P_{22} & 0 & 0 \\ P_{31} & 0 & P_{33} & 0 \\ 0 & 0 & 0 & P_{44} \end{bmatrix}. \qquad (7.21)$$

Note that the initial target velocity components in (7.20) are set to zero, while their variance is given by elements P_{22} and P_{44} of $\mathbf{P}_{0|0}$ in (7.21); that is,

$$P_{22} = P_{44} = \sigma_v^2. \qquad (7.22)$$

---

1 If the observer is moving along a straight line with a constant velocity (the same as the target), the calculated CRLBs are monotonically increasing. This confirms the observability requirement for nonzero acceleration.
2 The so-called "debiased consistent" conversion [8] is recommended for large values of σ_θ.

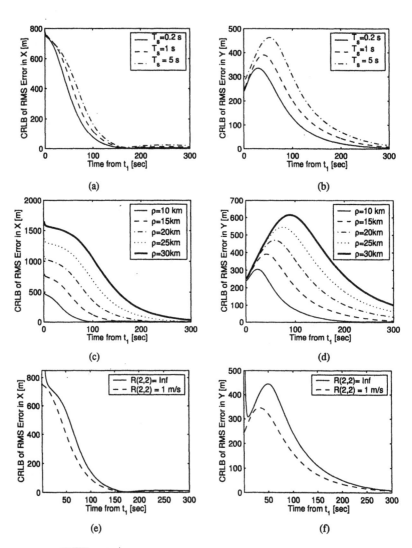

Figure 7.4 $\sqrt{\text{CRLB}}$ of position estimates in x and y direction shown as a function of T in (a) and (b); ρ in (c) and (d), and $\sigma_{\dot{r}}$ in (e) and (f).

The remaining elements of $\mathbf{P}_{0|0}$ are as follows [9, p. 155]:

$$P_{11} = r_0^2 \sigma_\theta^2 \cos^2 \theta_0 + \sigma_r^2 \sin^2 \theta_0$$
$$P_{33} = r_0^2 \sigma_\theta^2 \sin^2 \theta_0 + \sigma_r^2 \cos^2 \theta_0$$
$$P_{13} = P_{31} = (\sigma_r^2 - r_0^2 \sigma_\theta^2) \sin \theta_0 \cos \theta_0.$$

If the initial density at t_0 is assumed to be Gaussian, with covariance $\mathbf{P}_{0|0}$, we can start the recursions of (7.17) with $\mathbf{J}_1 = [\mathbf{F}_0^{-1}]^T \mathbf{P}_{0|0}^{-1} \mathbf{F}_0^{-1}$.

Figure 7.5 compares the $\sqrt{\text{CRLB}}$ curves obtained with and without the use of prior knowledge of the maximum target speed (x position only). The trajectories, measurement covariances, the sampling interval and the initial delay are the same as in Figure 7.3. The $\sqrt{\text{CRLB}}$ curves in Figure 7.5 are shown for the following values of σ_v: 5 m/s, 10 m/s, and 20 m/s. Figure 7.5 confirms that the CRLB which

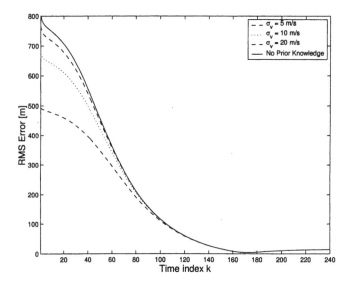

Figure 7.5 Comparison of $\sqrt{\text{CRLB}}$ curves (x position only) with and without the use of prior information.

uses prior knowledge is initially lower (hence the convergence is faster), although the significance of prior information diminishes very quickly over time and the bounds become indistinguishable for $k > 100$. The quality of prior information is measured by its variance; note that by increasing the value of σ_v, the CRLB with prior knowledge approaches the one with no prior knowledge. The practical value of σ_v for range-only tracking of surface vessels is approximately $\sigma_v \approx 10$ m/s (dotted line in Figure 7.5). Hence in this application, prior knowledge makes a measurable impact on the initial performance.

7.4 TRACKING ALGORITHMS

The Maximum Likelihood Estimator

For a deterministic state dynamic model as in (7.14), the MLE [10] is a suitable algorithm. As opposed to the tracking filters (described in Chapters 2 and 3), the MLE is a batch algorithm; it requires a block of accumulated measurements to estimate the target state $\mathbf{x}_k^t = [x_k^t \; \dot{x}_k^t \; y_k^t \; \dot{y}_k^t]^T$ by maximizing the likelihood function $p(\mathbf{Z}_k|\mathbf{x}_k^t)$:

$$\hat{\mathbf{x}}_k^t = \arg\max_{\mathbf{x}_k^t} p(\mathbf{Z}_k|\mathbf{x}_k^t). \tag{7.23}$$

The likelihood function can be expressed as:

$$p(\mathbf{Z}_k|\mathbf{x}_k^t) = \prod_{i=0}^{k} p(\mathbf{z}_i|\mathbf{x}_k^t), \tag{7.24}$$

where

$$p(\mathbf{z}_i|\mathbf{x}_k^t) = \begin{cases} \mathcal{N}(\mathbf{z}_0; \mathbf{h}_s(\mathbf{x}_0(\mathbf{x}_k^t)), \mathbf{R}_0) & \text{if } i = 0 \\ \mathcal{N}(\mathbf{z}_i; \mathbf{h}_1(\mathbf{x}_i(\mathbf{x}_k^t)), \mathbf{R}) & \text{if } i = 1, 2, \ldots, k \end{cases} \tag{7.25}$$

The notation $\mathbf{h}_s(\mathbf{x}_0(\mathbf{x}_k^t))$ and $\mathbf{h}_1(\mathbf{x}_i(\mathbf{x}_k^t))$ is intended to emphasize that the arguments of these two functions have to be expressed via $\mathbf{x}_k^t$. This means that

1. x_0 and y_0 in (7.13) should be expressed as

$$x_0(\mathbf{x}_k^t) = x_k^t - (t_k - t_0)\dot{x}_k^t - x_0^o \tag{7.26}$$
$$y_0(\mathbf{x}_k^t) = y_k^t - (t_k - t_0)\dot{y}_k^t - y_0^o \tag{7.27}$$

which follows from the assumption that target velocity is constant (i.e., both $\dot{x}_k^t$ and $\dot{y}_k^t$ take a fixed value due to the absence of process noise in the dynamic model). From (7.4) it follows that $t_k - t_0 = T_d + (k-1)T$.

2. x_i, y_i, $\dot{x}_i$, and $\dot{y}_i$ ($i = 1, 2, \ldots, k$), which according to (7.25) are to be used in (7.8) and (7.9), have to be expressed as:

$$x_i(\mathbf{x}_k^t) = x_k^t - (k - i) T \dot{x}_k^t - x_i^o \qquad (7.28)$$
$$y_i(\mathbf{x}_k^t) = y_k^t - (k - i) T \dot{y}_k^t - y_i^o \qquad (7.29)$$
$$\dot{x}_i(\mathbf{x}_k^t) = \dot{x}_k^t - \dot{x}_i^o \qquad (7.30)$$
$$\dot{y}_i(\mathbf{x}_k^t) = \dot{y}_k^t - \dot{y}_i^o. \qquad (7.31)$$

Equations (7.28)–(7.31) again follow from the fact that both $\dot{x}_k^t$ and $\dot{y}_k^t$ are fixed values.

Maximization of the likelihood function is equivalent to minimization of the negative log-likelihood function:

$$\lambda(\mathbf{x}_k^t) \triangleq -\log p(\mathbf{Z}_k | \mathbf{x}_k^t) \qquad (7.32)$$
$$= C + \frac{1}{2} \left\{ \mathbf{z}_0 - \mathbf{h}_s(\mathbf{x}_0(\mathbf{x}_k^t)) \right\}^T \mathbf{R}_0^{-1} \left\{ \mathbf{z}_0 - \mathbf{h}_s(\mathbf{x}_0(\mathbf{x}_k^t)) \right\}$$
$$+ \frac{1}{2} \sum_{i=1}^{k} \left\{ \mathbf{z}_i - \mathbf{h}_\mathrm{l}(\mathbf{x}_i(\mathbf{x}_k^t)) \right\}^T \mathbf{R}^{-1} \left\{ \mathbf{z}_i - \mathbf{h}_\mathrm{l}(\mathbf{x}_i(\mathbf{x}_k^t)) \right\}. \qquad (7.33)$$

Minimization of $\lambda(\mathbf{x}_k^t)$ is, in general, performed by numerical methods [11]. Our implementation is based on the MATLAB built-in routine *fminbnd*. Since the MLE is a batch algorithm, for on-line data processing applications (such as the one considered in this chapter), the MLE can be applied over a sliding window of accumulated measurements. It is well known [10, 12] that the MLE is an efficient estimator, and hence its variance meets the theoretical CRLB derived in Section 7.3.1 (with no use of prior knowledge).

Angle-Parameterized EKF

The recursive equations of the EKF were presented in Section 2.1. The goal is to evaluate the relative state estimate $\hat{\mathbf{x}}_{k+1|k+1}$ and its covariance matrix $\mathbf{P}_{k+1|k+1}$, given measurement $\mathbf{z}_{k+1}$ and the pair $(\mathbf{x}_k, \mathbf{P}_{k|k})$, for $k = 1, 2, \ldots$. The state prediction follows from (7.14):

$$\hat{\mathbf{x}}_{k+1|k} = \mathbf{F}_k \hat{\mathbf{x}}_{k|k} - \mathbf{U}_{k+1,k}, \qquad (7.34)$$

the covariance prediction is given by

$$\mathbf{P}_{k+1|k} = \mathbf{F}_k \mathbf{P}_{k|k} \mathbf{F}_k^T, \qquad (7.35)$$

and the measurement prediction by:

$$\hat{z}_{k+1|k} = h_1(\hat{x}_{k+1|k}). \tag{7.36}$$

The Kalman gain matrix can be evaluated as:

$$K_{k+1} = P_{k+1|k}\hat{H}_{k+1}^T \left[\hat{H}_{k+1}P_{k+1|k}\hat{H}_{k+1}^T + R\right]^{-1}, \tag{7.37}$$

where H_{k+1} is the Jacobian of (7.19), evaluated at the predicted state.
The updated state and its covariance matrix are given by

$$\hat{x}_{k+1|k+1} = \hat{x}_{k+1|k} + K_{k+1}[z_{k+1} - \hat{z}_{k+1|k}] \tag{7.38}$$

$$P_{k+1|k+1} = [I - K_{k+1}\hat{H}_{k+1}]P_{k+1|k} \tag{7.39}$$

where I is the 4×4 identity matrix.

If the initial azimuth measurement is characterized by a fairly large variance, the convergence of the EKF for range-only tracking can be improved by running in parallel a set of weighted EKFs, each with a different initial azimuth value. This type of a filter, in this chapter referred to as the angle-parameterized EKF (AP-EKF), in essence is the static MM estimator described in Section 2.3.1. The idea used in the development of the AP-EKF closely resembles that of the range-parameterized EKF for angle-only tracking [13, 14], described in Section 6.4.1.2.

Recall that the initial azimuth measurement is θ_0 with standard deviation σ_θ. We wish to construct N_f EKFs, each with an initial angular accuracy N_f times better than that of a single EKF. First define an interval in the angle domain with the limits $\theta_{\min} = \theta_0 - 3\sigma_\theta$ and $\theta_{\max} = \theta_0 + 3\sigma_\theta$. The interval is divided into N_f subintervals of width $\Delta\theta = (\theta_{\max} - \theta_{\min})/N_f$. The initial azimuth angle for EKF i is chosen to correspond the mid-point of the interval i; that is,

$$\theta_i = \theta_{\min} + (i - 0.5)\Delta\theta, \qquad i = 1, \ldots, N_f,$$

with the standard deviation set to $\sigma_{\theta_i} = \sigma_\theta/N_f$.

The next step is to calculate the weights associated with each EKF at discrete-time index $k + 1$. According to (2.23),

$$w_{k+1}^i = \frac{p(z_{k+1}|i)\, w_k^i}{\sum_{j=1}^{N_f} p(z_{k+1}|j)\, w_k^j} \tag{7.40}$$

where $p(z_{k+1}|i)$ is the likelihood of measurement z_{k+1}, given that the ith EKF is the correct one. Using a linear/Gaussian approximation, the likelihood $p(z_{k+1}|i)$

can be computed as:

$$p(z_{k+1}|i) = \frac{1}{|2\pi \mathbf{S}_{k+1}^i|^{\frac{1}{2}}} \exp\left[-\frac{1}{2}(z_{k+1} - \hat{z}_{k+1|k}^i)^T (\mathbf{S}_{k+1}^i)^{-1}(z_{k+1} - \hat{z}_{k+1|k}^i)\right]$$

where $\hat{z}_{k+1|k}^i$ is the predicted measurement vector for ith EKF with the covariance matrix given by:

$$\mathbf{S}_{k+1}^i = \hat{\mathbf{H}}_{k+1} \mathbf{P}_{k+1|k}^i \hat{\mathbf{H}}_{k+1}^T + \mathbf{R}. \tag{7.41}$$

Matrix $\mathbf{P}_{k+1|k}^i$ in (7.41) is the covariance matrix of the predicted state of the ith filter.

The initial weights w_0^i are computed assuming that the true initial azimuth obeys a Gaussian distribution with mean θ_0 and variance σ_θ^2; that is

$$w_0^i = \frac{1}{\sqrt{2\pi}\sigma_\theta} \int_{\theta_L^i}^{\theta_R^i} e^{-\frac{(\theta-\theta_0)^2}{2\sigma_\theta^2}} d\theta \tag{7.42}$$

where $\theta_L^i = \theta_{\min} + (i-1)\Delta\theta$ and $\theta_R^i = \theta_L^i + \Delta\theta$ are the lower and upper limits of the ith subinterval, respectively.

Suppose the updated state estimate of ith EKF at t_k is denoted by $\hat{\mathbf{x}}_{k|k}^i$. The combined AP-EKF state vector and its covariance are calculated using the Gaussian mixture formulas [15, p. 47] given by (2.26) and (2.27), respectively.

The improved tracking performance of the AP-EKF is achieved using N_f independent filters, each with a much smaller initial angular measurement variance than that of a single EKF. This improvement is achieved at the expense of an N_f-fold increase in the computational load if all the angle subintervals are processed throughout. However, it has been found that generally the weighting of some of the subintervals rapidly reduces to zero. In such cases, the corresponding filters can be removed from the tracking process without loss of accuracy, thereby reducing the processing requirement. Thus, a weighting threshold can be set and any filter corresponding to a subinterval with a weight less than the threshold may be removed from the tracking process.

Regularized Particle Filter

It was discussed at the end of Section 3.3 that using particle filters for deterministic dynamic models such as (7.14) is not entirely appropriate. Nevertheless we develop for this study a regularized PF described in Section 3.5.3. Improving particle diversity via regularization (or alternatively by applying the MCMC move step) is very important in the absence of process noise. Since $n_x = 4$ in this application, constant c_{n_x} that features in (3.50) is equal to $\pi^2/2$ [16, p. 96].

7.5 ALGORITHM PERFORMANCE AND COMPARISON

This section presents the error performance of three range-only tracking algorithms: the MLE, the AP-EKF, and the regularized PF. The trajectories, the measurement covariances $\mathbf{R}$ and $\mathbf{R}_0$, the sampling interval T, and the initial delay T_d have the same values as described in Section 7.3.2. The resulting error curves were compared to the theoretically derived Cramér-Rao lower bounds. The performance of range-only tracking algorithms was measured by the root mean square (RMS) error, which for component j of the state vector is defined as

$$\sigma_{\mathbf{e}_k[j]} = \sqrt{\mathrm{E}\{(\hat{\mathbf{x}}^t_{k|k}[j] - \mathbf{x}^t_k[j])^2\}}. \tag{7.43}$$

The expectation operator in (7.43) was carried out by averaging over 100 independent Monte Carlo runs.

Performance of the MLE

The RMS error performance of the MLE is shown in Figure 7.6 [x and y position only, i.e., $j = 1, 3$ in (7.43)]. Since the MLE is not a Bayesian estimator, its performance is compared to the CRLB that does not use prior knowledge. Figure 7.6 confirms (in the context of range-only tracking) the theoretical assertion that the MLE is an efficient estimator: it has no bias (not shown in the figure, but confirmed by simulations) and its RMS error agrees with the theoretical curve for $\sqrt{\mathrm{CRLB}}$.

The main disadvantage of the MLE is its computational complexity. Not only is the numerical minimization of the log-likelihood function a computationally demanding operation, but also at every time index k the MLE must operate on the accumulated set of all previous measurements[3] $\mathbf{Z}_k$. Using the same scenario as before, we compared the CPU time required by the MLE algorithm to that of the EKF algorithm. Both algorithms were implemented in MATLAB, and the comparison is performed at time index $k = 240$. It turns out that the MLE requires 580 times more CPU time than the EKF.

Another disadvantage of the MLE is that it is inappropriate for a stochastic model of target dynamics. Thus if a small amount of process noise were included in dynamic equation (7.14) to allow for gentle target maneuvers, both the EKF and the particle filter could still be applied directly, while the MLE would be inappropriate.

---

[3] By restricting the MLE to operate on the sliding window of previous measurements, the efficiency of the MLE is lost.

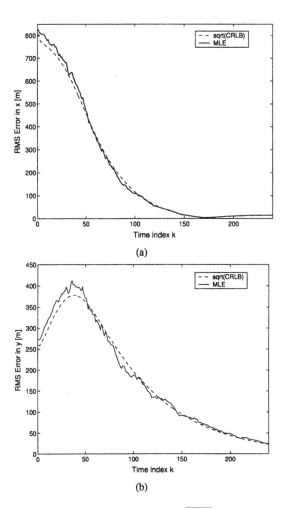

Figure 7.6 The error performance of the MLE against the $\sqrt{\text{CRLB}}$: (a) x position; and (b) y position.

Performance of the EKF and the AP-EKF

The RMS error performance of the EKF is shown in Figure 7.7 (x and y position only). The EKF is a Bayesian estimator, hence its performance is compared to the CRLB that uses prior knowledge of $\sigma_v = 10$ m/s. In comparison, the MLE performance curves are indicated by a dotted line in Figure 7.7. Observe that the EKF error is initially smaller than the MLE error, since the EKF uses prior knowledge about the maximum target velocity. Very quickly, though, the EKF departs from the $\sqrt{\text{CRLB}}$ curve, and in general demonstrates a slower convergence than the MLE. This is particularly pronounced in Figure 7.7(b). The suboptimal performance of the EKF is due to the approximation (local linearization) of the nonlinear measurement equation.

Angle parametrization effectively resolves this problem. The results, obtained using $N_f = 5$ parallel EKF filters, are shown in Figure 7.8. The performance of the AP-EKF is also compared to the CRLB (prior knowledge of $\sigma_v = 10$ m/s), and the MLE error curves (indicated by dotted lines for comparison). Initially, the AP-EKF error is below the MLE error (which is to be expected) and even below the $\sqrt{\text{CRLB}}$, which is unexpected. This result is due to the AP-EKF initialization, which is not exactly matched to the inverse of the initial information matrix used in the CRLB computation. As the time progresses, the performance of the AP-EKF is very close to that of the MLE. Overall, this represents a significant improvement over the EKF. The considered implementation of the AP-EKF required $N_f = 5$ times more CPU time than the EKF and $580/N_f = 116$ times less CPU time than the MLE.

It should be noted that in all simulations the value of $\sigma_\theta = 3°$ was used. If the accuracy of the initial angular measurement is higher (i.e., $\sigma_\theta < 1°$), it has been observed that the EKF actually attains its corresponding CRLB and hence the angle parametrization appears to be unnecessary.

Performance of the RPF

Figure 7.9 shows the RMS error performance of the RPF-based range-only tracking algorithm (x and y positions only) with $N = 2000$, 5000, and 10000 particles. In comparison, the CRLBs (using prior knowledge of $\sigma_v = 10$ m/s) and the EKF error curves are shown as well. Observe that as the number of particles N is increased, the performance of the RPF improves, and in the limit approaches the CRLBs. This improved accuracy of the RPF, however, is at the expense of the computational load. Thus the RPF with 2000, 5000, and 10000 particles requires 30, 70, and 134 times more CPU time respectively, than the EKF. In order to be fair, the RPF should be compared to the EKF, since the concept of angle parametrization can be equally applied to the particle filter. Nevertheless, the RPF appears to be computationally too demanding for this application. Overall, among the considered

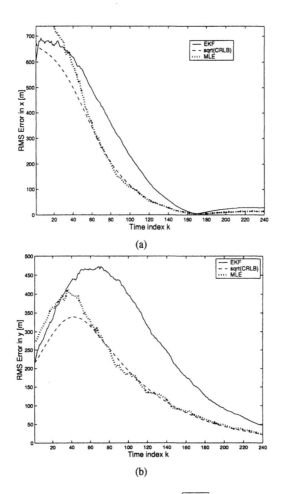

Figure 7.7 The error performance of the EKF against the $\sqrt{\text{CRLB}}$ (with prior knowledge of $\sigma_v = 10$ m/s): (a) x position and (b) y position. The dotted line is the MLE error curve.

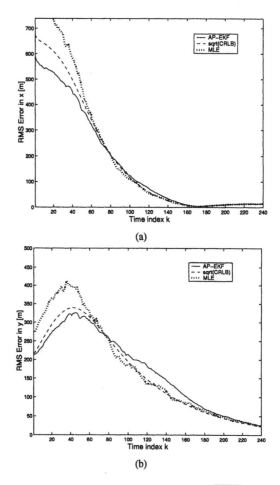

Figure 7.8 The error performance of the AP-EKF against the $\sqrt{\text{CRLB}}$ (with prior knowledge of $\sigma_v = 10$ m/s): (a) x position and (b) y position. The dotted line is the MLE error curve.

algorithms, the AP-EKF appears to be the most suitable for implementation in an operational system.

7.6 APPLICATION TO INGARA ISAR DATA

Inverse synthetic aperture radar is a technique for generating high-resolution radar images in range and Doppler (cross-range). This is achieved by sampling radar echoes from a target and coherently processing blocks of echoes to form ISAR images. The high resolution in range is achieved by using a stretch waveform while the high resolution in cross-range is achieved by processing a large aperture of data. A comprehensive treatment of the ISAR theory is given in [17]. Once the ISAR mode was developed for Ingara radar, a series of trials were conducted to build up a database of ISAR imagery of a representative collection of target types. During the trials, Ingara was installed on a Beechcraft KingAir 350 aircraft and the targets of opportunity were used in data collection. In processing real ISAR data for range-only target tracking, it was necessary to perform conversions between various coordinate systems. A review of coordinate transformations for target tracking is given in the Appendix. Note that the theoretical treatment of the range-only tracking problem (from Sections 7.2 to 7.5) was developed in the local tangential plane Cartesian coordinate system. The measurements of target range and range-rate (r_k and $\dot{r}_k$) were obtained by preprocessing the ISAR data; for details, see [6].

The geographic location of the relevant data collection scenario (in vicinity of Darwin, Australia) is shown in Figure 7.10. The initial observer position and the initial target location measurement are indicated by $\Diamond$ and $\star$, respectively. The reference point (the origin of the local coordinates) is shown as $\Box$. The observer trajectory after the time delay of $T_d = 65$ seconds is plotted by a solid line. The sampling interval (i.e. the time increment of the sliding window used for preprocessing range profiles) is $T_s = 1$ second. The observer height was constant at 210m.

The range and range rate measurements are displayed in Figure 7.11. Since the target range is approximately 20 km [see Figure 7.11(a)], in all our calculations we neglected the own-ship height (of 210 m) and assumed that the observer and the target are coplanar. The error in the measured range due to this approximation is less than 1m.

Since the targets of opportunity were used in data collection, we were only able to contract the "incomplete truth" target trajectory. This trajectory was built based on an accurate and independent measurement of target heading as $287°$ clockwise positive with respect to North. This "incomplete truth" trajectory is shown in Figure 7.12 by a dotted line. Part (a) of this figure also displays the estimated target trajectories using: the (progressive) MLE applied to $\mathbf{Z}_k$, $k = 1, \ldots, K$; the

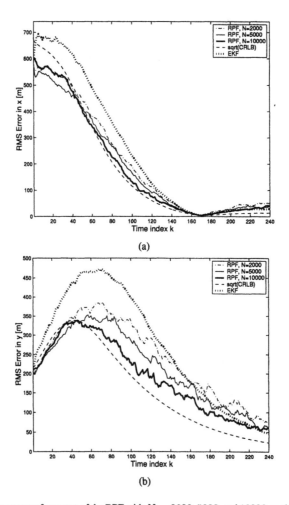

Figure 7.9 The error performance of the RPF with $N = 2000$, 5000, and 10000 particles. The CRLBs (using prior knowledge of $\sigma_v = 10$ m/s) are shown with dashed lines; the EKF curves are plotted with dotted lines; (a) x position and (b) y position.

Range-Only Tracking 175

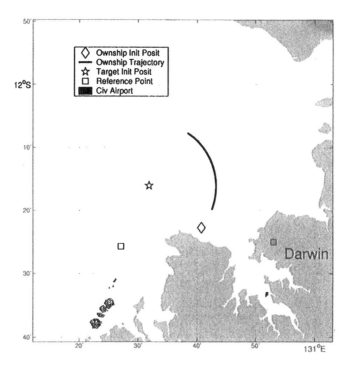

Figure 7.10 The geographic location of the data collection scenario.

angle-parameterized EKF (with $N_f = 5$ filters), and the RPF (with $N = 5000$ samples). The estimates were produced using $\mathbf{R} = \mathrm{diag}[(20\mathrm{m})^2; (3\mathrm{m/s})^2]$ and $\mathbf{R}_0 = \mathrm{diag}[(20\mathrm{m})^2; (1°)^2]$. Both the AP-EKF and the RPF were initialized using $\sigma_v = 10$ m/s. Note that all three algorithms fairly accurately follow the indicated target heading. The MLE and the AP-EKF converge towards the "incomplete truth" trajectory after approximately 2.5 minutes, while the RPF convergence is somewhat slower. Part (b) of Figure 7.12 displays the AP-EKF trajectory estimate, with overlaid 3-sigma ellipses of uncertainty. Observe the following characteristics of 3-sigma ellipses: (1) they are orthogonal to the line of sight (Figure 7.10 displays the observer trajectory); (2) they always include the "true" target position; and (3) they are large in angle (because the angle is not measured), but thin in range.

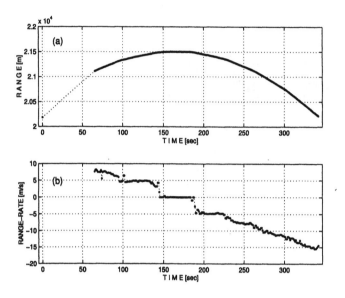

Figure 7.11 Target range and range-rate measurements.

7.7 SUMMARY

The chapter investigated the problem of target tracking using range and range-rate measurements. The problem is relevant for tracking surface vessels while collecting data in the ISAR mode of the Ingara multimode radar. The Cramér-Rao bounds were derived and three tracking algorithms were developed. The CRLBs enabled us to predict the best achievable error performance under various tracking conditions, such as the specific observer-target geometry, measurement accuracy, and measurement frequency. The considered algorithms included the maximum likelihood estimator, the extended Kalman filter (with and without angle parameterization), and the regularized particle filter. Considering both the statistical and computational performance of considered algorithms, this study has found the angle-parameterized EKF to be the preferred choice for implementation in an operational system.

The discussed algorithms have been applied successfully to a set of real ISAR data collected in the recent trials with the Ingara radar. For a typical scenario of interest (an extended ISAR data collection), the results indicate that tracking algorithms based on range and range-rate measurements can converge towards a steady state in less than 3 minutes.

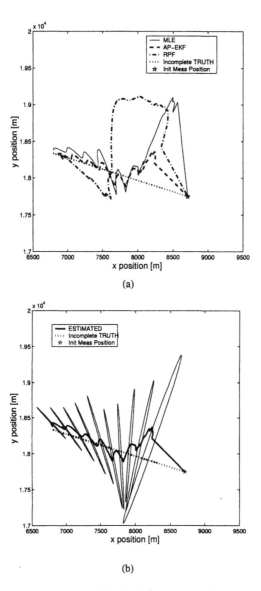

Figure 7.12 Estimated target trajectories in the local Cartesian coordinates: (a) MLE, AP-EKF, and RPF; and (b) AP-EKF with 3-sigma ellipses. Target moves with the heading of 287°.

References

[1] I. Bostock, "Ingara radar demonstrated by army," *Jane's International Defense Review*, vol. 33, p. 8, August 2000.

[2] C. Fantarella, "A Real-Time ISAR Mode for the Ingara Multi-Mode Radar," Tech. Rep. DSTO-TR-1036, Defence Science and Technology Organisation, Edinburgh, Australia, October 2000.

[3] T. L. Song, "Observability of target tracking with range-only measurements," *IEEE Journal of Oceanic Engineering*, vol. 24, pp. 383–387, July 1999.

[4] E. Fogel and M. Gavish, "Nth-order dynamics target observability from angle measurements," *IEEE Trans. Aerospace and Electronic Systems*, vol. AES-24, pp. 305–308, May 1988.

[5] S. Blackman and R. Popoli, *Design and Analysis of Modern Tracking Systems*. Norwood, MA: Artech House, 1999.

[6] B. Ristic, S. Arulampalam, and J. McCarthy, "Target motion analysis using range-only measurements: Algorithms, performance and application to ISAR data," *Signal Processing*, vol. 82, pp. 273–296, 2002.

[7] J. McCarthy, "Design and Implementation of Ingara Radar Scan Modes," Tech. Rep. DSTO-TR-1067, Defence Science and Technology Organisation, Edinburgh, Australia, 2001.

[8] D. Lerro and Y. Bar-Shalom, "Tracking with debiased consistent converted measurements versus EKF," *IEEE Trans. Aerospace and Electronic Systems*, vol. 29, pp. 1015–1022, July 1993.

[9] A. Farina and F. A. Studer, *Radar Data Processing*. New York: John Wiley, 1985.

[10] H. L. VanTrees, *Detection, Estimation and Modulation Theory*. New York: John Wiley & Sons, 1968.

[11] W. H. Press, B. P. Flannery, S. A. Teukolsky, and W. T. Vetterling, *Numerical Recipes in C*. Cambridge, U.K.: Cambridge University Press, 2nd ed., 1992.

[12] Y. Bar-Shalom, X. R. Li, and T. Kirubarajan, *Estimation with Applications to Tracking and Navigation*. New York: John Wiley & Sons, 2001.

[13] N. Peach, "Bearings-only tracking using a set of range-parametrised extended Kalman filters," *IEE Proc. Control Theory Appl.*, vol. 142, pp. 73–80, January 1995.

[14] S. Arulampalam and B. Ristic, "Comparison of the particle filter with range-parametrised and modified polar EKFs for angle-only tracking," in *Proc. SPIE, Signal and Data Processing of Small Targets*, vol. 4048, pp. 288–299, 2000.

[15] Y. Bar-Shalom and X. R. Li, *Estimation and Tracking*. Norwood, MA: Artech House, 1993.

[16] Y. Bar-Shalom and X. R. Li, *Multitarget-Multisensor Tracking: Principles and Techniques*. Storrs, CT: YBS Publishing, 1995.

[17] D. R. Wehner, *High-Resolution Radar*. Norwood, MA: Artech House, 2nd ed., 1995.

Chapter 8

Bistatic Radar Tracking

8.1 INTRODUCTION

Monostatic radars represent a well-established technology, but they are vulnerable to jamming and can be targeted by low-cost antiradiation missiles (ARMs). Bistatic radar systems [1, 2, 3, 4], characterized by a significant separation between the transmitter (illuminator) and the receiver, are becoming increasingly popular alternatives [5]. They are less vulnerable because the transmitter can be several tens or even hundreds of kilometers away from the forward line of troops. The receiver, being silent, is virtually undetectable by ordinary electromagnetic means. During the 1990s it was demonstrated that bistatic receivers could even exploit noncooperative illuminators such as TV and FM broadcasts [6]. Another advantage of the bistatic radar system is its capability of detecting low probability of intercept targets [1, 2]. The receiver of a bistatic system processes the forward scatter waves from the target. The forward scatter region is the angular region of enhanced radar cross section (RCS) that is located in the forward direction (i.e., 180 degrees away from the source of illumination). Within this region, the RCS can be more than 15 dB larger than backscatter RCS measured by conventional monostatic radar [2]. Since the magnitude of a target forward scatter return does not depend on material composition, a "forward scatter" radar is able to detect an object with a monostatic RCS specially reduced through shaping or by treatment with radar absorbing materials.

Consider the bistatic radar tracking problem, where the transmitter and receiver are well separated, typically of the order 40 to 100 km. The receiver, operating in a passive mode, processes the forward scatter EM waves corresponding to the transmitted signal that deflect off the aircraft. For a noncooperative transmitter, signal processing in the receiver leads to two forms of measurement processes: (1) target-receiver bearing, and (2) Doppler shift relative to the carrier frequency. For a cooperative transmitter, an additional third measurement component can be extracted, namely the bistatic range, which is proportional to the transit time of the scatter signal. This chapter, however, considers only the (more difficult) case where the bistatic range is unavailable. The bistatic radar measurements relate to the target state (typically consisting of position and velocity), in a nonlinear fashion. In addition, the problem is further complicated by the presence of spurious (or false) detections (clutter) that arise at the output of the signal processing chain. This chapter deals with tracking of position and velocity of a target using noise-corrupted measurements of angle and bistatic Doppler in the presence of clutter.

The described problem has received some attention in the recent past. Farina [7] proposed a tracking scheme for the bistatic radar tracking problem with all three measurement processes available to the tracker. A problem that closely resembles the one considered in this chapter was investigated in [8] where a noncooperative television transmitter was used as the illuminator of opportunity. Here a two-stage tracker was proposed. The first stage uses a nearest neighbor Kalman filter to track the angle and Doppler profiles in clutter, followed by an extended Kalman filter in the second stage that estimates the position and velocity of the aircraft.

This chapter first investigates the best achievable error performance for the problem of target tracking using bearings and bistatic Doppler measurements. This is done by means of the Cramér-Rao lower bounds. The bounds can be used to analyze the possible behavior of the tracker in various target-receiver geometries, sensor accuracies, sampling times, and transmitter-receiver separation. Next, the chapter proposes some tracking algorithms, analyzes their performance by Monte Carlo simulations, and compares this performance to the best achievable (the CRLB). For convenience, the developed trackers consist of two stages as in [8]. The algorithms used for the first stage are the probabilistic data association filter (PDAF) and the interacting multiple model PDAF (IMM-PDAF). The second stage consists of a nonlinear filter, which is designed either as the extended Kalman filter or as the particle filter. The chapter is based on the material presented in [9].

8.2 PROBLEM FORMULATION

Consider the two-dimensional bistatic geometry depicted in Figure 8.1, where the transmitter, located at coordinates $(0, L)$, illuminates the target. The receiver, located at the origin, operates in the passive mode by processing the forward scatter signal. The results of this processing are measurements of the target-receiver angle and the Doppler shift relative to the transmitter carrier signal. The basic problem is to estimate target kinematics parameters (position and velocity) from noise-corrupted angle and Doppler data in the presence of clutter.

Next we define the problem mathematically. The target, located at (x, y), is assumed to move with a nearly constant velocity vector $(\dot{x}, \dot{y})$, and is defined to

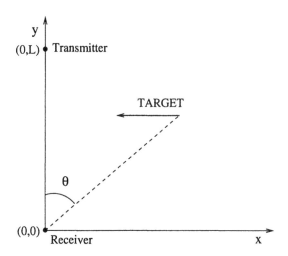

Figure 8.1 Typical target-receiver-transmitter geometry for bistatic tracking.

have the state vector

$$\mathbf{x} = [x \ \dot{x} \ y \ \dot{y}]^T. \tag{8.1}$$

The state equation can be written as:

$$\mathbf{x}_{k+1} = \mathbf{F}\mathbf{x}_k + \mathbf{v}_k, \tag{8.2}$$

where

$$\mathbf{F} = \begin{bmatrix} 1 & T & 0 & 0 \\ 0 & 1 & 0 & 0 \\ 0 & 0 & 1 & T \\ 0 & 0 & 0 & 1 \end{bmatrix}, \tag{8.3}$$

T is the sampling time, and $\mathbf{v}_k$ is zero-mean white Gaussian process noise vector with the covariance matrix

$$\mathbf{Q}_k = \begin{bmatrix} \Sigma_k & \mathbf{0}_2 \\ \mathbf{0}_2 & \Sigma_k \end{bmatrix}, \text{ where } \Sigma_k = q \begin{bmatrix} \frac{T^3}{3} & \frac{T^2}{2} \\ \frac{T^2}{2} & T \end{bmatrix}. \tag{8.4}$$

Parameter q represents the process noise intensity level [10, p. 269] and $\mathbf{0}_2$ denotes a 2×2 zero-matrix.

The target-originated measurements are available with a probability of detection P_D, and consist of noise-corrupted bearing θ and Doppler frequency δ. Here θ is the target-receiver bearing shown in Figure 8.1, and δ is the Doppler frequency shift relative to the carrier signal. This Doppler shift is related to the rate of change of the bistatic range. At time k the Doppler shift is defined as:

$$\delta_k \triangleq -\frac{1}{\lambda}\frac{d}{dt}(r_k^T + r_k^R)$$

$$= -\frac{1}{\lambda}\left\{\frac{x_k\dot{x}_k + y_k\dot{y}_k}{\sqrt{x_k^2 + y_k^2}} + \frac{x_k\dot{x}_k + (y_k - L)\dot{y}_k}{\sqrt{x_k^2 + (y_k - L)^2}}\right\} \quad (8.5)$$

where r_k^T and r_k^R are the target-transmitter and target-receiver distances, respectively, L is the distance between transmitter and receiver, and λ is the wavelength of the carrier signal. The measurement model for the target-originated measurement is then given by

$$\mathbf{z}_k = \mathbf{h}(\mathbf{x}_k) + \mathbf{w}_k \quad (8.6)$$

where

$$\mathbf{h}(\mathbf{x}_k) = \begin{bmatrix} h_\theta(\mathbf{x}_k) \\ h_\delta(\mathbf{x}_k) \end{bmatrix} \quad (8.7)$$

with

$$h_\theta(\mathbf{x}_k) = \arctan\left(\frac{x_k}{y_k}\right) \quad (8.8)$$

and $h_\delta(\mathbf{x}_k) = \delta_k$ defined by (8.5). Measurement noise $\mathbf{w}_k$ in (8.6) is modeled by a zero-mean Gaussian process with the covariance matrix

$$\mathbf{R} = \begin{bmatrix} \sigma_\theta^2 & 0 \\ 0 & \sigma_\delta^2 \end{bmatrix}. \quad (8.9)$$

The clutter measurements (false alarms) are uniformly distributed over the observation space, and their number obeys a Poisson distribution with parameter ρ.

The measurement process of the bistatic radar is characterized by the regions of blind Doppler for Doppler shifts close to zero. In this region, the forward scatter signal from the target interferes with the carrier signal and this leads to the suppression of target-originated measurements. This effect can be modeled as

$$|\delta_k| < \delta_{TH} \Rightarrow P_D = 0, \quad (8.10)$$

where δ_{TH} is referred to as the blind Doppler limit[1] (threshold).

Let us denote a set of all $n_k \geq 0$ measurements (target-originated and clutter) available at time k as $\mathbf{Z}^k = \{\mathbf{z}_k^i\}_{i=1}^{n_k}$. Given a set of measurements $\mathcal{Z}_k = \{\mathbf{Z}^1, \ldots, \mathbf{Z}^k\}$, the bistatic tracking problem is to obtain an estimate of the target state vector $\mathbf{x}_k$. The estimates are required for every $k = 1, 2, \ldots,$ as measurements are received sequentially.

---

[1] Chapter 9 considers in more detail the problem of tracking a target through the blind Doppler.

There are three salient features to note about the described bistatic tracking problem. First, the measurement processes are nonlinear functions of the target state. This necessitates the use of nonlinear filtering algorithms for this problem. Second, the availability of Doppler information, in addition to angle measurements, leads to an observable problem (in contrast to the bearings-only tracking, considered in Chapter 6, where the target is observable only if the receiver "outmaneuvers" the target [11]). Third, note that the contours of the constant bistatic range are ellipses whose focii are the transmitter and receiver positions [3]. Hence, for a given bistatic range, the Doppler shift is maximum when the target motion is perpendicular to the corresponding ellipse, and is zero (the so-called blind Doppler) when its motion is tangential to the ellipse.

8.3 CRAMÉR-RAO BOUNDS

This section develops the Cramér-Rao lower bounds for sequential estimators of the target state for the bistatic tracking problem. Bounds relevant to both Bayesian and non-Bayesian estimators are derived. For simplicity of derivation, we restrict the problem to $P_D = 1$, the clutter-free case with zero process noise and $\delta_{TH} = 0$. It has been recently shown [12, 13] that the effect of clutter and $P_D < 1$ is to increase the CRLB by an *information reduction factor*, which depends on sensor accuracy, false alarm rate, P_D, and the field of sensor observation (for gating). In the clutter-free case, this factor amounts to $1/P_D$.

8.3.1 Derivations

The problem is very similar to those treated in Chapters 6 and 7. Since the state equation (8.2) is linear and the measurement equation (8.6) is nonlinear, the CRLBs can be computed recursively using a simplified form of (4.50) given by:

$$\mathbf{J}_{k+1} = \left[\mathbf{F}^{-1}\right]^T \mathbf{J}_k \mathbf{F}^{-1} + \tilde{\mathbf{H}}_{k+1}^T \mathbf{R}^{-1} \tilde{\mathbf{H}}_{k+1}. \tag{8.11}$$

Here $\tilde{\mathbf{H}}_{k+1}$ is the Jacobian of $\mathbf{h}(\mathbf{x}_k)$ evaluated at the true state $\mathbf{x}_k$; that is,

$$\tilde{\mathbf{H}}_{k+1} = \left[\nabla_{\mathbf{x}_k} \mathbf{h}^T(\mathbf{x}_k)\right]^T \tag{8.12}$$

$$= \begin{bmatrix} \frac{\partial h_\theta}{\partial x_k} & \frac{\partial h_\theta}{\partial \dot{x}_k} & \frac{\partial h_\theta}{\partial y_k} & \frac{\partial h_\theta}{\partial \dot{y}_k} \\ \frac{\partial h_\delta}{\partial x_k} & \frac{\partial h_\delta}{\partial \dot{x}_k} & \frac{\partial h_\delta}{\partial y_k} & \frac{\partial h_\delta}{\partial \dot{y}_k} \end{bmatrix}. \tag{8.13}$$

Let us denote the receiver-target and transmitter-target ranges and their derivatives at time k by

$$r_k^R = \sqrt{x_k^2 + y_k^2} \tag{8.14}$$

$$r_k^T = \sqrt{x_k^2 + (y_k - L)^2} \tag{8.15}$$

$$\dot{r}_k^R = \frac{x_k \dot{x}_k + y_k \dot{y}_k}{\sqrt{x_k^2 + y_k^2}} \tag{8.16}$$

$$\dot{r}_k^T = \frac{x_k \dot{x}_k + (y_k - L)\dot{y}_k}{\sqrt{x_k^2 + (y_k - L)^2}}. \tag{8.17}$$

The analytic expressions for the elements of $\tilde{\mathbf{H}}_{k+1}$ are then given by:

$$\frac{\partial h_\theta}{\partial x_k} = \frac{y_k}{x_k^2 + y_k^2} \tag{8.18}$$

$$\frac{\partial h_\theta}{\partial \dot{x}_k} = 0 \tag{8.19}$$

$$\frac{\partial h_\theta}{\partial y_k} = -\frac{x_k}{x_k^2 + y_k^2} \tag{8.20}$$

$$\frac{\partial h_\theta}{\partial \dot{y}_k} = 0 \tag{8.21}$$

$$\frac{\partial h_\delta}{\partial x_k} = -\frac{1}{\lambda}\left\{\frac{\dot{x}_k r_k^R - x_k \dot{r}_k^R}{x_k^2 + y_k^2} + \frac{\dot{x}_k r_k^T - x_k \dot{r}_k^T}{x_k^2 + (y_k - L)^2}\right\} \tag{8.22}$$

$$\frac{\partial h_\delta}{\partial \dot{x}_k} = -\frac{1}{\lambda}\left\{\frac{x_k r_k^R}{x_k^2 + y_k^2} + \frac{x_k r_k^T}{x_k^2 + (y_k - L)^2}\right\} \tag{8.23}$$

$$\frac{\partial h_\delta}{\partial y_k} = -\frac{1}{\lambda}\left\{\frac{\dot{y}_k r_k^R - y_k \dot{r}_k^R}{x_k^2 + y_k^2} + \frac{\dot{y}_k r_k^T - y_k \dot{r}_k^T}{x_k^2 + (y_k - L)^2}\right\} \tag{8.24}$$

$$\frac{\partial h_\delta}{\partial \dot{x}_k} = -\frac{1}{\lambda}\left\{\frac{y_k r_k^R}{x_k^2 + y_k^2} + \frac{y_k r_k^T}{x_k^2 + (y_k - L)^2}\right\}. \tag{8.25}$$

From these equations we note that the information matrix $\mathbf{J}_k$ depends on target trajectory, sensor accuracy (determined by the covariance matrix $\mathbf{R}$), sampling interval T, and the transmitter-receiver separation L. The CRLBs of the components of $\mathbf{x}_k$ are calculated as the diagonal elements of the inverse of the information matrix (see Chapter 4).

The CRLB for non-Bayesian estimators (such as the maximum likelihood estimator) is initialized with $\mathbf{J}_0 = \mathbf{0}$, that using (8.11) leads to $\mathbf{J}_1 = \tilde{\mathbf{H}}_1^T \mathbf{R}^{-1} \tilde{\mathbf{H}}_1$.

The Bayesian estimators (such as the EKF or the particle filter) use prior knowledge about the target state (if available). Suppose the true initial target range, considered as a random variable, has a mean $\bar{R}$ and standard deviation σ_R. Furthermore, assume that the initial velocity of the target in each of the x and y directions is zero mean with standard deviations σ_{v_x} and σ_{v_y}, respectively. Then, we can construct an initial density for the state vector at time $k = 1$ using the first angle measurement θ_1 as

$$\mathbf{x}_1 \sim \mathcal{N}(\mathbf{x}_1; \bar{\mathbf{x}}_1, \mathbf{P}_1) \tag{8.26}$$

where

$$\bar{\mathbf{x}}_1 = \begin{bmatrix} \bar{R}\sin(\theta_1) \\ 0 \\ \bar{R}\cos(\theta_1) \\ 0 \end{bmatrix} \quad \mathbf{P}_1 = \begin{bmatrix} P_{11} & 0 & P_{13} & 0 \\ 0 & P_{12} & 0 & 0 \\ P_{31} & 0 & P_{33} & 0 \\ 0 & 0 & 0 & P_{44} \end{bmatrix} \tag{8.27}$$

with

$$\begin{aligned} P_{11} &= \bar{R}^2 \sigma_\theta^2 \cos^2\theta_1 + \sigma_r^2 \sin^2\theta_1 \\ P_{33} &= \bar{R}^2 \sigma_\theta^2 \sin^2\theta_1 + \sigma_r^2 \cos^2\theta_1 \\ P_{13} &= P_{31} = (\sigma_r^2 - \bar{R}^2 \sigma_\theta^2)\sin\theta_1 \cos\theta_1 \\ P_{22} &= \sigma_{v_x}^2 \\ P_{44} &= \sigma_{v_y}^2. \end{aligned} \tag{8.28}$$

The recursions in (8.11) for Bayesian estimators then start up with $\mathbf{J}_1 = \mathbf{P}_1^{-1}$.

8.3.2 Analysis

Consider again the bistatic geometry shown in Figure 8.1, where the receiver is located at the origin, while the transmitter is at $(0, 60 \text{ km})$. The receiver operates in a passive mode by processing the forward scatter signal whose frequency is set to be 429 MHz. The target, initially located at $(3 \text{ km}, 5.2 \text{ km})$, is moving in a westerly direction with a speed of 45 m/s. The angle and Doppler measurement accuracies are set to be $\sigma_\theta = 3°$ and $\sigma_\delta = 3$ Hz, respectively. A sampling time $T = 0.4$ s is used and measurements are collected for a period of 2 minutes.

Figures 8.2 to 8.6 show the square-root values of the theoretical CRLBs for non-Bayesian unbiased estimators of target position coordinates. Since the magnitudes of errors, as predicted by the CRLB, are large in the initial interval of 30 seconds, the bounds are plotted only for $t > 30$ seconds.

Figure 8.2 presents the error bounds as a function of bearings measurement standard deviation σ_θ. As expected, the errors decrease with increased measurement accuracy. For the nominal values of $\sigma_\theta = 3°$ and $\sigma_\delta = 3$ Hz, the best achievable position accuracy in 2 minutes of tracking is about 120 m. Note that the initial $\sqrt{\text{CRLB}}$ value for the position error in the x direction is smaller than that for the

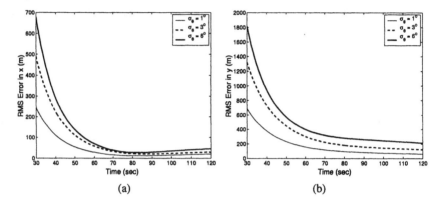

Figure 8.2 $\sqrt{\text{CRLB}}$ curves for various bearings accuracies, $\sigma_\theta = \{1°, 3°, 6°\}$: (a) x position; and (b) y position.

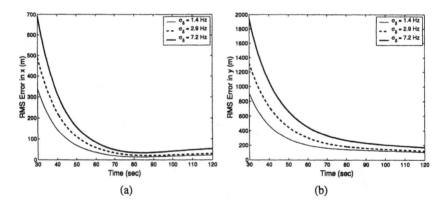

Figure 8.3 $\sqrt{\text{CRLB}}$ curves for various Doppler accuracies, $\sigma_\delta = \{1.4, 2.9, 7.2\}$ Hz: (a) x position; and (b) y position.

y direction. This can be explained as follows. Intuitively, the tracking process can be viewed as a two-stage activity. In the first stage, the Doppler measurements are integrated to obtain the bistatic range that gives an ellipse of constant bistatic range on which the target could possibly lie. The intersection of the line of sight (angle measurement) with the ellipse defines the position. For the considered geometry, this ellipse has a major axis in the vertical direction. Thus, errors in angle manifest as large errors in the y direction and small errors in the x direction.

Figure 8.3 presents the RMS error bounds for three values of σ_δ. Again, as expected, the error bound decreases with an increased accuracy of Doppler measurements.

Figure 8.4 shows a comparison of CRLBs for different sampling times. For a given observation period of 120 seconds, small values of T imply more frequent measurements and thus we can expect greater accuracy in tracking, as reflected by this graph. Comparing the values of the curves corresponding to $T = 0.1$ and $T = 0.4$, we observe that for a four-fold decrease in T, the position accuracy is improved by a factor of 2. Note, however, that this accuracy comes at the expense of increased computational load.

Two interesting features can be observed from Figure 8.5, which displays $\sqrt{\text{CRLB}}$ curves for target speed of $v_t = \{45, 150, 250\}$ m/s. First, note that the error bounds for the faster targets are initially smaller, but with time this trend reverses (at least in x coordinate). The explanation is as follows. The higher the rate of change of angle, the greater is the information content due to angle measurements. In the initial period, when the target is relatively close to the receiver, the rate of change of angle is greater for the fast targets and so their CRLBs are below that of the slow target. As time progresses, the target-receiver distance increases for the fast targets. Due to the geometry of the scenario, the rate of change of angle is smaller for the fast targets at long ranges compared to the slow target at a given instant in time. In addition, the position errors due to angle measurement error are greater at long ranges compared to short ranges. This explains the first feature of Figure 8.5.

The second feature is that for fast targets there appears to be negligible difference in performance as time progresses (for $t > 60$ seconds). The explanation is that for $t > 60$ seconds and for the considered geometry, the rate of change of angle is minimal for fast targets, and thus the new information from angular measurements is very marginal. Likewise, for the same reasons, for $t > 60$ seconds the rate of change of Doppler is minimal, leading to minimal new information due to Doppler measurements. These two effects account for the relatively stable performance depicted by the CRLB for fast targets in the interval $t > 60$ seconds.

Similar reasons account for the behavior of the CRLBs for various initial starting range values of the target, see Figure 8.6. At large ranges, the position errors are large due to: (1) smaller rate of change of angle, and (2) errors in angle measurements leading to large position errors.

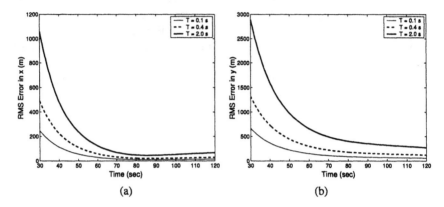

Figure 8.4 $\sqrt{\text{CRLB}}$ curves for sampling interval $T = \{0.1, 0.4, 2\}$ seconds: (a) x position; (b) y position.

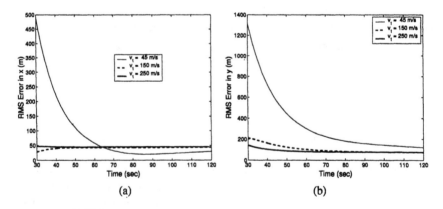

Figure 8.5 $\sqrt{\text{CRLB}}$ curves for target speeds $v_t = \{45, 150, 250\}$ m/s: (a) x position; (b) y position.

The influence of prior knowledge on CRLBs for Bayesian estimators is discussed next. The scenario is still the same that we have used so far, but the initial values of target range and speed are set to 6 km and 45 m/s, respectively. The standard deviations of velocity components are $\sigma_{v_x} = 5$ m/s, and $\sigma_{v_y} = 0.01$ m/s. Figure 8.7 shows a comparison of $\sqrt{\text{CRLB}}$ curves for various values of standard deviation of initial range, σ_R. Note that as the information content of prior knowledge of initial range is better (small σ_R), position errors are smaller. Also note from the same figure that the difference in performance between the various cases of σ_R is significant only during the initial period. The importance of prior knowledge becomes less significant as time progresses and more measurements are available.

8.4 TRACKING ALGORITHMS

The tracking approach adopted for this problem is a two-stage process depicted in Figure 8.8. At each scan, the input to the tracking algorithm consists of a set of angle and Doppler measurements, denoted by $\{\theta_k^i\}_{i=1}^{n_k}$ and $\{\delta_k^i\}_{i=1}^{n_k}$, respectively, where n_k is the number of measurements at scan k. Stage 1 consists of a tracker with data association capability that filters the angle and Doppler measurements in the measurement space to produce estimates $\hat{\theta}_k$ and $\hat{\delta}_k$. The algorithms used for this stage are: (1) probabilistic data association filter (PDAF), and (2) interacting multiple model PDAF (IMM-PDAF). The estimates from Stage 1, together with their covariances, are then passed as measurements to a nonlinear filter in Stage 2, which processes them to output the target-state estimates $\hat{\mathbf{x}}_{k|k}$. The Stage 2 processing is implemented as the EKF or the particle filter.

Two remarks are in order here. First, the estimates of bearings and Doppler that are fed into Stage 2 are, strictly speaking, correlated over time (meaning that the "measurement noise" for Stage 2 is not white, but colored). However, since the purpose of Stage 1 is primarily to perform data association, we incorporate into the PDAF large amounts of process noise, so that the described approximation has a very small effect on the overall performance.[2] Second, Stage 1 (data association) can alternatively be implemented using one of the sequential Monte Carlo techniques; see, for example, [15, 16] and Chapter 12. We adopt here the conventional techniques for data association (the PDAF) for two reasons: (1) to show that a particle filter can be easily combined with conventional data association methods; and (2) as stated in Section 3.7, particle filters should be used only for hard nonlinear/non-Gaussian problems[3] (which is not the case here).

---

2 A possible alternative would be to use equivalent measurements [14, p. 684].
3 Note, however, that the problem of joint data association and filtering using a sequential Monte Carlo method is considered in Chapter 12, in the context of group tracking.

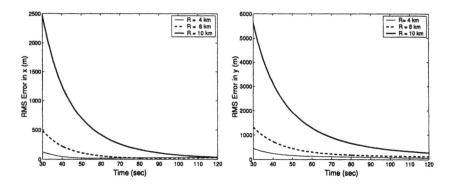

Figure 8.6 $\sqrt{\text{CRLB}}$ curves for various initial target range $\bar{R} = \{4, 6, 10\}$ km: (a) x position; and (b) y position.

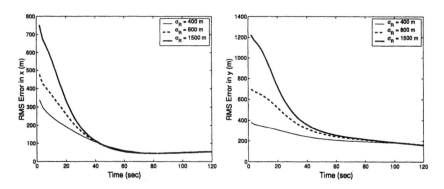

Figure 8.7 The influence of prior knowledge of target range, expressed by $\sigma_R = \{0.4, 0.8, 1.5\}$ km, on $\sqrt{\text{CRLB}}$: (a) x position; and (b) y position.

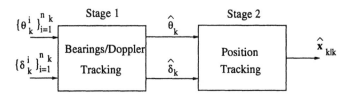

Figure 8.8 The two stages of the bistatic tracking algorithm.

8.4.1 Stage 1 of Tracker

Target and Measurement Models

To track the measurements in the bearings-Doppler space, we define a new state vector to be $\mathbf{s} = \begin{bmatrix} \theta & \dot\theta & \delta & \dot\delta \end{bmatrix}^T$, where $\dot\theta$ and $\dot\delta$ denote the rate of change of bearings and Doppler, respectively, with respect to time. Assuming a constant velocity model for the evolution of this state vector, the dynamics of the process can be written as

$$\mathbf{s}_{k+1} = \mathbf{F}\mathbf{s}_k + \mathbf{u}_k, \qquad (8.29)$$

where transition matrix $\mathbf{F}$ is given by (8.3) and $\mathbf{u}_k$ is a zero-mean white Gaussian process noise vector with the covariance matrix

$$\mathbf{Q}_u = \begin{bmatrix} T^3 q_\theta/3 & T^2 q_\theta/2 & 0 & 0 \\ T^2 q_\theta/2 & T q_\theta & 0 & 0 \\ 0 & 0 & T^3 q_\delta/3 & T^2 q_\delta/2 \\ 0 & 0 & T^2 q_\delta/2 & T q_\delta \end{bmatrix}. \qquad (8.30)$$

Here q_θ and q_δ are process noise intensity parameters for the θ and δ dimensions, respectively.

The available target originated measurement, detected with probability P_D, consists of noise corrupted angle and Doppler, modeled as

$$\mathbf{z}_k = \mathbf{H}\mathbf{s}_k + \mathbf{w}_k \qquad (8.31)$$

where

$$\mathbf{H} = \begin{bmatrix} 1 & 0 & 0 & 0 \\ 0 & 0 & 1 & 0 \end{bmatrix}$$

and $\mathbf{w}_k$ is zero-mean white Gaussian with covariance matrix given by (8.9). As described in Section 8.2, in addition to the target originated measurements, there is a possibility of false measurements. Furthermore, the target is not detected in the blind Doppler region, defined by (8.10).

The PDAF and IMM-PDAF

The probabilistic data association filter (PDAF) [17, 18] is an algorithm that estimates the state of a dynamic process in which there exists measurement origin uncertainty. In this framework, when state and measurement equations are linear (the case we have in Stage 1), the posterior state probability density function is a mixture of Gaussian pdfs [17]. The PDAF approximates this mixture by a single Gaussian density having the same mean and covariance as the mixture. Thus, at each scan, estimation is built upon a Gaussian predicted density and converted to a Gaussian mixture posterior, which is then forced back to a single Gaussian for the succeeding scan.

Suppose the target state up to time $k-1$, based on measurements up to $k-1$, is estimated as $\hat{\mathbf{s}}_{k-1|k-1}$ with associated covariance $\mathbf{P}_{k-1|k-1}$. The set of available measurements at time k was denoted as $\mathbf{Z}^k = \{\mathbf{z}_k^i\}_{i=1}^{n_k}$. Then, the operation of the PDAF based on this set of measurements can be summarized as follows:

1. Predict the target state to scan k from the prior at scan $k-1$

$$\hat{\mathbf{s}}_{k|k-1} = \mathbf{F}\hat{\mathbf{s}}_{k-1|k-1}.$$

2. Compute the innovation covariance for the predicted measurement $\hat{\mathbf{z}}_{k|k-1} = \mathbf{H}\hat{\mathbf{s}}_{k|k-1}$ as

$$\mathbf{S}_k = \mathbf{H}\mathbf{P}_{k|k-1}\mathbf{H}^T + \mathbf{R}$$

where

$$\mathbf{P}_{k|k-1} = \mathbf{F}\mathbf{P}_{k-1|k-1}\mathbf{F}^T + \mathbf{Q}_u.$$

3. A validation "gate" centered around the predicted measurement is set up to select the set of measurements to be associated probabilistically to the target. The validation region satisfies

$$(\mathbf{z}_k - \hat{\mathbf{z}}_{k|k-1})^T \mathbf{S}_k^{-1} (\mathbf{z}_k - \hat{\mathbf{z}}_{k|k-1}) \leq \gamma,$$

where γ determines the size of the gate. Suppose m_k of the n_k measurements fall in the gate. Denote this set by $\mathcal{Z}^k = \{\mathbf{z}_k^{i_j}\}_{j=1}^{m_k}$, and form innovations as

$$\tilde{\mathbf{z}}_k^j = \mathbf{z}_k^{i_j} - \hat{\mathbf{z}}_{k|k-1}, \quad j = 1, \ldots, m_k.$$

4. Calculate the association probabilities β_k^j, $j = 1, \ldots, m_k$, where β_k^j is the posterior probability that measurement $\mathbf{z}_k^{i_j}$ is from the target,

$$\beta_k^j = \begin{cases} c \dfrac{1}{\sqrt{|2\pi\mathbf{S}_k|}} e^{-[\tilde{\mathbf{z}}_k^j]^T \mathbf{S}_k^{-1} \tilde{\mathbf{z}}_k^j}, & j = 1, \ldots, m_k \\ c \dfrac{(1-P_D P_G)m_k}{P_D V_G}, & j = 0, \end{cases}$$

where β_k^0 is the posterior probability that all measurements at this scan are clutter originated, and c is a constant which ensures that $\sum_{j=0}^{m_k} \beta_k^j = 1$. Here P_G is the probability that the target-originated measurement falls in the validation region (determined by γ), and

$$V_G = \gamma^{n_z/2} c_{n_z} |\mathbf{S}_k|$$

is the volume of the validation gate, where c_{n_z} is the volume of the unit hypersphere of dimension n_z, the dimension of the measurement z_k. In our case $n_z = 2$ and $c_{n_z} = \pi$ [18, p. 96].

5. Use the above association probabilities β_k^j to update the state estimate as

$$\hat{\mathbf{s}}_{k|k} = \hat{\mathbf{s}}_{k|k-1} + \mathbf{K}_k \tilde{\mathbf{z}}_k$$

where $\mathbf{K}_k = \mathbf{P}_{k|k-1} \mathbf{H}^T \mathbf{S}_k^{-1}$ is the Kalman gain, and

$$\tilde{\mathbf{z}}_k = \sum_{j=1}^{m_k} \beta_k^j \tilde{\mathbf{z}}_k^j$$

is the aggregate innovation.

6. The covariance associated with the updated state is computed as

$$\mathbf{P}_{k|k} = \mathbf{P}_{k|k-1} - (1 - \beta_k^0) \mathbf{K}_k \mathbf{S}_k \mathbf{K}_k^T + \tilde{\mathbf{P}}_k,$$

where $\tilde{\mathbf{P}}_k$ is a measurement dependent term, called the "spread of the innovations,"

$$\tilde{\mathbf{P}}_k = \mathbf{K}_k \left[\sum_{j=1}^{m_k} \beta_k^j \tilde{\mathbf{z}}_k^j [\tilde{\mathbf{z}}_k^j]^T - \tilde{\mathbf{z}}_k \tilde{\mathbf{z}}_k^T \right] \mathbf{K}_k^T.$$

The above steps describe the operation of the PDAF at a particular scan. As measurements are received sequentially, the algorithm proceeds in a recursive manner with the same steps as outlined above.

The IMM estimator, described in Section 2.3.2, is a popular Bayesian filtering technique for tracking targets with switching modes of operation. The IMM-PDAF method [18] combines elements of the IMM estimator with the PDAF described above. The resulting filter is a solution for multiple-model tracking using measurements of uncertain origin. The key idea is to use a bank of PDAFs as the mode-matched filters in the IMM.

For the bistatic tracking problem, we assume a nonmaneuvering target, and use the following three target models in the IMM: (1) "no target" model, with

$P_D = 0$, (2) a constant velocity model with $P_D = 0$ for a target in blind Doppler zone (BDZ), and (3) a constant velocity model with $P_D \neq 0$ for a target outside BDZ. For the first two models, the updated state and covariance are essentially the predicted state and covariance, respectively. The state dependent transition probability matrix associated with the evolution of the models is of the form

$$\Pi_{k+1} = \begin{bmatrix} p_{11}(k+1) & p_{12}(k+1) & p_{13}(k+1) \\ p_{21}(k+1) & p_{22}(k+1) & p_{23}(k+1) \\ p_{31}(k+1) & p_{32}(k+1) & p_{33}(k+1) \end{bmatrix}, \qquad (8.32)$$

where the elements $p_{ij}(k+1)$ can be computed as follows. Suppose the probabilities of a target appearing and disappearing are given by

$$P(\text{target appearing}) = P(\text{target disappearing}) = \epsilon \qquad (8.33)$$

Then, the first row of Π_{k+1} can be set to

$$p_{12}(k+1) = p_{13}(k+1) = \epsilon/2, \quad p_{11}(k+1) = 1 - \epsilon. \qquad (8.34)$$

Now consider the computation of the elements of the second row of Π_{k+1}. The transition probability to model 1 is essentially the target disappearing probability, hence $p_{21}(k+1) = \epsilon$. By defining the event

$$\bar{D} \triangleq \text{"target does not disappear"},$$

the element $p_{22}(k+1)$ can be computed as

$$\begin{aligned} p_{22}(k+1) &= P(\text{target in BDZ at } k+1 | \text{target in BDZ at } k) & (8.35) \\ &= P(\bar{D})P(\text{target in BDZ at } k+1 | \text{target in BDZ at } k, \bar{D}) & (8.36) \\ &= (1-\epsilon)\tilde{p}_{22}(k+1) & (8.37) \end{aligned}$$

where

$$\begin{aligned} \tilde{p}_{22}(k+1) &= P(\text{target in BDZ at } k+1 | \text{target in BDZ at } k, \bar{D}) & (8.38) \\ &= \frac{1}{\sqrt{2\pi\sigma_{\hat{\delta}_{k+1|k}}^2}} \int_{-\delta_{TH}}^{\delta_{TH}} e^{-\frac{(\delta_{k+1} - \hat{\delta}_{k+1|k})^2}{2\sigma_{\hat{\delta}_{k+1|k}}^2}} d\delta_{k+1}. & (8.39) \end{aligned}$$

where $\hat{\delta}_{k+1|k}$ and $\sigma_{\hat{\delta}_{k+1|k}}$ are filter calculated predicted estimates of δ_{k+1} and its standard deviation, respectively. Similarly,

$$\begin{aligned} p_{23}(k+1) &= (1-\epsilon)\tilde{p}_{23}(k+1) & (8.40) \\ &= (1-\epsilon)(1 - \tilde{p}_{22}(k+1)). & (8.41) \end{aligned}$$

The computation of the third row of Π_{k+1} follows the same lines of argument as above.

The IMM-PDAF with the above models can be used to initiate and terminate tracks as well as to maintain tracks through the blind Doppler region[4] [18].

Track Initiation and Termination

At each scan, any measurement that does not fall within the gate of any of the current tracks is stored to form possible associations with unassociated measurements in the next scan. Those associations that are plausible form "tentative" tracks. The plausible associations are those with the rate of Doppler change $\dot{\delta}$ and bearing change $\dot{\theta}$ such that:

$$|\dot{\delta}| \leq \delta_{\max} \quad \text{and} \quad |\dot{\theta}| \leq \theta_{\max}. \tag{8.42}$$

Each tentative track is propagated and updated according to the PDAF equations given above. For a standard PDAF filter applied to the bistatic tracking problem, the track status is promoted to "confirmed" if the detections in the following scans satisfy the $2/2 \times r/s$ cascaded logic [18]. That is, it requires 2 detections in consecutive scans (counting the initial detection as the first detection), followed by r detections out of s scans to be promoted. If the requirement is not satisfied, the track is deleted. A confirmed track is terminated according to the r/s logic. That is, a track is terminated if r missed detections are reported in a sliding window of s scans.

The explicit initiation and termination logics discussed above are not required for the IMM-PDAF algorithm. This filter uses the true target probability (TTP) [18, p. 208], given by

$$\begin{align} \text{TTP} &= P(\text{"target is present"}) \tag{8.43} \\ &= w_k^2 + w_k^3 \tag{8.44} \end{align}$$

to initiate and terminate tracks. Coefficients w_k^2 and w_k^3 in (8.44) denote IMM model 2 and 3 probabilities (or weights), respectively. When TTP exceeds a predefined threshold, a track is initiated; likewise, when TTP falls below a threshold, the track is terminated.

Track Maintenance in the Blind Doppler

For a standard PDAF filter applied to this problem, the track maintenance in the blind Doppler zone is done as follows. A target is deemed to be in the blind Doppler

---

[4] Note that a more efficient PDAF based algorithm (requiring only one model) could be designed for this problem, using the target existence (visibility) concept [19, 20]. This concept is implemented using particles in Chapter 11.

region if its estimate of the Doppler frequency $\hat{\delta}_k$ at time k satisfies

$$|\hat{\delta}_k| < \delta_{TH}. \qquad (8.45)$$

Thereafter, the track corresponding to this target is maintained by extrapolation according to the target dynamics. Specifically, if $\hat{s}_{k-1|k-1}$ and $\mathbf{P}_{k-1|k-1}$ denote the target state estimate and covariance at $k-1$, they are updated according to

$$\hat{s}_{k|k} = \mathbf{F}\hat{s}_{k|k-1} \qquad (8.46)$$
$$\mathbf{P}_{k|k} = \mathbf{F}\mathbf{P}_{k-1|k-1}\mathbf{F}^T + \mathbf{Q}_u. \qquad (8.47)$$

The condition (8.45) is tested in subsequent scans to determine if the target has exited the blind Doppler zone. If so, the tracker resumes normal operation and the track is updated according to standard PDAF equations.

For the IMM-PDAF, there is no explicit logic required for track maintenance in the Doppler zone. Model 2, which corresponds to a target in the Doppler zone, takes care of this implicitly.

8.4.2 Stage 2 of Tracker

The filtered estimate (the output of Stage 1), $\hat{s}_{k|k} = \begin{bmatrix} \hat{\theta}_k, & \hat{\delta}_k \end{bmatrix}^T$, is passed as the "measurement" to Stage 2. Its covariance $\mathbf{P}_{k|k}$ is supplied as the measurement covariance. Stage 2 of the tracking algorithm computes the estimate of the target state vector defined by (8.1). The target dynamics and measurement models applicable at this stage are identical to those described in Section 8.2 except for one difference: there is no measurement origin uncertainty because it has been resolved in Stage 1 of the tracker. The output of Stage 2 is denoted by $\hat{\mathbf{x}}_{k|k}$.

The general recursive equations of the EKF were presented in Section 2.1. For the special case where state dynamics is linear and the measurement equation is nonlinear, relationships (7.34)–(7.39) are applicable (note however that the bistatic radar receiver is not moving, hence $\mathbf{U}_{k+1,k}$ in (7.34) is zero). The Jacobian required in the EKF formulation is given in Section 8.3.1. This Jacobian is evaluated at the predicted state $\hat{\mathbf{x}}_{k|k-1}$.

The particle filter was implemented as the regularized PF (RPF) described in Section 3.5.3. Since $n_x = 4$ in this application, constant c_{n_x} that features in (3.50) is equal to $\pi^2/2$ [18, p. 96].

8.5 ALGORITHM PERFORMANCE

This section studies the comparative performance of the tracking algorithms described in Section 8.4. The performance measure is the RMS error, which for the

jth component of the state vector is specified as:

$$\sigma_{\mathbf{e}_k[j]} = \sqrt{\mathrm{E}\{(\hat{\mathbf{x}}_{k|k}[j] - \mathbf{x}_k[j])^2\}}, \qquad (8.48)$$

where $\mathbf{x}_k$ is the true target state vector. The expectation operator in (8.48) is carried out by averaging over 100 independent Monte Carlo runs.

The tracking scenario is identical to the one shown in Figure 8.1 and considered in Section 8.3.2: The target moves in a westerly direction with the speed of 45 m/s. The initial range of the target is distributed as $R \sim \mathcal{N}(R; \bar{R}, \sigma_R^2)$, where $\bar{R} = 6$ km and $\sigma_R = 1$ km. Bistatic measurements of Doppler and bearing are collected at the receiver at a sampling rate $T = 2$ sec for a period of 2 minutes. The detection probability and the Doppler cutoff frequency for the blind Doppler zone are set to be $P_D = 0.95$, and $\delta_{TH} = 10$ Hz, respectively. In addition, clutter measurements were generated with a density of 8.3×10^{-5} deg^{-1}Hz^{-1}. All other parameters are set according to Section 8.3.2.

The r/s track initiation and termination parameters for the PDAF were set to be $r_{init} = 4$, $s_{init} = 5$, $r_{term} = 10$, and $s_{term} = 10$. Likewise, the initiation and termination true target probabilities for the IMM-PDAF were set to be 0.9 and 0.02, respectively. In addition, both filters require the following parameters that relate to the dynamics of the evolution of the measurement process: the process noise intensity parameters $q_\theta = 1 \times 10^{-3}$, $q_\delta = 8 \times 10^{-3}$; the maximum allowable rates of change in track formation $\dot{\theta}_{\max} = 0.6$ deg/s and $\dot{\delta}_{\max} = 0.5$ Hz/s. Parameter ϵ of (8.33) was set to 0.04.

A typical output of Stage 1 of the tracker is shown in Figure 8.9. The solid thick line shows the true trajectory in the measurement space, while the thin line shows the estimated trajectory. The crosses represent measurements, which are either target or clutter originated. The blind Doppler zone is also indicated, where no measurements are received from the target. In this region, the tracker maintains tracks by extrapolation, based on assumed target dynamics.

Figure 8.10 shows the RMS error curves in bearing and Doppler space for PDAF and IMM-PDAF. Both filters exhibit comparable performance for the considered scenario.[5] The RMS errors grow in the middle of the observation period due to the lack of measurements – during this period the target is in the blind Doppler region.

Consider next the performance of two Bayesian filters (EKF and RPF) for Stage 2 of tracking. Figure 8.11 shows the RMS errors and the corresponding posterior CRLB for the EKF and RPF trackers. These trackers were initialized according to (8.26)–(8.28), and the number of particles for the RPF was set to be $N = 5000$. For this particular scenario, both EKF and RPF exhibit comparable

---

[5] We could expect an improved performance using the IMM-PDAF when tracking maneuvering targets, but this would require additional dynamic models in the IMM.

198   Beyond the Kalman Filter: Particle Filters for Tracking Applications

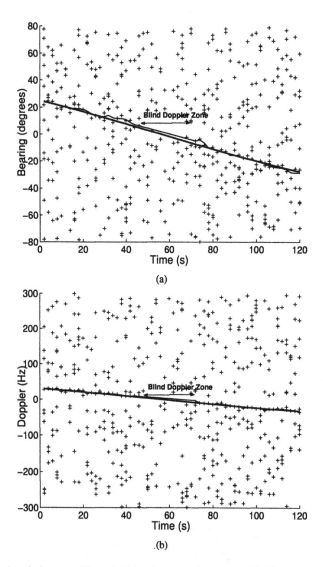

Figure 8.9 A typical output of Stage 1: (a) bearing versus time plot; and (b) Doppler versus time plot.

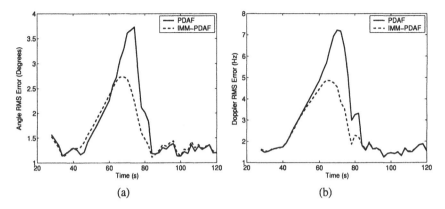

Figure 8.10 Stage 1 RMS errors: (a) bearing, and (b) Doppler.

performance, although the RPF is marginally better. From Figure 8.11 it is also evident that both EKF and RPF are very close to the CRLB corresponding to the Bayesian estimators. Although the EKF shows comparable performance to the RPF, it must be noted that this comparable performance was achieved only for this particular case, where both filters were initialized with a tight initial density. When the initial pdf is vague, the performance of the EKF degrades as will be seen in the next figure.

The last set of numerical results deals with the overall performance of the tracker, with both Stages 1 and 2 combined. For this experiment, we deliberately set bad initialization parameters for the nonlinear filters to test the extent to which they can cope with a bad initial density. In particular, the mean of the initial range to the filter was set to be 24 km away from the true mean, with a standard deviation of 8 km. Figure 8.12 compares the performances of the IMM-PDAF-EKF and IMM-PDAF-RPF for this example. When the initial pdf is poor, it can be seen that the IMM-PDAF-EKF shows a tendency to diverge, while the estimation errors for the IMM-PDAF-RPF continue to decrease with time.

The better performance of the RPF, however, comes at the expense of increased computation time. We found that this particular implementation of the particle filter requires about 100 times more CPU time than the EKF, for a single run of the algorithm.

8.6 SUMMARY

The chapter presented an investigation of tracking algorithms and their performance for the bistatic radar using the measurements of target bearings and Doppler only.

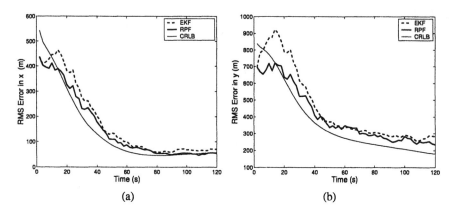

Figure 8.11 Performance of EKF and RPF in Stage 2: (a) RMS error in x direction; (b) RMS error in y direction.

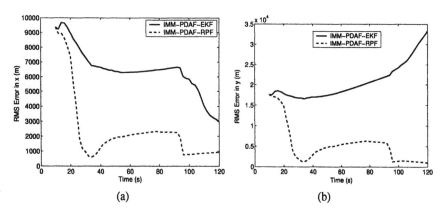

Figure 8.12 Overall performance of the tracker with poor initialization: (a) RMS error in x direction; (b) RMS error in y direction. The particle filter-based solution (dashed line) converges, while the EKF-based tracker (solid line) does not.

This is an important problem for air surveillance, as bistatic radars are becoming increasingly popular due to their better detection capabilities of stealth targets and reduced vulnerability to jamming.

The two considered nonlinear filters (the EKF and the particle filter) demonstrated comparable error performance when the available initial density was "tight." Moreover, this performance is in agreement with the theoretical CRLB. When the initial density is vague, however, the EKF showed a tendency to diverge, while the particle filter exhibited good performance. The described tracking algorithms have been recently applied to a set of experimental real data [9] and the superiority of the particle filter has been confirmed.

References

[1] E. Hanle, "Survey of bistatic and multistatic radar," *IEE Proceedings, Part F*, vol. 133, pp. 587–595, December 1986.

[2] J. I. Glaser, "Fifty years of bistatic and multistatic radar," *IEE Proceedings, Part F*, vol. 133, pp. 596–603, December 1986.

[3] M. C. Jackson, "The geometry of bistatic radar systems," *IEE Proceedings, Part F*, vol. 133, pp. 604–612, December 1986.

[4] N. J. Willis, *Bistatic Radar*. Norwood, MA: Artech House, 1991.

[5] N. J. Willis, "Bistatic radars and their third resurgence: Passive coherent location." *IEEE Radar Conference Tutorial Notes*, April 2002.

[6] J. M. Hawkins, "An opportunistic bistatic radar," in *Proc. Radar 97 Conf.*, October 1997. IEE Publ. No. 449.

[7] A. Farina, "Tracking function in bistatic and multistatic radar systems," *IEE Proceedings, Part F*, vol. 133, pp. 630–637, December 1986.

[8] P. E. Howland, "Target tracking using television-based bistatic radar," *IEE Proceedings, Radar, Sonar and Navigation*, vol. 146, no. 3, pp. 166–174, 1999.

[9] M. S. Arulampalam, N. Gordon, S. Maskell, and W. J. Fitzgerald, "Bistatic Radar Tracking Using Bearings and Doppler Measurements," Tech. Rep. DSTO-TR-1303, Defence Science and Technology Organisation, Edinburgh, Australia, April 2002.

[10] Y. Bar-Shalom, X. R. Li, and T. Kirubarajan, *Estimation with Applications to Tracking and Navigation*. New York: John Wiley & Sons, 2001.

[11] J. M. Passerieux and D. V. Cappel, "Optimal observer maneuver for bearings-only tracking," *IEEE Trans. Aerospace Electron. Syst.*, vol. 34, no. 3, pp. 777–788, 1998.

[12] M. L. Hernandez, A. D. Marrs, N. J. Gordon, S. R. Maskell, and C. M. Reed, "Cramér-Rao bounds for non-linear filtering with measurement origin uncertainty," in *Proc. 5th Int. Conf. Information Fusion*, vol. 1, (Annapolis, MD), pp. 18–25, 2002.

[13] M. Hernandez, B. Ristic, A. Farina, and L. Timmoneri, "A comparison of two Cramér-Rao bounds for nonlinear filtering with $P_d < 1$," *IEEE Trans. Signal Processing*. (to be published in 2004).

[14] S. Blackman and R. Popoli, *Design and Analysis of Modern Tracking Systems*. Norwood, MA: Artech House, 1999.

[15] C. Hue, J. L. Cadre, and P. Pérez, "Tracking multiple objects with particle filtering," *IEEE Trans. Aerospace and Electronic Systems*, vol. 38, pp. 791–812, July 2002.

[16] M. Morelande, S. Challa, and N. J. Gordon, "Study of the application of particle filters to single target tracking problems," in *Proc. SPIE, Signal and Data Processing of Small Targets*, vol. 5204, August 2003.

[17] Y. Bar-Shalom and T. E. Fortmann, *Tracking and Data Association*. Boston, MA: Academic Press, 1988.

[18] Y. Bar-Shalom and X. R. Li, *Multitarget-Multisensor Tracking: Principles and Techniques*. Storrs, CT: YBS Publishing, 1995.

[19] S. B. Colegrove, A. W. Davis, and J. K. Ayliffe, "Track initiation and nearest neighbours incorporated into probabilistic data association," *Journal of Electrical and Electronics Engineers, Australia*, vol. 6, pp. 191–198, September 1986.

[20] D. Musicki, R. Evans, and S. Stankovic, "Integrated probabilistic data association," *IEEE Trans. Automatic Control*, vol. 39, pp. 1237–1240, June 1994.

Chapter 9

Tracking Targets Through the Blind Doppler

9.1 INTRODUCTION

Most combat aircraft are equipped with some form of electronic warfare intelligence system, such as the electronic support measures (ESM) sensor or a radar warning receiver (RWR). These systems in general detect nearby RF emissions, process them in real time, and report to the pilot and/or to the mission computer: (1) where an RF emission comes from (its angle of arrival) and (2) what is the likely identification (ID) of its source [1, Ch. 36]. In a hostile environment, there are various electronic protection (EP) measures available to the pilot that can hide the aircraft's true position from the enemy radar. The aim is to protect the aircraft from threat radar systems that might deploy radar guided weapons against the platform. This can be achieved by preventing or delaying the formation of a track on the target or by causing a track loss if the track has been initiated. Typical EP measures are noise jamming, deceptive jamming, and chaff [1, 2]. However, one of the simplest and most effective EP measures against any CW or pulse Doppler radar is to hide in what is commonly referred to as the radar blind-Doppler zone (BDZ).

The blind Dopplers are the bands of Doppler frequency falling within the rejection notches covering the regions around zero Doppler and the integer multiples of the pulse repetition frequency (PRF). The function of these notch filters is to remove the ground clutter and the ground-moving vehicles from radar echoes. The notch filter cutoff frequency can be up to ± 80 knots ($\approx \pm 150$ km/h) in airborne radars [3, p. 314].

Hiding in the blind Doppler is aided by the on-board ESM or the RWR, since either system indicates to the pilot the direction of the enemy radar. The pilot can then maneuver to reduce the aircraft's radial velocity by flying tangentially with respect to the enemy radar. This reduction in radial velocity is often combined with the release of chaff. While hiding in the blind Doppler is only a temporary measure, it can often cause a loss of track. Once the target comes out of the blind Doppler a new track has to be initiated and the target has to be identified. By the time the new track is established, the target could be lost again because it hides once again in the blind Doppler. In this "cat and mouse" game, track continuity is the most important performance measure, since the delays in track/ID initiation can be fatal.

With the standard Kalman type filters it is difficult to exploit prior knowledge of the geographical or sensor limitations, such as the position of mountains or the extent of the blind Doppler zones. These effects introduce gross nonlinearities in the form of "hard edges" on probability distributions. These hard constraints however, can be fairly easily incorporated into the framework of sequential Monte Carlo (SMC) estimation techniques. In this chapter we compare two filters for tracking an airborne target that temporarily hides in the blind Doppler. The available radar measurements are target range, azimuth, and range rate (obtained from Doppler). Two filters are used for comparison: the extended Kalman filter (EKF) and the particle filter. Both filters assume a constant velocity target motion with a fair amount of process noise to handle the possible maneuvers. The EKF is tracking the target using only radar measurements. The particle filter, in addition to the radar measurements, is exploiting prior knowledge of the blind Doppler zone limits. This chapter is based on the material presented in [4]. The main results have been independently verified in [5].

9.2 PROBLEM FORMULATION

Let the true target state at discrete time t_k be defined as $\mathbf{x}_k = [x_k \ \dot{x}_k \ y_k \ \dot{y}_k]^T$, where x_k and y_k denote the position in Cartesian coordinates and $\dot{x}_k$ and $\dot{y}_k$ are velocities. The polar measurements of the target location at t_k (range r_k and azimuth θ_k) obtained from a radar are conveniently converted into the Cartesian system using the debiasing transformations of [6]; no such linearization is necessary for the particle filter but we use the same procedure for convenience.

The target motion model is the constant velocity (CV) model, with an appropriate level of process noise to deal with possible maneuvers. In order to simplify analysis we assume a static tracking radar platform. We adopt the following model of target state dynamics:

$$\mathbf{x}_{k+1} = \mathbf{F}_k \mathbf{x}_k + \mathbf{G}_k \mathbf{v}_k \tag{9.1}$$

where

$$\mathbf{F}_k = \begin{bmatrix} 1 & T_k & 0 & 0 \\ 0 & 1 & 0 & 0 \\ 0 & 0 & 1 & T_k \\ 0 & 0 & 0 & 1 \end{bmatrix}, \quad \mathbf{G}_k = \begin{bmatrix} T_k^2/2 & 0 \\ T_k & 0 \\ 0 & T_k^2/2 \\ 0 & T_k \end{bmatrix}, \tag{9.2}$$

$T_k = t_{k+1} - t_k$, $\mathbf{v}_k$ is a 2×1 white Gaussian process noise vector with covariance matrix $\mathbf{Q} = \text{diag}[\sigma_v^2, \sigma_v^2]$.

The measurement equation is given by:

$$\mathbf{z}_k = \mathbf{h}(\mathbf{x}_k) + \mathbf{w}_k \tag{9.3}$$

where $\mathbf{z}_k = [X_k \ Y_k \ \dot{r}_k]^T$ is the measurement vector at time t_k. It consists of the Cartesian target position components $X_k = \lambda^{-1} \cdot r_k \cdot \cos\theta_k$, $Y_k = \lambda^{-1} \cdot r_k \cdot \sin\theta_k$, $\lambda = exp(-\sigma_\theta^2/2)$ and the target range rate measurement $\dot{r}_k$. For a moving radar platform, the range rate would have to be compensated for the platform motion. The error statistics of radar measurements are given in terms of the range standard deviation σ_r, range rate standard deviation $\sigma_{\dot{r}}$, and the azimuth standard deviation σ_θ. With these statistics, the position variances in the respective directions and their crosscovariance are as follows [6]

$$\sigma_{X_k}^2 = (\lambda^{-2} - 2)r_k^2 \cos^2\theta_k + (r_k^2 + \sigma_r^2)(1 + \lambda^4 \cos 2\theta_k)/2 \tag{9.4}$$

$$\sigma_{Y_k}^2 = (\lambda^{-2} - 2)r_k^2 \sin^2\theta_k + (r_k^2 + \sigma_r^2)(1 - \lambda^4 \cos 2\theta_k)/2 \tag{9.5}$$

$$\sigma_{X_k Y_k}^2 = (\lambda^{-2} - 2)r_k^2 \cos\theta_k \sin\theta_k + (r_k^2 + \sigma_r^2)\lambda^4 \sin 2\theta_k/2. \tag{9.6}$$

The nonlinear measurement function $\mathbf{h}(\mathbf{x}_k)$ is defined as:

$$\mathbf{h}(\mathbf{x}_k) = \begin{bmatrix} x_k \\ y_k \\ \frac{x_k \dot{x}_k + y_k \dot{y}_k}{\sqrt{x_k^2 + y_k^2}} \end{bmatrix} = \begin{bmatrix} \mathbf{x}_k[1] \\ \mathbf{x}_k[3] \\ \frac{\mathbf{x}_k[1]\mathbf{x}_k[2] + \mathbf{x}_k[3]\mathbf{x}_k[4]}{\sqrt{\mathbf{x}_k^2[1] + \mathbf{x}_k^2[3]}} \end{bmatrix} \tag{9.7}$$

where as usual $\mathbf{x}_k[i]$ denotes the ith component of the state vector. Measurement noise $\mathbf{w}_k$ in (9.3) is a 3×1 zero-mean Gaussian noise vector with covariance matrix

$$\mathbf{R}_k = \begin{bmatrix} \sigma_{X_k}^2 & \sigma_{X_k Y_k}^2 & 0 \\ \sigma_{X_k Y_k}^2 & \sigma_{Y_k}^2 & 0 \\ 0 & 0 & \sigma_{\dot{r}}^2 \end{bmatrix}. \tag{9.8}$$

The assumption is that $\mathbf{v}_k$ and $\mathbf{w}_k$ are independent.

A simple model of a 2D radar is assumed for the purpose of studying the problem of tracking targets occasionally hidden in the blind Doppler. The probability of detection according to this model is:

$$P_D(\mathbf{x}_k) = \begin{cases} P_d, & \text{if } \left|\frac{x_k \dot{x}_k + y_k \dot{y}_k}{\sqrt{x_k^2 + y_k^2}}\right| \geq L_0 \\ 0, & \text{otherwise} \end{cases} \tag{9.9}$$

where L_0 is the limit of the blind Doppler (compensated for the platform motion) and P_d is a positive constant less or equal to unity. The sampling interval (frame time) of the radar, T, is assumed to be constant.

In this study we are concerned only with the track maintenance aspect of the tracking filter. Hence we ignore the track initiation intricacies, and make an additional assumption that the track on the target has been established.

The problem is to investigate the performance of a standard EKF type filter (as an example of a filter that ignores the existence of blind Doppler zones) against another more sophisticated filter that uses prior knowledge of L_0 in (9.9). The latter was developed using the SMC techniques mainly because any nonstandard form of information (such as prior knowledge of L_0) can be easily incorporated into the framework of the SMC estimation (particle filters). An alternative design can be based on the Gaussian sum filter [7, 8].

To illustrate the problem further we show in Figure 9.1 a typical scenario of interest. The target trajectory (solid line), the radar location (indicated by a triangle), and a set of radar measurements (indicated by little squares) are all shown in Figure 9.1(a). The target motion is generated without process noise. The target is at first moving radially towards the radar at a speed of 800 km/h. Then it makes a $3g$ turn ($g = 9.81$ m/s^2 is acceleration due to gravity) to its right and continues the tangential motion with respect to the radar for about 25 seconds. Finally the target makes another $3g$ turn and moves again towards the radar with a high range rate. The target range rate and the corresponding measurements are shown in Figure 9.1(b) as the function of time. This figure also indicates the BDZ limits. Note that while the target is in the BDZ (the second leg of target trajectory), there are no target measurements. Figure 9.1(c) shows the time evolution of the target "inverse time-to-go" (TTG^{-1}), defined as the ratio between the target range rate and range. TTG^{-1} is often used in tracking systems to prioritize targets according to the potential threat [9].

9.3 EKF-BASED TRACK MAINTENANCE

A simple tracker based on the standard extended Kalman filter (EKF), which ignores the existence of BDZ, is developed as follows. Although the sampling interval T is constant, the state vector will be updated only when the target is detected and the measurement is gated. Let us denote the time when the state vector was updated for the last time as t_{k_0}. Then the predicted state at time $t_k > t_{k_0}$ is:

$$\hat{\mathbf{x}}_{k|k_0} = \mathbf{F}_k \hat{\mathbf{x}}_{k_0|k_0} \qquad (9.10)$$

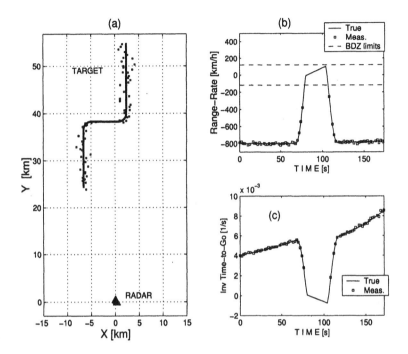

Figure 9.1 A typical tracking scenario for a target hiding in the blind Doppler: (a) target trajectory; (b) range rate versus time; and (c) TTG^{-1} versus time.

and the associated predicted state covariance is

$$\mathbf{P}_{k|k_0} = \mathbf{F}_k \mathbf{P}_{k_0|k_0} \mathbf{F}_k^T + \mathbf{G}_k \mathbf{Q} \mathbf{G}^T, \quad (9.11)$$

where $\mathbf{F}_k$ and $\mathbf{G}_k$ are the functions of $T_k = t_k - t_0$, which is a multiple of T. Measurements are gated using ellipsoidal gates in x and y directions only (i.e., range rate is not used for gating). If a measurement is detected and gated at time t_k, the state vector will be updated. For this we use the EKF equations following [10]. The predicted measurement is:

$$\hat{\mathbf{z}}_{k|k_0} = \mathbf{h}(\hat{\mathbf{x}}_{k|k_0}) \quad (9.12)$$

and its associated innovation covariance matrix is

$$\mathbf{S}_k = \mathbf{H}_k \mathbf{P}_{k|k_0} \mathbf{H}_k^T + \mathbf{R}_k \quad (9.13)$$

where $\mathbf{H}_k$ is the Jacobian of $\mathbf{h}(\mathbf{x}_k)$ evaluated as [10]:

$$\mathbf{H}_k = \begin{bmatrix} 1 & 0 & 0 & 0 \\ 0 & 0 & 1 & 0 \\ l_{k|k_0}\sin\theta_{k|k_0} & \cos\theta_{k|k_0} & -l_{k|k_0}\cos\theta_{k|k_0} & \sin\theta_{k|k_0} \end{bmatrix} \quad (9.14)$$

with

$$\theta_{k|k_0} = \arctan\left(\frac{\hat{\mathbf{x}}_{k|k_0}[3]}{\hat{\mathbf{x}}_{k|k_0}[1]}\right) \quad (9.15)$$

$$l_{k|k_0} = \frac{\hat{\mathbf{x}}_{k|k_0}[2]\sin\theta_{k|k_0} - \hat{\mathbf{x}}_{k|k_0}[4]\cos\theta_{k|k_0}}{\sqrt{\hat{\mathbf{x}}_{k|k_0}^2[1] + \hat{\mathbf{x}}_{k|k_0}^2[3]}}. \quad (9.16)$$

The state estimate is updated using:

$$\hat{\mathbf{x}}_{k|k} = \hat{\mathbf{x}}_{k|k_0} + \mathbf{K}_k \nu_k \quad (9.17)$$

where $\mathbf{K}_k$ is the filter gain given by

$$\mathbf{K}_k = \mathbf{P}_{k|k_0} \mathbf{H}_k^T \mathbf{S}_k^{-1} \quad (9.18)$$

and $\nu_k = \mathbf{z}_k - \hat{\mathbf{z}}_{k|k_0}$ is the innovation. The covariance matrix associated with $\hat{\mathbf{x}}_{k|k}$ is given by

$$\mathbf{P}_{k|k} = (\mathbf{I} - \mathbf{K}_k \mathbf{H}_k)\mathbf{P}_{k|k_0}. \quad (9.19)$$

If no measurement is detected or gated, the EKF will report the predicted state and its associated covariance matrix, but will not change the state estimate at k_0.

9.4 PARTICLE FILTER-BASED SOLUTION

This section outlines a particle filter that allows information about the limits of the Doppler blind zone to be directly utilized in the tracking process. By doing this we have a situation where receiving no measurement actually conveys useful information about the probable target location and velocity. A string of periods with no measurement is then strongly indicative of a target hiding in the blind zone of the radar and the particles spread out to cover the possible trajectories and locations where this is possible and effectively wait for the target to appear out of the blind zone. Hence we expect increased probability of track maintenance. The filter we create draws on stratified sampling ideas from [11] and the multiple-model PF ideas described in Section 3.5.5. The basic idea is to have one set of particles restricted to the motion within the Doppler blind zone and the other set able to move unrestricted. The two sets of particles are used to hedge between the two possible motions.

We require the following notation. The unrestricted sample set is denoted $\{\mathbf{x}_k^U(i) : i = 1, \ldots, N_U\}$ and the blind zone set by $\{\mathbf{x}_k^{BZ}(i) : i = 1, \ldots, N_{BZ}\}$. The overall probability weight attached to each set is denoted $p_k(U)$ and $p_k(BZ)$. The indicator function $I_{BZ}(\mathbf{x})$ is defined as

$$I_{BZ}(\mathbf{x}) = \begin{cases} 1 & \text{if } |\dot{r}(\mathbf{x})| \leq L_0 \\ 0 & \text{otherwise} \end{cases} \quad (9.20)$$

with $\dot{r}(\mathbf{x})$ as defined in (9.3) and (9.7). The pseudocode of the particle filter for tracking targets through the blind Doppler is as follows:

Initialization: $k = 1$

- For $i = 1, \ldots, N_U$ draw sample $\mathbf{x}_1^U(i)$ from $p(\mathbf{x}_1)$;
- Set $p_1(U) = 1, p_1(BZ) = 0$.

For $k = 2, 3, \ldots$ (main loop)

Prediction and gating

- For $i = 1, \ldots, N_U$, sample $\tilde{\mathbf{x}}_k^U(i) \sim p(\mathbf{x}_k \mid \mathbf{x}_{k-1}^U(i))$
- If $p_{k-1}(BZ) > 0$, for $i = 1, \ldots, N_{BZ}$, sample

$$\tilde{\mathbf{x}}_k^{BZ}(i) \sim p(\mathbf{x}_k \mid \mathbf{x}_{k-1}^{BZ}(i))$$

subject to the constraint $I\left(\tilde{\mathbf{x}}_k^{BZ}(i)\right) = 1$

- Use sample sets for gating

Information update

If measurement $\mathbf{z}_k$ is available

- Set $p_k(U) = 1, p_k(BZ) = 0$
- Assign sample weight $w_k^U(i) \propto p(\mathbf{z}_k \mid \tilde{\mathbf{x}}_k^U(i))$ to $\tilde{\mathbf{x}}_k^U(i)$

else if no measurement available and $p_{k-1}(BZ) = 0$

- Use auxiliary sequential importance sampling [12] to generate $\{\tilde{\mathbf{x}}_k^{BZ}(i)\}$ from $\{\mathbf{x}_{k-1}^U(i)\}$

- Attach sample weight $w_k^{BZ}(i) = N_{BZ}^{-1}$ to $\tilde{\mathbf{x}}_k^{BZ}(i)$
- Set $p_k(BZ) = P_d$ and $p_k(U) = 1 - P_d$

else if no measurement is available and $p_{k-1}(BZ) > 0$

- Assign sample weight $w_k^{BZ}(i) = N_{BZ}^{-1}$ to $\tilde{\mathbf{x}}_k^{BZ}(i)$
- Assign sample weight $w_k^U(i) = N_U^{-1}$ to $\tilde{\mathbf{x}}_k^U(i)$
- Set $p_k(U) = (1 - P_d)p_{k-1}(U)$ and $p_k(BZ) = 1 - p_k(U)$.

Selection

- For each stratum, multiply/discard particles with respect to high/low normalized importance weights to obtain N_U particles $\{\mathbf{x}_k^U(i)\}$ and N_{BZ} particles $\{\mathbf{x}_k^{BZ}(i)\}$.

Output

- Output the weighted mean

$$\bar{\mathbf{x}}_k = p_k(U) \sum_{i=1}^{N_U} w_k^U(i)\mathbf{x}_k^U(i) + p_k(BZ) \sum_{i=1}^{N_{BZ}} w_k^{BZ}(i)\mathbf{x}_k^{BZ}(i)$$

End of main loop

9.5 SIMULATION RESULTS

The performance of two tracking algorithms will be measured by the probability of track maintenance. In order to define a track loss, we adopt a simple scheme for calculating *track score S*. At a discrete time t_k track score is computed by the following two-step procedure:

$$S_k' = \begin{cases} S_{k-1} + \delta^+(T_k) & \text{if target detected and } z_k \text{ gated} \\ S_{k-1} - \delta^-(T_k) & \text{otherwise} \end{cases} \quad (9.21)$$

$$S_k = \min(S_k', 1). \quad (9.22)$$

Parameters $\delta^+(T_k)$ and $\delta^-(T_k)$ are the score increment and decrement, respectively. When track score S_k falls below a certain threshold, the track is declared lost (deleted).

Both the EKF and the particle filter use the same process noise, gating probability, and track score calculation parameters. Parameter σ_v of the process noise covariance matrix $\mathbf{Q}$ is adopted as $\sigma_v = gT_k^{-1/2}$. The gating probability is set to $P_g = 0.995$. The track score parameters are: $\delta^+ = 0.2 \cdot T_k$, $\delta^- = 0.03 \cdot T_k$ and the track deletion threshold is 0.05.

Figure 9.2 illustrates one run of the EKF-based tracker on the scenario described in Section 9.2 (see Figure 9.1). The parameters of the radar model used in simulating both Figures 9.1 and 9.2 are as follows: the sampling interval is $T = 2$ s; the error statistics for radar measurements are $\sigma_r = 250$ m, $\sigma_{\dot{r}} = 3$ m/s, and $\sigma_\theta = 1°$; the BDZ limit is $L_0 = 100$ km/h; the probability of detection is $P_d = 0.9$. Note that the track score calculation parameters are chosen in such a way that if the first measurement after the reappearance of the target from the BDZ is not gated, the track is lost. In the Monte Carlo run displayed in Figure 9.2, the EKF has lost the track.

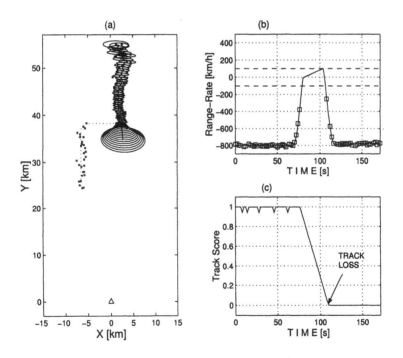

Figure 9.2 The EKF in action: (a) state estimates with covariance ellipsoids; (b) the range rate; and (c) track score S_k.

Figure 9.3 illustrates the results obtained using the particle-filter-based tracker on the same scenario with the *same* measurements. Because of the proper shape of the validation region, the first measurement after the reappearance of the target from the BDZ falls inside the gate of the particle filter. The shape of this gate, just before the crucial moment (at time 108 seconds) is indicated by a cloud of particles in Figure 9.3(a). The shape of this cloud, which resembles an arc, accurately represents the uncertainty associated with the target position. As a result, the particle filter does not lose the track – the track score S_k, after dipping to the value of 0.1, recovers quickly to 1.

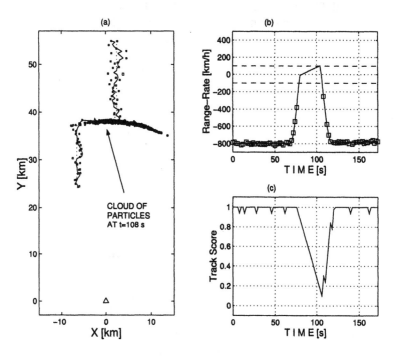

Figure 9.3 The particle filter in action: (a) state estimates (solid line) with a cloud of particles at $t = 108$ seconds; (b) the range rate; and (c) track score S_k.

Next we show the probability of track maintenance, for both the EKF and the particle filter, as a function of the BDZ limit L_0. The probability of track maintenance is estimated by Monte Carlo simulations, as the ratio between the number of runs in which the track was not lost and the total number of Monte Carlo runs. The results, obtained by running 100 Monte Carlo trials, are shown in Figure 9.4 for the sampling interval: (a) $T = 2$ seconds and (b) $T = 4$ seconds. The particle filter used 1000 particles for the CV motion and 2000 particles for the blind Doppler. Figure 9.4 indicates a far superior performance of the particle filter, which almost never loses the track. This improvement in track continuity, however,

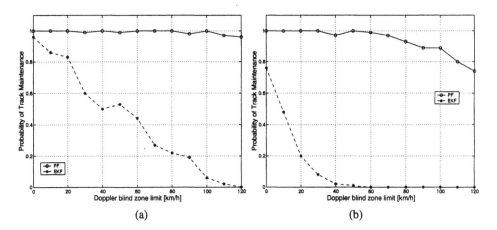

Figure 9.4 Probability of track maintenance: EKF $(-*-)$; particle filter $(-o-)$. The sampling interval: (a) $T = 2$ s; (b) $T = 4$ s.

is paid by a significant increase in the computational load: the particle filter requires approximately 100 times more CPU time than the EKF.

9.6 SUMMARY

The problem of tracking a target occasionally hidden in the blind Doppler of a radar has been investigated. It has been demonstrated that by using prior knowledge of the limits of the blind Doppler zone, one can design a tracker that will perform better, in terms of track continuity, than a tracker that ignores this prior information. The concept has been proven using a particle filter, mainly because any nonstandard information (such as blind Doppler) can be easily incorporated into the framework of sequential Monte Carlo estimation. An EKF with multiple dynamic models (a type of a Gaussian sum filter) could possibly achieve a similar effect, although its implementation would be very complicated (e.g., how to implement hard edges on probability distributions). Another operationally viable solution could be a hybrid tracker that can switch between the EKF and the particle filter depending on the tracking conditions. Such a filter builds upon the ideas of [13]. In a more realistic environment other effects such as false detections (due to clutter or decoys) and multiple targets can further complicate the tracking process. These effects can be also dealt within the particle filtering framework; see for example Chapter 12 and [13, 14, 15].

References

[1] G. W. Stimson, *Introduction to Airborne Radar*. Mendham, NJ: SciTech Publishing, 2nd ed., 1998.

[2] B. J. Slocumb and P. D. West, "ECM modeling for multitarget tracking and data association," in *Multisensor-Multitarget Tracking: Applications and Advances* (Y. Bar-Shalom and W. D. Blair, eds.), vol. III, ch. 8, Norwood, MA: Artech House, 2000.

[3] B. Edde, *Radar: Principles, Technology, Applications*. Englewood Hills, NJ: Prentice-Hall, 1993.

[4] N. J. Gordon and B. Ristic, "Tracking airborne targets occasionally hidden in the blind Doppler," *Digital Signal Processing*, vol. 12, pp. 383–393, April/July 2002.

[5] D. A. Zaugg, A. A. Samuel, D. E. Waagen, and H. A. Schmitt, "Combined particle/Kalman filter for improved tracking of beam aspect targets," in *Proc. IEEE Workshop on Statistical Signal Processing*, (St. Louis, MI), Sepember/October 2003.

[6] L. Mo, X. Song, Y. Zhou, Z. K. Sun, and Y. Bar-Shalom, "Unbiased converted measurements for tracking," *IEEE Trans. Aerospace and Electronic Systems*, vol. 34, pp. 1023–1027, July 1998.

[7] K. Ito and K. Xiong, "Gaussian filters for nonlinear filtering problems," *IEEE Trans. Automatic Control*, vol. 45, pp. 910–927, May 2000.

[8] W. Koch and R. Klemm, "Ground target tracking with STAP radar," *IEEE Proc. - Radar, Sonar Navig.*, vol. 148, no. 3, pp. 173–185, 2001.

[9] S. Blackman and R. Popoli, *Design and Analysis of Modern Tracking Systems*. Norwood, MA: Artech House, 1999.

[10] T. Kirubarajan, Y. Bar-Shalom, K. R. Pattipati, and I. Kadar, "Ground target tracking with variable structure IMM estimator," *IEEE Trans. Aerospace and Electronic Systems*, vol. 36, pp. 26–45, January 2000.

[11] P. Fearnhead, *Sequential Monte Carlo Methods in Filter Theory*. PhD thesis, University of Oxford, 1998.

[12] M. Pitt and N. Shephard, "Filtering via simulation: Auxiliary particle filters," *Journal of the American Statistical Association*, vol. 94, no. 446, pp. 590–599, 1999.

[13] A. Doucet, N. Gordon, and V. Krishnamurthy, "Particle filters for state estimation of jump Markov linear systems," *IEEE Trans. Signal Processing*, vol. 49, pp. 613–624, March 2001.

[14] C. Hue, J. L. Cadre, and P. Pérez, "Sequential Monte Carlo methods for multiple target tracking and data fusion," *IEEE Trans. Signal Processing*, vol. 50, pp. 309–325, February 2002.

[15] C. Hue, J. L. Cadre, and P. Pérez, "Tracking multiple objects with particle filtering," *IEEE Trans. Aerospace and Electronic Systems*, vol. 38, pp. 791–812, July 2002.

Chapter 10

Terrain-Aided Tracking

10.1 INTRODUCTION

In standard tracking problems, the only inputs available to the tracker are sensor measurements obtained through one or more sensors. In certain applications, however, there may be some additional prior information available, which could be exploited in the Bayesian estimation framework. For instance, we may have some knowledge of the environment in which the target is being tracked or some limitations in the dynamic motion of the target (such as speed or acceleration constraints). In the context of airborne surveillance of ground-moving vehicles using the ground moving target indicator (GMTI) radar, one may have some prior information of the terrain, road maps, and visibility conditions. This chapter is primarily devoted to such an application. Since the Gulf War in 1991, the GMTI radar has become an extremely useful sensor for military surveillance [1] and during the last few years a number of large-scale programs have been devoted to GMTI tracking of ground targets [2, 3, 4, 5]. Similar problems to those of GMTI tracking have been encountered in the context of airport surface traffic management, where the objective is to ensure safe, orderly, and expeditious movement of aircraft and support vehicles (buses, cars) along the airport runway network [6]. The ultimate goal in all these applications was to exploit prior nonstandard information (speed constraints, road networks, and so forth) in the tracker to produce better (sequential) estimates of the target state.

It turns out, however, that incorporating such nonstandard information within the conventional Kalman filtering framework is not an easy task. The reason is that, in general, nonstandard information leads to highly non-Gaussian posterior densities that are difficult to represent accurately using the conventional techniques. The most common approach to the treatment of such nonstandard information within the Kalman filtering framework is based on the variable structure interacting multiple-model (VS-IMM) algorithm [2]. The VS-IMM uses a modified version of the standard IMM, where the number and structure of multiple models that are active at any particular time are allowed to vary. The various models may represent motions under different conditions of visibility, road constraints, and target speeds. Although the VS-IMM has been shown to produce better results than methods that discard such nonstandard information, it still has some major shortcomings. In particular, the nonstandard information available to the tracker, which leads to highly non-Gaussian posterior pdfs, is approximated by a finite mixture of Gaussian densities. Moreover, the VS-IMM does not have a mechanism to incorporate hard constraints on position and speed. Because of these weaknesses, the use of the VS-IMM has only resulted in a modest improvement in accuracy over methods that do not use such nonstandard information.

In this chapter we propose a new algorithm based on SMC methods, which we term variable structure multiple-model particle filter (VS-MMPF) [7]. The basic principle is to use particles (random samples) to represent the posterior density of the state of a target in a dynamic state estimation framework where nonstandard information is utilized. Since particle filtering methods have no restrictions on the type of models, including the noise distributions used, one can choose rather complex models to represent ground vehicle motion in a GMTI context. In particular, the nonstandard information available through road maps, speed constraints, and so forth, is modeled by a generalized jump Markov system with constraints on the state. In addition, the transition probabilities of the Markov process are designed to be state dependent, thus allowing for realistic characterization of ground vehicles. The proposed algorithm is tested on simulated data and compared with the performance of the VS-IMM.

The organization of the chapter is as follows. Section 10.2 describes the GMTI tracking problem and its mathematical formulation for both VS-IMM and VS-MMPF. Section 10.3 reviews the VS-IMM algorithm followed by Section 10.4 which presents the VS-MMPF algorithm. Finally, simulation results are presented in Section 10.5.

10.2 PROBLEM DESCRIPTION AND FORMULATION

10.2.1 Problem Description

This section describes the problem of GMTI tracking with nonstandard information. We consider the problem of tracking ground targets from measurements obtained using a single sensor. The surveillance region includes road networks and varying terrain conditions, such as hills, tunnels, open fields, and so forth. Depending on the target present location, its motion is influenced by these external factors. For example, a target on a particular road has a high probability of continuing its motion constrained along that road. Similarly, an off-road target traveling in the open field is free to move in any direction; however, it may enter a road only at certain locations due to the terrain features such as rivers or hills. Likewise, an on-road target at a junction can continue only in one of the roads meeting at the junction. Thus, road networks and terrain conditions result in constrained target motion capabilities. The target motion can also be constrained by speed restrictions (due to the weather conditions, a type of vehicle, and so forth) that may be known a priori.

The terrain conditions, however, not only constrain the target motion (as described above), but also influence the performance of a sensor responsible for the acquisition of target measurements. Thus, depending on target location, the terrain features such as hills and tunnels may hide the target from the sensor field of view. The varying target obscuration conditions therefore need to be taken into consideration in the tracking filter design.

A typical road map is illustrated in Figure 10.1 with four roads (AJ, BJ, CJ, and DJ) meeting at junction J. The road segments represented by solid lines allow entry to or exit from the roads. The road segments shown by broken lines (eg., TU and BJ) indicate that targets on the road cannot get off or those off the road cannot get on the road. In a typical GMTI tracking problem, such restrictions are generally determined by the surrounding terrain conditions, for example, rivers, open fields, or ditches. The road segment TU represents a tunnel with zero target detection probability (while BJ could be a bridge).

The complete road map with terrain conditions can be specified as follows. Each road segment is represented by a pair of waypoints, as shown in Table 10.1. These waypoints determine the direction, location, and length of each road segment. The visibility can be defined as a binary-valued probability, and the entry/exit condition is given by a Boolean variable. The entire topographic information (topography is a description on a map of all natural and artificial features of a certain region) is summarized in Table 10.1 and in the sequel will be denoted by $\mathcal{T}$.

Figure 10.1 also shows a sample target trajectory through the road network described above. A target starts at point I, merges with road AJ at M, then proceeds to junction J through tunnel TU, branches off to road JD, exits road JD at E, and terminates at F in the open field.

The GMTI tracking problem can now be summarized as follows: given the standard GMTI measurements of target range, azimuth, and Doppler, and given the (nonstandard) topographic information (described above), the objective is to estimate sequentially the target state vector. The key feature of this problem is to exploit the nonstandard information to yield an enhanced tracker performance.

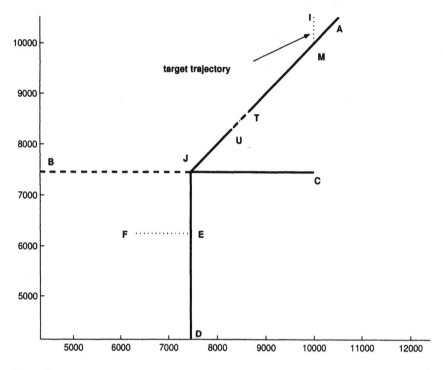

Figure 10.1 An example of a road network with a target trajectory: target starts at point I and terminates at point F.

Table 10.1

Specification of a Road Map

Road Segments	Waypoints	Visibility	Allow Entry/Exit
1	AT	A and T	1.0	True
2	TU	T and U	0.0	False
3	UJ	U and J	1.0	True
4	BJ	B and J	1.0	False
5	CJ	C and J	1.0	True
6	DJ	D and J	1.0	True

10.2.2 Dynamics and Measurement Models for VS-IMM

The problem of target tracking with road constraints is handled by the VS-IMM [2] using the concept of *directional* process noise. To explain this, note that the standard motion models assume that the target can move in any direction with equal probability, and therefore use equal process noise variance in both x and y directions (i.e., $\sigma_x^2 = \sigma_y^2$). For on-road targets, however, the road constraints mean more uncertainty along the road than orthogonal to it. Suppose σ_a^2 and σ_o^2 are the variances *along* and *orthogonal* to the road, respectively, and the road direction is specified by the angle ψ (measured clockwise from y-axis). Then, the process noise covariance matrix corresponding to on-road motion is given by

$$\mathbf{Q} = \rho_\psi \cdot \begin{bmatrix} \sigma_o^2 & 0 \\ 0 & \sigma_a^2 \end{bmatrix} \cdot \rho_\psi^T \tag{10.1}$$

where

$$\rho_\psi = \begin{bmatrix} \cos \psi & \sin \psi \\ -\sin \psi & \cos \psi \end{bmatrix} \tag{10.2}$$

is the (clockwise) rotation matrix. With the above directional process noise matrix concept, the GMTI tracking problem is formulated in a jump Markov system framework (see Section 1.4). Let the state vector at discrete-time k consist of position and velocity components of the target in the Cartesian coordinates; that is $\mathbf{x}_k = \begin{bmatrix} x_k & y_k & \dot{x}_k & \dot{y}_k \end{bmatrix}^T$. Then, the discrete time kinematic model for the problem can be expressed by

$$\mathbf{x}_k = \mathbf{F}\mathbf{x}_{k-1} + \mathbf{G}\mathbf{v}_{k-1}(r_k), \tag{10.3}$$

where

$$\mathbf{F} = \begin{bmatrix} 1 & 0 & T & 0 \\ 0 & 1 & 0 & T \\ 0 & 0 & 1 & 0 \\ 0 & 0 & 0 & 1 \end{bmatrix}, \quad \mathbf{G} = \begin{bmatrix} T^2/2 & 0 \\ 0 & T^2/2 \\ T & 0 \\ 0 & T \end{bmatrix}. \tag{10.4}$$

Here T is the sampling time, discrete variable r_k represents the mode (regime) in effect in the interval $(k-1, k]$, and $\mathbf{v}_{k-1}(r_k)$ is a mode-dependent white Gaussian process noise sequence with covariance $\mathbf{Q}(r_k)$. Physically, r_k represents the type of motion (off-road, on-road in a specific direction, and so forth) of the target. This variable is modeled as a Markov process whose states belong to a variable set defined by the road network. Suppose there are s road segments in the network and let $r_k \in \mathcal{S}^a$, where $\mathcal{S}^a \triangleq \{0, 1, \ldots, s\}$ is the set of all possible motion models. Here we adopt the following convention: $r_k = r \in \{1, \ldots, s\}$ implies a motion model corresponding to road segment r (see Table 10.1) and $r_k = 0$ refers to the off-road motion. Now let $\mathcal{S}_k \subseteq \mathcal{S}^a$ denote the set of modes active in the interval $(k-1, k]$. Then, the Markov process r_k is characterized by the following transition probabilities that depend on $\mathcal{S}_{k-1}$, $\mathcal{S}_k$, and $\mathbf{x}_{k-1}$; that is,

$$p_{ij}(\mathcal{S}_{k-1}, \mathcal{S}_k, \mathbf{x}_{k-1}) = \mathrm{P}\{r_k = j \in \mathcal{S}_k | r_{k-1} = i \in \mathcal{S}_{k-1}, \mathbf{x}_{k-1}\}. \tag{10.5}$$

The measurement equation applicable to this problem is given by[1]

$$\mathbf{z}_k = \mathbf{h}(\mathbf{x}_k, r_k) + \mathbf{w}_k \tag{10.6}$$

where

$$\mathbf{h}(\mathbf{x}_k) = \begin{bmatrix} \rho_k \\ \theta_k \\ \dot{\rho}_k \end{bmatrix} = \begin{bmatrix} \sqrt{x_k^2 + y_k^2} \\ \tan^{-1}(x_k/y_k) \\ \frac{x_k \dot{x}_k + y_k \dot{y}_k}{\sqrt{x_k^2 + y_k^2}} \end{bmatrix}, \tag{10.7}$$

with $\mathbf{w}_k \sim \mathcal{N}(\mathbf{0}, \mathbf{R}_k)$. Here $\mathbf{R}_k$ is a 3×3 diagonal measurement covariance matrix with elements equal to the variances in range ρ, azimuth θ, and Doppler $\dot{\rho}$ that are σ_ρ^2, σ_θ^2, and $\sigma_{\dot{\rho}}^2$, respectively. For model r, this measurement is received with detection probability P_D^r, which is chosen to be either 0 or 1 depending on its visibility specified in Table 10.1.

Equation (10.6) is nonlinear and thus each model-matched filter of the VS-IMM would require an EKF. To minimize the required linearization in the measurement function, we adopt elements of the converted measurement Kalman filter [8] for the position component (ρ_k, θ_k) of the measurements. That is, we form the pseudomeasurement

$$\mathbf{z}'_k = \mathbf{h}'(\mathbf{x}_k) + \mathbf{w}'_k, \tag{10.8}$$

where

$$\mathbf{h}'(\mathbf{x}_k) = \begin{bmatrix} \rho_k \sin \theta_k \\ \rho_k \cos \theta_k \\ \dot{\rho}_k \end{bmatrix} = \begin{bmatrix} x_k \\ y_k \\ \dot{\rho}_k \end{bmatrix}, \tag{10.9}$$

with $\mathbf{w}'_k \sim \mathcal{N}(\mathbf{0}, \mathbf{R}'_k)$,

$$\mathbf{R}'_k = \begin{bmatrix} \sigma_x^2(\mathbf{R}'_k) & \sigma_{xy}(\mathbf{R}'_k) & 0 \\ \sigma_{yx}(\mathbf{R}'_k) & \sigma_y^2(\mathbf{R}'_k) & 0 \\ 0 & 0 & \sigma_{\dot{\rho}}^2 \end{bmatrix}, \tag{10.10}$$

---

[1] In this study, for convenience, we assume a static GMTI sensor, located at the origin of the x-y plane.

$$\sigma_x^2(\mathbf{R}_k') = \rho_k^2 \sigma_\theta^2 \cos^2 \theta_k + \sigma_\rho^2 \sin^2 \theta_k, \tag{10.11}$$

$$\sigma_y^2(\mathbf{R}_k') = \rho_k^2 \sigma_\theta^2 \sin^2 \theta_k + \sigma_\rho^2 \cos^2 \theta_k, \tag{10.12}$$

$$\sigma_{xy}(\mathbf{R}_k') = \sigma_{yx}(\mathbf{R}_k') = (\sigma_\rho^2 - \rho_k^2 \sigma_\theta^2) \sin \theta_k \cos \theta_k. \tag{10.13}$$

The Jacobian of the measurement function $\mathbf{h}'(\mathbf{x}_k)$ required for the EKF is given by

$$\begin{aligned}\mathbf{H}_k' &= \nabla_{\mathbf{x}_k} \mathbf{h}'(\mathbf{x}_k) \\ &= \begin{bmatrix} 1 & 0 & 0 & 0 \\ 0 & 1 & 0 & 0 \\ \frac{\partial \mathbf{h}'[3]}{\partial x_k} & \frac{\partial \mathbf{h}'[3]}{\partial y_k} & \frac{\partial \mathbf{h}'[3]}{\partial \dot{x}_k} & \frac{\partial \mathbf{h}'[3]}{\partial \dot{y}_k} \end{bmatrix} \end{aligned} \tag{10.14}$$

where $\mathbf{h}'[3]$ stands for the third component of $\mathbf{h}'(\mathbf{x}_k)$. By differentiation we obtain:

$$\frac{\partial \mathbf{h}'[3]}{\partial x_k} = \frac{\dot{x}_k \rho_k - x_k \dot{\rho}_k}{x_k^2 + y_k^2} \tag{10.15}$$

$$\frac{\partial \mathbf{h}'[3]}{\partial y_k} = \frac{\dot{y}_k \rho_k - y_k \dot{\rho}_k}{x_k^2 + y_k^2} \tag{10.16}$$

$$\frac{\partial \mathbf{h}'[3]}{\partial \dot{x}_k} = \frac{x_k}{r_k} \tag{10.17}$$

$$\frac{\partial \mathbf{h}'[3]}{\partial \dot{y}_k} = \frac{y_k}{r_k}. \tag{10.18}$$

For the jump Markov system defined by (10.3) and (10.8), given a sequence of measurements $\mathbf{Z}_k = \{\mathbf{z}_1, \ldots, \mathbf{z}_k\}$ and topographic information $\mathcal{T}$, the problem is to estimate the target state $\hat{\mathbf{x}}_{k|k} = \mathbb{E}[\mathbf{x}_k | \mathbf{Z}_k, \mathcal{T}]$.

10.2.3 Dynamic Models for VS-MMPF

Modeling the motion dynamics of a target under terrain constraints is a complex process. To do so, we have to capture various inherent features of motion dynamics due to road restrictions, the behavior of transiting from on-road to off-road motion and vice versa, transiting to different roads at junctions and modeling detection probabilities under the varying visibility conditions. Next, we present a mathematical model for target dynamics with terrain constraints that will be utilized by the VS-MMPF algorithm. This model is more complex but also more accurate than the one described in Section 10.2.2, simply because the VS-MMPF can be designed according to this model, while the VS-IMM cannot.

As in Section 10.2.2, let us assume that there are s road segments in the network and let $r_k \in \{0, 1, \ldots, s\}$ denote the type of motion dynamics applicable in the interval $(k-1, k]$. The evolution of the mode (regime) variable r_k is modeled as a state and topography-dependent Markov process with transition probabilities

$$p_{ij}(\mathbf{x}_{k-1}, \mathcal{T}) = P\{r_k = j | r_{k-1} = i, \mathbf{x}_{k-1}, \mathcal{T}\}. \quad (10.19)$$

The generic dynamics model for a target is then given by

$$\mathbf{x}_k = \mathbf{f}(\mathbf{x}_{k-1}, r_{k-1}, r_k, \mathbf{v}_{k-1}(\mathbf{x}_{k-1}, r_k), u(r_{k-1}, r_k)), \quad (10.20)$$

where the parameters on the right hand side of (10.20) are defined as follows:

- process noise vector

$$\mathbf{v}_{k-1}(\mathbf{x}_{k-1}, r_k) \sim \begin{cases} \tilde{\mathcal{N}}(\mathbf{0}, \mathbf{Q}(r_k); \mathbf{x}_{k-1}), & \text{if } r_k \neq 0, \\ \mathcal{N}(\mathbf{0}, \mathbf{Q}(0)), & \text{if } r_k = 0. \end{cases} \quad (10.21)$$

Here $\tilde{\mathcal{N}}(\mathbf{0}, \mathbf{Q}(r_k); \mathbf{x}_{k-1})$ denotes the truncated version of Gaussian density $\mathcal{N}(\mathbf{0}, \mathbf{Q}(r_k))$ that ensures that $\mathbf{x}_k$ remains on road r_k. Process noise covariance matrix for off-road motion, $\mathbf{Q}(0)$, is a special case of (10.1) with $\sigma_a = \sigma_o$.

- $u(r_{k-1}, r_k)$ is a variable that models the possibility of the target turning left or right as it enters the road (as it will be further explained later in the text). This variable is defined as

$$u(r_{k-1}, r_k) \triangleq \begin{cases} 0, & \text{if } r_{k-1} = r_k = 0 \\ & \text{or } r_{k-1} \neq 0, r_k \neq 0 \\ -1 \text{ or } 1 \text{ with prob} = \frac{1}{2}, & \text{otherwise.} \end{cases} \quad (10.22)$$

Thus if only one of r_{k-1} and r_k is zero, then $u(r_{k-1}, r_k)$ is a binary random variable with values -1 or 1.

Depending on the current and the previous mode (r_k and r_{k-1}), the form of the target dynamics given by (10.20) will vary. The transition probabilities of the Markov chain and the form of the dynamics (10.20) will now be specified in detail.

10.2.3.1 Transition Probabilities of the Markov Chain

First, some terminology and functions need to be defined. A junction j is defined as a point where two or more roads meet. Let $\mathcal{J}_r$ denote the set of junctions that are directly connected to road r. Similarly, let $\mathcal{R}_j$ denote the set of roads that are directly connected to junction j. Suppose the predicted state of the target at time k is denoted by

$$\mathbf{x}_k^* = \mathbf{F}\mathbf{x}_{k-1}, \qquad (10.23)$$

where $\mathbf{F}$ is the standard transition matrix given in (10.4). If the mode at $k-1$ corresponds to off-road motion (i.e., if $r_{k-1} = 0$), the transition $\mathbf{x}_{k-1} \to \mathbf{x}_k^*$ may have resulted in the crossing of some road. We define a function $c(\mathbf{x}_{k-1}, r)$ as an indicator function for this; that is

$$c(\mathbf{x}_{k-1}, r) = \begin{cases} 1 & \text{if } \mathbf{x}_{k-1} \to \mathbf{x}_k^* \text{ crosses road } r \\ 0 & \text{otherwise.} \end{cases} \qquad (10.24)$$

Likewise, if the mode at $k-1$ is $r_{k-1} = r \neq 0$ (on-road motion), the transition $\mathbf{x}_{k-1} \to \mathbf{x}_k^*$ may have resulted in the crossing of some junction. An indicator function for this is defined to be

$$\bar{c}(\mathbf{x}_{k-1}, j) = \begin{cases} 1 & \text{if } \mathbf{x}_{k-1} \to \mathbf{x}_k^* \text{ crosses junction } j \\ 0 & \text{otherwise.} \end{cases} \qquad (10.25)$$

The state and topography dependent transition probabilities can now be defined as follows. Suppose the mode at $k-1$ is $r_{k-1} = i$ with $i \neq 0$ (i.e., the target is on road i). The mode transition probabilities to the next time are defined as

$$p_{ii}(\mathbf{x}_{k-1}, \mathcal{T}) = \begin{cases} \bar{p}, & \text{if } \bar{c}(\mathbf{x}_{k-1}, j) = 0, \; \forall j \in \mathcal{J}_i \\ 0, & \text{if } \bar{c}(\mathbf{x}_{k-1}, j) = 1 \text{ for some } j \in \mathcal{J}_i \end{cases} \qquad (10.26)$$

$$p_{it}(\mathbf{x}_{k-1}, \mathcal{T}) = \begin{cases} \bar{p}/n_i^j, & \text{if } \bar{c}(\mathbf{x}_{k-1}, j) = 1 \text{ for some } j \in \mathcal{J}_i \\ 0, & \text{if } \bar{c}(\mathbf{x}_{k-1}, j) = 0, \; \forall j \in \mathcal{J}_i \end{cases} \qquad (10.27)$$

where $i = 1, \ldots, s$ and $t \in \mathcal{R}_j \setminus \{i\}$ (notation $\setminus$ stands for the set difference). Parameter $0 < \bar{p} < 1$ is a filter design parameter and n_i^j is the cardinality of $\mathcal{R}_j \setminus \{i\}$; that is the number of roads (other than road i) meeting road i at junction j. Finally, the probability of transition to an off-road motion mode is then given by

$$p_{i0}(\mathbf{x}_{k-1}, \mathcal{T}) = 1 - \bar{p}, \qquad (10.28)$$

so that $\sum_{m=0}^{s} p_{im} = 1$.

Now suppose the mode at $k-1$ is $r_{k-1} = 0$ (i.e., the target is doing off-road motion). Let $d(\mathbf{x}_k^*, r)$ denote the shortest distance from $\mathbf{x}_k^*$ to road r. Then, the transition probabilities for this off-road target transiting to one of the roads is given by

$$p_{0i}(\mathbf{x}_{k-1}, T) = \begin{cases} p^* & c(\mathbf{x}_{k-1}, i) = 1 \\ 0 & c(\mathbf{x}_{k-1}, i) = 0, \ d(\mathbf{x}_k^*, i) > \tau \\ p^*(\tau - d(\mathbf{x}_k^*, i))/\tau & \text{otherwise} \end{cases} \quad (10.29)$$

where $i \in S^a \backslash \{0\}$, $0 < p^* < 1$ is a design parameter and τ is a threshold chosen according to the acceleration capabilities of the target. Finally, assuming that an off-road target can potentially transit to only one of the roads in the network, the probability of remaining in off-road motion in given by

$$p_{00}(\mathbf{x}_{k-1}, T) = 1 - p_{0i}(\mathbf{x}_{k-1}, T). \quad (10.30)$$

Next we describe all possible state dynamic models.

10.2.3.2 State Dynamics Models

Recall that depending on the current and previous modes (r_k and r_{k-1}), the form of the state dynamics model (10.20) will vary. In this section we present models for all possible combinations of mode sequences r_{k-1} and r_k. Specifically, we can model the dynamics by considering four different cases:

Case 1: $r_{k-1} = 0$, $r_k = 0$;

Case 2: $r_{k-1} = 0$, $r_k = r \neq 0$;

Case 3: $r_{k-1} = r \neq 0$, $r_k = 0$; and

Case 4: $r_{k-1} = r \neq 0$, $r_k = r' \neq 0$.

We now proceed to describe the dynamics corresponding to each of the above cases. In the following, $\mathbf{f}_j(\cdot)$ will denote the dynamics function corresponding to case $j = 1, 2, 3, 4$.

<u>Case 1: $r_{k-1} = 0$, $r_k = 0$</u>

This corresponds to an off-road motion target remaining in the same mode. Thus, the motion dynamics for this case is given by the standard constant velocity (CV) motion model:

$$\mathbf{f}_1(\mathbf{x}_{k-1}, r_{k-1}, r_k, \mathbf{v}_{k-1}(\mathbf{x}_{k-1}, r_k), u(r_{k-1}, r_k)) \\ = \mathbf{F}\mathbf{x}_{k-1} + \mathbf{G}\mathbf{v}_{k-1}(\mathbf{x}_{k-1}, r_k) + u(r_{k-1}, r_k) \quad (10.31)$$

where we note from (10.22) that for this case $u(r_{k-1}, r_k) = 0$ and from (10.21) that $\mathbf{v}_{k-1}(\mathbf{x}_{k-1}, r_k)$ is distributed according to (untruncated) Gaussian density $\mathcal{N}(\mathbf{0}, \mathbf{Q}(0))$.

Case 2: $r_{k-1} = 0$, $r_k = r \neq 0$

This corresponds to the situation where a target in the off-road motion enters one of the roads. For this case let the predicted state of the target be denoted by $\mathbf{x}_k^*$ as in (10.23). The new state of the target is defined as follows: the x and y components of the new state are defined to be the shortest distance point from $\mathbf{x}_k^*$ to road r. The magnitude of the velocity component of the new state remains the same as that of $\mathbf{x}_{k-1}$. However, the direction of motion is in line with road r. Mathematically, this can be written as

$$\mathbf{f}_2(\mathbf{x}_{k-1}, r_{k-1}, r_k, \mathbf{v}_{k-1}(\mathbf{x}_{k-1}, r_k), u(r_{k-1}, r_k))$$
$$= \begin{bmatrix} d^x(\mathbf{x}_k^*, r) \\ d^y(\mathbf{x}_k^*, r) \\ u(r_{k-1}, r_k) v(\mathbf{x}_{k-1}) \sin \psi_r \\ u(r_{k-1}, r_k) v(\mathbf{x}_{k-1}) \cos \psi_r \end{bmatrix} \quad (10.32)$$

where $d^x(\mathbf{x}_k^*, r)$ and $d^y(\mathbf{x}_k^*, r)$ are the x and y components of the shortest distance point from $\mathbf{x}_k^*$ to road r, $v(\mathbf{x}_{k-1})$ is the speed corresponding to state $\mathbf{x}_{k-1}$, and angle ψ_r is the direction of road r measured clockwise from the y-axis. Note that the binary random variable $u(r_{k-1}, r_k)$ models the possibility of the target turning left or right as it enters the road.

Case 3: $r_{k-1} = r \neq 0$, $r_k = 0$

This corresponds to the case where an on-road target is getting off the road. For this case, the new state is obtained by applying standard CV motion transition to a modified target state whose velocity is perpendicular to the road. To define this mathematically, we form an intermediate state

$$\mathbf{x}'_{k-1} = \gamma(\mathbf{x}_{k-1}) \triangleq (x_{k-1}, y_{k-1}, \dot{x}'_{k-1}, \dot{y}'_{k-1})^T, \quad (10.33)$$

where $(x_{k-1}, y_{k-1})^T$ is the position component of the original state $\mathbf{x}_{k-1}$ and $(\dot{x}'_{k-1}, \dot{y}'_{k-1})^T$ is the rotated velocity vector defined as

$$\begin{bmatrix} \dot{x}'_{k-1} \\ \dot{y}'_{k-1} \end{bmatrix} = \rho_\psi \cdot \begin{bmatrix} \dot{x}_{k-1} \\ \dot{y}_{k-1} \end{bmatrix}, \quad (10.34)$$

with the rotation angle

$$\psi = 90^\circ \cdot u(r_{k-1}, r_k). \tag{10.35}$$

In (10.34), vector $[\dot{x}_{k-1}, \dot{y}_{k-1}]^T$ denotes the velocity component of $\mathbf{x}_{k-1}$ and ρ_ψ is the rotational transformation matrix defined in (10.2). With the above definition, we can now write the state dynamics corresponding to case 3 as

$$\mathbf{f}_3(\mathbf{x}_{k-1}, r_{k-1}, r_k, \mathbf{v}_{k-1}(\mathbf{x}_{k-1}, r_k), u(r_{k-1}, r_k))$$
$$= \mathbf{F}\gamma(\mathbf{x}_{k-1}) + \mathbf{G}\mathbf{v}_{k-1}(\gamma(\mathbf{x}_{k-1}), r_k). \tag{10.36}$$

Note from (10.35) that with the binary random variable $u(r_{k-1}, r_k)$ we model the possibility the target making a left or a right turn exit out of the road.

Case 4: $r_{k-1} = r \neq 0$, $r_k = r' \neq 0$

This corresponds to the situation where an on-road target remains an on-road target. There are two possibilities (subcases) here: (a) $r' = r$, which means that the target remains on the same road, and (b) $r' \neq r$ which corresponds to the case where the target enters another road via some junction j. To describe the motion dynamics for both subcases, first define r^* to be an imaginary road formed by extending road r beyond junction j, and define the function

$$\phi(\mathbf{x}_{k-1}, r_{k-1}, r_k) = \mathbf{F}\mathbf{x}_{k-1} + \mathbf{G}\mathbf{v}_{k-1}(\mathbf{x}_{k-1}, r_k) + u(r_{k-1}, r_k). \tag{10.37}$$

Furthermore, let $\rho^r_{r'}$ be a 4×4 rotational matrix corresponding to a rotation about junction j that places target with state $\mathbf{x}^*_k = \phi(\mathbf{x}_{k-1}, r_{k-1}, r^*)$ on road r'. Then, a motion dynamics corresponding to case 4 can be written as

$$\mathbf{f}_4(\mathbf{x}_{k-1}, r_{k-1}, r_k, \mathbf{v}_{k-1}(\mathbf{x}_{k-1}, r_k), u(r_{k-1}, r_k))$$
$$= \begin{cases} \phi(\mathbf{x}_{k-1}, r_{k-1}, r_k), & r' = r \\ \rho^r_{r'} \cdot [\phi(\mathbf{x}_{k-1}, r_{k-1}, r^*) - \mathbf{x}^j] + \mathbf{x}^j, & r' \neq r \end{cases} \tag{10.38}$$

where $\mathbf{x}^j = \begin{bmatrix} x^j & y^j & 0 & 0 \end{bmatrix}^T$ is the state vector of the junction j joining roads r and r'. Essentially, for $r' = r$ the above dynamics implies that the target executes a constrained CV motion model (constrained to road r). When $r' \neq r$ the above dynamics ensures that the target is constrained to road r' with new velocity vector that favors motion along road r'. Note that the 4×4 rotation matrix $\rho^r_{r'}$ rotates both the position and the velocity component of the target state.

10.3 VARIABLE STRUCTURE IMM

A general formulation of the VS-IMM algorithm is described in [9]. A specific implementation developed for GMTI tracking is presented in [2]. Here we briefly review its basic features.

Recall that $\mathcal{S}_k$ denotes the set of modes active in the interval $(k-1, k]$. Then, the probability of mode $r \in \mathcal{S}_k$ is defined as

$$\mu_k^r \triangleq P\{r_k = r \in \mathcal{S}_k | \mathbf{Z}_k\}. \tag{10.39}$$

The mode-conditioned state estimate (mode being $r \in \mathcal{S}_k$) and its associated covariance matrix are denoted by $\hat{\mathbf{x}}_{k|k}^r$ and $\mathbf{P}_{k|k}^r$, respectively. Furthermore, note that in the VS-IMM described in [2], the transitional probabilities $p_{ij}(\mathcal{S}_{k-1}, \mathcal{S}_k, \mathbf{x}_{k-1})$ that we defined in (10.5) are determined on the basis of mode sojourn times and do not explicitly depend on $\mathbf{x}_{k-1}$. Hence, we will use notation $p_{ij}(\mathcal{S}_{k-1}, \mathcal{S}_k)$ for transitional probabilities of the VS-IMM.

With the above notation, the five steps of a single cycle of the VS-IMM are as follows.

Step 1: *Mode set update*
Based on the state estimate at $k-1$ and the topography, the mode set of the IMM is updated as

$$\begin{aligned}\mathcal{S}_k &= \{t \in \mathcal{S}^a | \mathcal{S}_{k-1}, \mathcal{T}, \mathbf{Z}_{k-1}\} \\ &= \left\{t \in \mathcal{S}^a | \mathcal{S}_{k-1}, \mathcal{T}, \{\hat{\mathbf{x}}_{k-1|k-1}^r, \mathbf{P}_{k-1|k-1}^r, r \in \mathcal{S}_{k-1}\}\right\}.\end{aligned}$$

This step is described in more detail in Section 10.3.1.

Step 2: *Mode interaction or mixing*
The mode-conditioned state estimates and the associated covariances from the previous scan are combined to obtain the initial condition for the mode-matched filters. The initial condition in scan k for the filter module $t \in \mathcal{S}_k$ is computed using

$$\hat{\mathbf{x}}_{k-1|k-1}^{0t} = \sum_{r \in \mathcal{S}_{k-1}} \mu_{k-1|k-1}^{r|t} \hat{\mathbf{x}}_{k-1|k-1}^r, \tag{10.40}$$

where

$$\begin{aligned}\mu_{k-1|k-1}^{r|t} &= P\{r_{k-1} = r | r_k = t, \mathbf{Z}_{k-1}\} \\ &= \frac{p_{rt}(\mathcal{S}_{k-1}, \mathcal{S}_k) \mu_{k-1}^r}{\sum_{\ell \in \mathcal{S}_{k-1}} p_{\ell t}(\mathcal{S}_{k-1}, \mathcal{S}_k) \mu_{k-1}^\ell},\end{aligned} \tag{10.41}$$

are the mixing probabilities. The covariance matrix associated with the estimate of (10.40) is given by

$$\mathbf{P}_{k-1|k-1}^{0t} = \sum_{r \in S_{k-1}} \mu_{k-1|k-1}^{r|t} \left(\mathbf{P}_{k-1|k-1}^{r} + \tilde{\mathbf{d}}_{k-1}^{rt} \cdot (\tilde{\mathbf{d}}_{k-1}^{rt})^T \right)$$

(10.42)

where $\tilde{\mathbf{d}}_{k-1}^{rt} = \hat{\mathbf{x}}_{k-1|k-1}^{r} - \hat{\mathbf{x}}_{k-1|k-1}^{0t}$.

Step 3: *Mode-conditioned filtering*

An extended Kalman filter is used for each module to compute the mode-conditioned state estimate and covariance, given the initial conditions (10.40) and (10.42). In addition, depending on whether a measurement was received and based on the detection probability P_D^t corresponding to module t, we compute the likelihood function Λ_k^t for each module t as

$$\Lambda_k^t = \begin{cases} \mathcal{N}(\nu_k^t; 0, \mathbf{S}_k^t), & \text{if } z_k \text{ received and } P_D^t = 1 \\ 0, & \text{if } z_k \text{ received and } P_D^t = 0 \\ 1, & \text{no } z_k \text{ and } P_D^t = 0 \\ 0, & \text{no } z_k \text{ and } P_D^t = 1 \end{cases}$$

(10.43)

where ν_k^t and $\mathbf{S}_k^t$ are the innovation and its covariance, respectively, of the measurement z_k in module t.

Step 4: *Mode probability update*

The mode probabilities are updated based on the likelihood of each mode using

$$\mu_k^t = \frac{\Lambda_k^t \sum_{\ell \in S_{k-1}} p_{\ell t}(S_{k-1}, S_k) \mu_{k-1}^{\ell}}{\sum_{r \in S_k} \Lambda_k^r \sum_{\ell \in S_{k-1}} p_{\ell r}(S_{k-1}, S_k) \mu_{k-1}^{\ell}}$$

(10.44)

Step 5: *State combination*

The mode-conditioned estimates and their covariances are combined to give the overall state estimate and covariance using the Gaussian mixture formulas (2.26) and (2.27), repeated here for convenience:

$$\hat{\mathbf{x}}_{k|k} = \sum_{t \in S_k} \mu_k^t \hat{\mathbf{x}}_{k|k}^t$$

(10.45)

$$\mathbf{P}_{k|k} = \sum_{t \in S_k} \mu_k^t \left(\mathbf{P}_{k|k}^t + [\hat{\mathbf{x}}_{k|k}^t - \hat{\mathbf{x}}_{k|k}][\hat{\mathbf{x}}_{k|k}^t - \hat{\mathbf{x}}_{k|k}]^T \right).$$

(10.46)

10.3.1 Model Set Update

The VS-IMM adaptively updates the set of active modes S_k based on the current estimate, its covariance, and the topography. We present a brief review next, although details can be found in [2]. Let $\mathcal{L}_\ell \in \mathcal{T}$ denote the ℓth road in the network. Then, a model corresponding to this road is included in S_k by testing whether any segment of this road lies within a certain neighborhood ellipse centered at the predicted location $(\hat{x}_{k|k-1}, \hat{y}_{k|k-1})$. The ellipse $\mathcal{E}_k$ is the region of the x-y plane that satisfies the following condition:

$$\mathcal{E}_k \triangleq \left\{ \begin{bmatrix} x \\ y \end{bmatrix} : \begin{bmatrix} x - \hat{x}_{k|k-1} \\ y - \hat{y}_{k|k-1} \end{bmatrix}^T \left[\mathbf{P}^{pos}_{k|k-1}\right]^{-1} \begin{bmatrix} x - \hat{x}_{k|k-1} \\ y - \hat{y}_{k|k-1} \end{bmatrix} \leq \alpha \right\} \quad (10.47)$$

where α is the "gate threshold" and

$$\mathbf{P}^{pos}_{k|k-1} = \begin{bmatrix} \mathbf{P}^{11}_{k|k-1} & \mathbf{P}^{12}_{k|k-1} \\ \mathbf{P}^{21}_{k|k-1} & \mathbf{P}^{22}_{k|k-1} \end{bmatrix} \quad (10.48)$$

is the position submatrix of the prediction covariance $\mathbf{P}_{k|k-1}$. A road is deemed *validated* if any segment of it belongs to $\mathcal{E}_k$.

The model set update proceeds as follows. First, every junction $j \in \mathcal{T}$ is subjected to the above test. If $j \in \mathcal{E}_k$, then a model corresponding to each road meeting j is added to S_k. Next, all remaining road segments (that were not tested during the junction test) are tested. If any point on the road satisfies the test, a model corresponding to that road is added to S_k. Finally, the baseline models that correspond to off-road motion are added if any one of the following criteria are satisfied: (a) a road (with entry/exit) is validated; (b) a road (with no entry/exit) is validated and one of its endpoints from which the target can get into the open field is also validated; and (c) none of the road segments is validated.

10.4 VARIABLE STRUCTURE MULTIPLE-MODEL PARTICLE FILTER

The concept of multiple-model particle filter (MMPF) was described in Section 3.5.5, and its application to bearings-only tracking of maneuvering targets was presented in Chapter 6. Recall that the MMPF assumed a fixed number of modes and a time homogeneous Markov chain for the entire observation period. In this section we present a modified version of the multiple-model particle filter that is suited to terrain-aided tracking. The resulting algorithm is termed variable structure multiple-model particle filter (VS-MMPF). The key feature of the VS-MMPF algorithm is its constant adaptation of:

- The active model set $\mathcal{S}_k$;
- Transition probabilities p_{ij}

based on the current state and the topographic information.

To explain the operation of the VS-MMPF, consider a set of particles $\{(\mathbf{x}_{k-1}^i, r_{k-1}^i)\}_{i=1}^N$ that represents the posterior pdf $p(\mathbf{x}_{k-1}, r_{k-1}|\mathbf{Z}_{k-1})$ of the state and mode at $k-1$. Now, suppose at time k we receive some measurement $\mathbf{z}_k$. As usual it is required to construct a sample $\{(\mathbf{x}_k^i, r_k^i)\}_{i=1}^N$ that characterizes the posterior pdf $p(\mathbf{x}_k, r_k|\mathbf{Z}_k)$ at k. This is carried out in two steps: prediction and update, which are described in detail next. Following this, the initialization step required for the algorithm is briefly described. One cycle of the algorithm is summarized in Table 10.2.

10.4.1 Prediction Step

Here we describe the prediction phase of a specific particle i, denoted as a pair $(\mathbf{x}_{k-1}^i, r_{k-1}^i)$. This phase consists of six substeps. First, the state- and topography-dependent transition probabilities are computed to give $P(r_k|r_{k-1}^i, \mathbf{x}_{k-1}^i, \mathcal{T})$ where $r_k \in \{0, 1, \ldots, s\}$. Second, a mode r_k^{*i} is drawn according to the computed transition probabilities. Third, the case $\delta^i \in \{1, 2, 3, 4\}$ corresponding to transition (r_{k-1}^i, r_k^{*i}) is determined. Fourth, depending on the combination of (r_{k-1}^i, r_k^{*i}) the value of $u(r_{k-1}^i, r_k^{*i})$ is set. Fifth, depending on the value of r_k^{*i}, the process noise vector $\mathbf{v}_{k-1}^i$ is drawn either from an untruncated ($r_k^{*i} = 0$) or a truncated ($r_k^{*i} \neq 0$) Gaussian density. Finally, the particle $(\mathbf{x}_{k-1}^i, r_{k-1}^i)$ is predicted to $(\mathbf{x}_k^{*i}, r_k^{*i})$, where

$$\mathbf{x}_k^{*i} = \mathbf{f}_{\delta^i}(\mathbf{x}_{k-1}^i, r_{k-1}^i, r_k^{*i}, \mathbf{v}_{k-1}^i, u(r_{k-1}^i, r_k^{*i})), \quad (10.49)$$

and $\mathbf{f}_{\delta^i}(\ )$ is the state dynamic function corresponding to case δ^i, defined in Section 10.2.3.2.

10.4.2 Update Step

This step involves updating the predicted particles $\{(\mathbf{x}_k^{*i}, r_k^{*i})\}_{i=1}^N$ with the new measurement $\mathbf{z}_k$. Note that $\mathbf{z}_k$ can also be absent (or null), which indicates no measurements due to obscurations. To update the particles, each is assigned a weight proportional to its likelihood function:

$$\tilde{w}_k^i = p(\mathbf{z}_k|\mathbf{x}_k^{*i}) = \begin{cases} \mathcal{N}(\mathbf{z}_k; \mathbf{h}'(\mathbf{x}_k^{*i}), \mathbf{R}_k') & \mathbf{z}_k \neq \emptyset, P_D^i = 1 \\ 0 & \mathbf{z}_k \neq \emptyset, P_D^i = 0 \\ 1 & \mathbf{z}_k = \emptyset, P_D^i = 0 \\ 0 & \mathbf{z}_k = \emptyset, P_D^i = 1 \end{cases} \quad (10.50)$$

where P_D^i is the detection probability corresponding to mode r_k^{*i} and $z_k = \emptyset$ denotes the null measurement. Each particle i, $(i = 1, \ldots, N)$ then receives the normalized weight, $w_k^i = \frac{\tilde{w}_k^i}{\sum_{j=1}^N \tilde{w}_k^j}$. Next, we resample N times with replacement from the set $\{(\mathbf{x}_k^{*i}, r_k^{*i})\}_{i=1}^N$ with weights $\{w_k^i\}_{i=1}^N$ to obtain a new set of particles $\{(\mathbf{x}_k^i, r_k^i)\}_{i=1}^N$ such that $P\{(\mathbf{x}_k^i, r_k^i) = (\mathbf{x}_k^{*j}, r_k^{*j})\} = w_k^j$.

Initialization

To initialize the VS-MMPF we use a two-point differencing [10, p. 247] to obtain $\mathbf{x}_0$ and $\mathbf{P}_0$. Then, we sample N times from $\mathcal{N}(\mathbf{x}_0, \mathbf{P}_0)$ to obtain $\{\mathbf{x}_0^i\}_{i=1}^N$. Also, the initial mode is set to off-road mode $r_0^i = 0$ for $i = 1, \ldots N$. Thus we have $\{(\mathbf{x}_0^i, r_0^i)\}_{i=1}^N$, which characterizes $p(\mathbf{x}_0, r_0 | \mathbf{Z}_0)$ where $\mathbf{Z}_0$ refers to "prior" information.

10.5 SIMULATION RESULTS

This section compares the performance of the two trackers described in Sections 10.3 and 10.4. The scenario used for the simulations is identical to the one shown in Figure 10.1. A ground target, initially at $(10, 10.5)$ km travels at speed 12 m/s along the path indicated by the dotted line. The width of the road network is 8 meters and it includes a tunnel in segment TU where no measurements are received. The receiver, for the purpose of simulation, was considered static and located at the origin. Measurements of range, azimuth, and Doppler are received at a sampling rate of $T = 5$ seconds for a total of 110 scans. The sensor accuracies of range and azimuth measurements are $\sigma_\rho = 20$ m, $\sigma_\theta = 0.5°$. The Doppler measurements are not used in simulations.

The parameters of the VS-IMM algorithm are as follows. This filter is initialized with two second-order white noise acceleration models with equal process noise levels in both the x and y directions. For these baseline models, $\sigma_x = \sigma_y$ were set to 0.6 m/s² and 2.4 m/s², which correspond to constant velocity and maneuver modes, respectively. In addition, six road motion models (one corresponding to each segment in Table 10.1) were used with parameters $\sigma_a = 0.6$ m/s² and $\sigma_0 = 0.0001$ m/s². The sojourn times of the baseline constant velocity and maneuver models are 100 seconds and 7.1 seconds, respectively, while that of road models is 100 seconds. These sojourn times are used in the computation of the Markov chain transition matrix [8].

The VS-MMPF used $s = 7$ modes: one for each road segment and one constant velocity baseline mode with $\sigma_x = \sigma_y = 0.6$ m/s² for off-road motion.

Table 10.2
VS-MMPF Algorithm for Terrain-Aided Tracking

$[\{\mathbf{x}_k^i, r_k^i\}_{i=1}^N] = $ VS-MMPF $[\{\mathbf{x}_{k-1}^i, r_{k-1}^i\}_{i=1}^N, \mathbf{z}_k, \mathcal{T}]$

1. Predict particles to next step to get $\{\mathbf{x}_k^{*i}, r_k^{*i}\}_{i=1}^N$:
   - FOR $i = 1 : N$
     - compute $P\{r_k = j | r_{k-1}^i, \mathbf{x}_{k-1}^i, \mathcal{T}\}$, $j = 0, 1, \ldots, s$
     - draw r_k^{*i} according to $P\{r_k = j | r_{k-1}^i, \mathbf{x}_{k-1}^i, \mathcal{T}\}$
     - determine the case δ^i corresponding to transition (r_{k-1}^i, r_k^{*i})
     - IF $(r_{k-1}^i = r_k^{*i} = 0)$ or $(r_{k-1}^i \neq 0 \text{ and } r_k^{*i} \neq 0)$
       * Set $u(r_{k-1}^i, r_k^{*i}) = 0$
     - ELSE
       * Set $u(r_{k-1}^i, r_k^{*i}) = -1$ or 1 with equal probability
     - END IF
     - IF $(r_k^{*i} = 0)$
       * draw $\mathbf{v}_{k-1}^i \sim \mathcal{N}(\mathbf{0}, \mathbf{Q}(0))$
     - ELSE
       * draw $\mathbf{v}_{k-1}^i \sim \tilde{\mathcal{N}}(\mathbf{0}, \mathbf{Q}(r_k^{*i}); \mathbf{x}_{k-1}^i)$
     - END IF
     - $\mathbf{x}_k^{*i} = \mathbf{f}_{\delta^i}(\mathbf{x}_{k-1}^i, r_{k-1}^i, r_k^{*i}, \mathbf{v}_{k-1}^i, u(r_{k-1}^i, r_k^{*i}))$.
   - END FOR

2. Update the particles
   - FOR $i = 1 : N$
     - Compute weights $\tilde{w}_k^i = p(\mathbf{z}_k | \mathbf{x}_k^{*i})$ using (10.50).
   - END FOR
   - Calculate total weight: $t = \text{SUM}[\{\tilde{w}_k^i\}_{i=1}^N]$
   - FOR $i = 1 : N$
     - Normalize: $w_k^i = t^{-1} \tilde{w}_k^i$
   - END FOR
   - Resample using the algorithm in Table 3.2:
     $[\{[\mathbf{x}_k^i, r_k^i]^T, -, -\}_{i=1}^N] = \text{RESAMPLE}\,[\{[\mathbf{x}_k^{*i}, r_k^{*i}]^T, w_k^i\}_{i=1}^N]$

The process noise covariance matrix $\mathbf{Q}(r)$ for $r \neq 0$ is identical to that of VS-IMM. In addition, the parameter $\bar{p}$ and p^* are set to $\bar{p} = p^* = 0.98$. The threshold τ required in (10.29) is set to

$$\tau = 1.25 \, T^2 \sigma_x \qquad (10.51)$$

such that $\mathrm{P}\{d > \tau\} \approx 0$, where d is the distance traveled due to process noise with variance σ_x^2, and T is the sampling time. The number of particles for the VS-MMPF was chosen to be $N = 1000$.

Figures 10.2, 10.3, 10.4, and 10.5 display the output of the VS-MMPF at scans $k = 68, 70, 73$, and 95, respectively. The left-hand side in this set of figures always shows the distribution of particles, with their mean value represented by "■." The right-hand side presents the probabilities of modes and should be compared with Table 10.1. For example, at $k = 68$ (Figure 10.2), the target is on the road segment UJ (number 3 in Table 10.1), and indeed the probability of mode 3 in Figure 10.2(b) is close to 1. Observe from Figure 10.2(a) that most of the particles are distributed along the road in this case. At scan $k = 70$ the target has reached the junction J and the particles are spread along four possible directions of this intersection in Figure 10.3(a). The probabilities of modes 3, 4, 5, and 6, according to Figure 10.3(b), are similar in value. Figure 10.4 corresponds to $k = 73$ when the target is moving along the road segment DJ (number 6 in Table 10.1). Finally, Figure 10.5 shows the results at $k = 95$, when the target is moving off the road (segment EF in Figure 10.1). Observe from Figure 10.5 that particles are more evenly spread for the off-road motion (no hard constraints) and that mode 0 is the dominant mode.

Figure 10.6 shows a comparison of the RMS error performance for the two trackers averaged over 100 MC runs. The dotted vertical lines in the graph correspond to mode transition times. The VS-MMPF shows a much reduced position error compared to VS-IMM, particularly in the region where the target is moving along the road network. Specifically, the average RMS error for the on-road target is about 70 m and 25 m for the VS-IMM and VS-MMPF, respectively. This implies that VS-MMPF provides a 65% improvement over VS-IMM. Observe that while the target is within the tunnel (scan number k is between 40 and 50) the RMS error increases (for both VS-IMM and VS-MMPF) due to the lack of measurements. Also, notice that while the target is off-road ($k \leq 10$ and $k > 90$) the performance of the two filters is comparable. For this region, since no explicit hard constraints are introduced, the posterior pdf will tend to be nearly Gaussian, which is handled equally well by both filters.

Next, we compare the RMS error for the case where the tracker has prior knowledge of speed constraints. Suppose we know that the speed of the target is in the range $6 \leq |v| \leq 17$ m/s. Such information can be incorporated in the VS-MMPF, which ensures that particles not satisfying the above constraints are rejected.

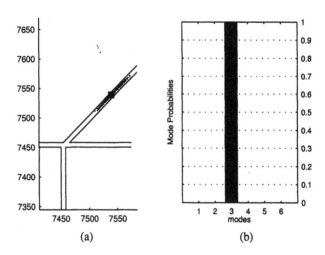

Figure 10.2 A "snapshot" of the particle filter at $k = 68$: (a) the particle cloud; and (b) the mode probabilities.

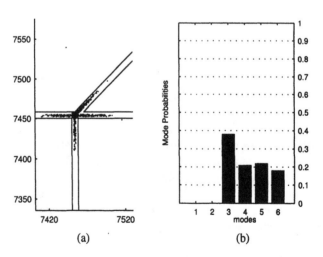

Figure 10.3 A "snapshot" of the particle filter at $k = 70$: (a) the particle cloud; and (b) the mode probabilities.

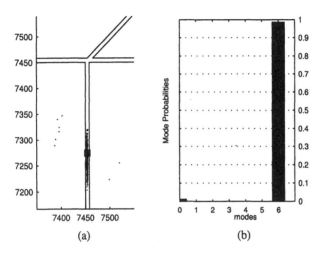

Figure 10.4 A "snapshot" of the particle filter at $k = 73$: (a) the particle cloud; and (b) the mode probabilities.

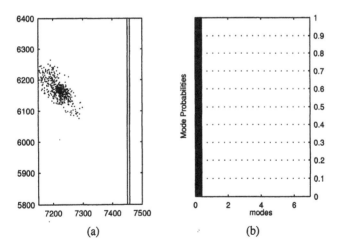

Figure 10.5 A "snapshot" of the particle filter at $k = 95$: (a) the particle cloud; and (b) the mode probabilities.

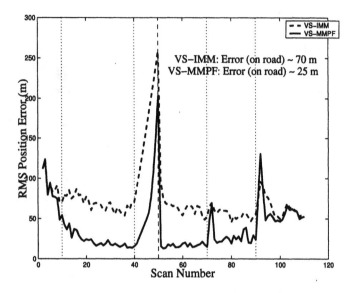

Figure 10.6 Comparison of RMS error performance.

On the contrary, it is difficult to incorporate such nonstandard information in the VS-IMM framework. Figure 10.7 shows a comparison of RMS error performance for this case in which this speed constraint information is used only by the VS-MMPF algorithm. We notice that the VS-MMPF performance is further improved here. Specifically, the average error on road is reduced from 25 to 16 meters leading to a 77% improvement over the VS-IMM.

10.6 CONCLUSIONS

In this chapter we presented a particle-filter-based algorithm for GMTI tracking with nonstandard information. The algorithm is referred to as the variable structure multiple-model particle filter (VS-MMPF), since it adaptively selects a subset of modes that are active at a particular time. The RMS error performance of the VS-MMPF is compared with that of the VS-IMM algorithm. For the scenario considered, the VS-MMPF outperformed the VS-IMM, giving 65% improvement in error performance for an on-road target. The VS-MMPF performance is further improved with other nonstandard information such as speed constraints. In particular, for a case with speed constraint knowledge, the VS-MMPF showed a 77% improvement in RMS error compared to the VS-IMM.

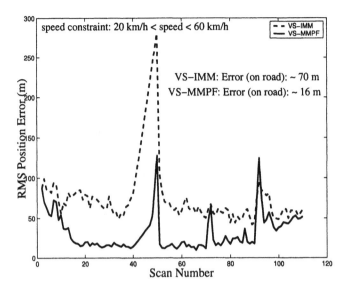

Figure 10.7 Comparison of RMS error performance: case with speed constraints $6 \leq |v| \leq 17$ m/s.

This superiority of the VS-MMPF over the VS-IMM is due to the better dynamics models adopted by the VS-MMPF. These models capture the motion dynamics with terrain information in an intricate but accurate manner. While the particle filter (VS-MMPF) is able to incorporate such complex models into its framework, the VS-IMM is unable to do so. The dynamics models adopted by the VS-IMM are simpler and less accurate in capturing the motion of targets in a road network. Future work on terrain-aided tracking should explore more realistic models of target visibility (with probability $0 \leq P_D^r \leq 1$) and consider road networks with possibly winding roads.

References

[1] J. N. Entzminger, C. A. Fowler, and W. J. Kenneally, "JointSTARS and GMTI: Past, present and future," *IEEE Trans. Aerospace and Electronic Systems*, vol. 35, pp. 748–761, April 1999.

[2] T. Kirubarajan, Y. Bar-Shalom, K. R. Pattipati, and I. Kadar, "Ground target tracking with variable structure IMM estimator," *IEEE Trans. Aerospace and Electronic Systems*, vol. 36, pp. 26–45, January 2000.

[3] T. Kirubarajan and Y. Bar-Shalom, "Tracking evasive move-stop-move targets with a MTI radar using a VS-IMM estimator," in *Proc. SPIE, Signal and Data Processing of Small Targets*, vol. 4048, pp. 236–246, April 2000.

[4] K. Kastella, C. Kreucher, and M. A. Pagels, "Nonlinear filtering for ground target applications," in *Proc. SPIE, Signal and Data Processing of Small Targets*, vol. 4048, pp. 266–276, April 2000.

[5] W. Koch and R. Klemm, "Ground target tracking with STAP radar," *IEEE Proc. - Radar, Sonar Navig.*, vol. 148, no. 3, pp. 173–185, 2001.

[6] A. Farina, L. Ferranti, and G. Golino, "Constrained tracking filters for A-SMGCS," in *Proc. 6th Int. Conf. Information Fusion (Fusion 2003)*, (Cairns, Australia), July 2003.

[7] M. S. Arulampalam, N. J. Gordon, M. Orton, and B. Ristic, "A variable structure multiple model particle filter for GMTI tracking," in *Proc. 5th Int. Conf. Information Fusion*, (Annapolis, MD), July 2002.

[8] Y. Bar-Shalom and X. R. Li, *Multitarget-Multisensor Tracking: Principles and Techniques*. Storrs, CT: YBS Publishing, 1995.

[9] X. R. Li, "Engineer's guide to variable-structure multiple-model estimation for tracking," in *Multitarget-Multisensor Tracking: Applications and Advances* (Y. Bar-Shalom and W. D. Blair, eds.), vol. III, ch. 10, Norwood, MA: Artech House, 2000.

[10] Y. Bar-Shalom, X. R. Li, and T. Kirubarajan, *Estimation with Applications to Tracking and Navigation*. New York: John Wiley & Sons, 2001.

Chapter 11

Detection and Tracking of Stealthy Targets

11.1 INTRODUCTION

The conventional approach to target tracking is based on target measurements (position, range rate, and so forth) that are extracted by thresholding the output of a signal processing unit of a surveillance sensor [1]. The primary role of thresholding is to reduce the data flow and thus simplify tracking. For a target of a certain signal-to-noise ratio (SNR), the choice of the detection threshold determines the probability of target detection and the density of false alarms. The false alarm rate, on the other hand, affects the complexity of the data association problem in the tracking system – in general, higher densities of false alarms require more sophisticated data association algorithms.

The undesirable effect of thresholding the sensor data, however, is that in restricting the data flow, it also throws away potentially useful information. For high SNR targets this loss of information is of little concern because one can achieve good probability of detection with a small false alarm rate. Recent developments of stealthy military aircraft and cruise missiles have emphasized the need to detect and track low SNR targets. For these dim (stealthy) targets, there is a considerable advantage in using the unthresholded data for simultaneous detection and track initiation [2, 3]. Depending on the type of sensor in use, the unthresholded data can be a sequence of range-Doppler maps (for a radar), bearing-frequency distributions (for a passive sonar), or gray-scale images (for a staring TV or an infrared camera).

The concept of simultaneous detection and tracking using unthresholded data is known in literature as *track-before-detect* (TBD) approach. Typically TBD is implemented as a batch algorithm using the Hough transform [4], dynamic programming [2, 3], or maximum likelihood estimation [5]. These methods operate on several scans of data, prohibit or penalize deviations from the straight-line motion, and in general require enormous computational resources. As an alternative,

Stone et al. proposed a recursive Bayesian track-before-detect, implemented using the numerical techniques referred to by authors as "sample-field, cell-based and Kalman like" [6, Chs. 6 and 7].

In this chapter we also develop a recursive Bayesian TBD estimator, however, our formulation and implementation are based on the particle filter. The basic concept follows the approach presented by Salmond and coworkers [7, 8]. The advantages of the proposed recursive Bayesian TBD are manifold:

- There is no need to store and process multiple scans of data;
- Target motion is modeled by a stochastic dynamic equation (and therefore the method is not restricted to the straight-line motion);
- The approach is valid for non-Gaussian and structured background noise;
- The effects of unknown and fluctuating target intensity (or SNR), point spread functions and extended objects are accommodated;
- Target presence and absence (due to occlusions for example) are explicitly modeled.

This chapter differs from the rest of the book in two important details. First, the sensor is assumed to produce a sequence of images characterized by *finite resolution* (i.e., the target is of a subpixel size). Finite sensor resolution, often ignored by the research community, is an important practical issue for target tracking [9]. Second, the concept of *target existence* is introduced in order to detect targets and initiate tracks. This concept, which has its roots in [10, 11], enables us to compute the probability of target presence (existence) directly from the filter.

The chapter is organized as follows. Section 11.2 introduces the target and sensor models. Section 11.3 formulates the TBD approach as a nonlinear filtering problem and describes the conceptual recursive Bayesian solution. The implementation of this solution using a particle filter is presented in Section 11.4. A numerical example is presented in Section 11.5, while the statistical performance of the PF is analyzed in Section 11.6.

11.2 TARGET AND SENSOR MODELS

11.2.1 Target Model

Consider a target of a certain intensity level (brightness) moving in the x-y plane according to a known discrete-time dynamic model of the form

$$\mathbf{x}_{k+1} = \mathbf{f}_k(\mathbf{x}_k, \mathbf{v}_k), \tag{11.1}$$

where k is the discrete-time index, $\mathbf{v}_k$ is the process noise sequence, and $\mathbf{x}_k$ is the state vector defined as

$$\mathbf{x}_k = [x_k \ \dot{x}_k \ y_k \ \dot{y}_k \ I_k]^T. \tag{11.2}$$

Here (x, y), $(\dot{x}, \dot{y})$, and I_k denote the position, velocity, and the intensity of the target, respectively. The specification of (11.1) is equivalent to knowledge of the transitional density $p(\mathbf{x}_{k+1}|\mathbf{x}_k)$.

The target can be present or absent from the surveillance region (or the field of view) at a discrete-time k. Target presence (existence) variable E_k is modeled by a two-state Markov chain, that is $E_k \in \{0, 1\}$. Here 0 denotes the event that a target is not present, while 1 denotes the opposite [10, 11]. Furthermore, we assume that transitional probabilities of target "birth" (P_b) and "death" (P_d), defined as

$$P_b \triangleq P\{E_k = 1|E_{k-1} = 0\} \tag{11.3}$$

$$P_d \triangleq P\{E_k = 0|E_{k-1} = 1\}, \tag{11.4}$$

are known. The other two transitional probabilities of this Markov chain, the probability of staying alive and the probability of remaining absent, are given by $1 - P_d$ and $1 - P_b$, respectively. In summary the transitional probability matrix (TPM) is given by:

$$\Pi = \begin{bmatrix} 1 - P_b & P_b \\ P_d & 1 - P_d \end{bmatrix}. \tag{11.5}$$

The initial target existence probability (at time $k = 1$), denoted as $\mu_1 = P\{E_1 = 1\}$, is also assumed to be known.

11.2.2 Sensor Model

The sensor provides a sequence of two-dimensional images (maps or frames) of the surveillance region, each image consisting of $n \times m$ resolution cells (pixels). A resolution cell corresponds to a rectangular region of dimensions $\Delta_x \times \Delta_y$ so that the center of each cell (i, j) is defined to be at $(i\Delta_x, j\Delta_y)$ for $i = 1, \ldots, n$ and $j = 1, \ldots, m$.

Measured images are recorded at discrete instants k with the sampling interval T. At each resolution cell (i, j) the measured intensity is denoted as $z_k^{(i,j)}$ and modeled as:

$$z_k^{(i,j)} = \begin{cases} h_k^{(i,j)}(\mathbf{x}_k) + w_k^{(i,j)} & \text{if target present} \\ w_k^{(i,j)} & \text{if target absent} \end{cases} \tag{11.6}$$

where

- $h_k^{(i,j)}(\mathbf{x}_k)$ is the target contribution to intensity level in the resolution cell (i,j) and

- $w_k^{(i,j)}$ is measurement noise in the resolution cell (i,j), assumed to be independent from pixel to pixel and from frame to frame. For simplicity we will adopt a homogeneous and Gaussian model of background noise; that is, $w_k^{(i,j)} \sim \mathcal{N}(w; 0, \sigma^2)$. We point out, however, that the recursive Bayesian solution (to be presented in the next section) is applicable even to non-homogeneous and non-Gaussian measurement noise, as long as the statistics of $w_k^{(i,j)}$ are known for each cell (i,j).

Target contribution intensity, $h_k^{(i,j)}(\mathbf{x}_k)$, can be either due to target extent (for large targets) or to the sensor point spread function. We concentrate on the latter; in particular, we consider a point target and the sensor point spread function approximated by a two-dimensional Gaussian density with circular symmetry. Thus for a point target of intensity I_k at position (x_k, y_k), the contribution to pixel (i,j) is approximated as:

$$h_k^{(i,j)}(\mathbf{x}_k) \approx \frac{\Delta_x \Delta_y I_k}{2\pi \Sigma^2} \exp\left\{-\frac{(i\Delta_x - x_k)^2 + (j\Delta_y - y_k)^2}{2\Sigma^2}\right\} \quad (11.7)$$

where Σ is a known parameter that represents the amount of blurring introduced by the sensor. The expression in (11.7) approximates the integral of the point spread function over the resolution cell and the sampling interval. For a more accurate model of an optical sensor, the reader is referred to [5].

The complete measurement recorded at time k is an $n \times m$ matrix denoted as

$$\mathbf{z}_k = \{z_k^{(i,j)} : i = 1, \ldots, n, j = 1, \ldots, m\} \quad (11.8)$$

while the set of complete measurements collected up to time k is denoted as usual: $\mathbf{Z}_k = \{\mathbf{z}_i, i = 1, \ldots, k\}$.

11.3 CONCEPTUAL SOLUTION IN THE BAYESIAN FRAMEWORK

The problem of TBD can now be formulated in the framework of recursive Bayesian estimation as follows. Given the joint posterior pdf of target state and target existence at time $k-1$, denoted as $p(\mathbf{x}_{k-1}, E_{k-1}|\mathbf{Z}_{k-1})$, and given the latest available frame $\mathbf{z}_k$, the goal is to construct the joint posterior pdf at time k, $p(\mathbf{x}_k, E_k|\mathbf{Z}_k)$. The posterior probability of target existence at time k

$$P_k \triangleq \mathrm{P}\{E_k = 1|\mathbf{Z}_k\} \quad (11.9)$$

is then computed as the marginal of $p(\mathbf{x}_k, E_k = 1|\mathbf{Z}_k)$. The problem is conceptually one of hybrid estimation described in Section 1.4. This time, however, the discrete-valued variable is different: instead of regime r_k, we deal with target existence E_k.

As described by (1.40) and (1.41), the formal recursive Bayesian solution can be presented as a two-step procedure, consisting of *prediction* and *update*. If $E_k = 0$, the target state is not defined. For $E_k = 1$, similarly to (1.40), the prediction step can be expressed as:

$$p(\mathbf{x}_k, E_k = 1|\mathbf{Z}_{k-1}) =$$
$$\int p(\mathbf{x}_k, E_k = 1|\mathbf{x}_{k-1}, E_{k-1} = 1, \mathbf{Z}_{k-1}) p(\mathbf{x}_{k-1}, E_{k-1} = 1|\mathbf{Z}_{k-1}) d\mathbf{x}_{k-1}$$
$$+ \int p(\mathbf{x}_k, E_k = 1|\mathbf{x}_{k-1}, E_{k-1} = 0, \mathbf{Z}_{k-1}) p(\mathbf{x}_{k-1}, E_{k-1} = 0|\mathbf{Z}_{k-1}) d\mathbf{x}_{k-1}$$
$$(11.10)$$

where

$$p(\mathbf{x}_k, E_k = 1|\mathbf{x}_{k-1}, E_{k-1} = 1, \mathbf{Z}_{k-1}) =$$
$$p(\mathbf{x}_k|\mathbf{x}_{k-1}, E_k = 1, E_{k-1} = 1) \, \mathrm{P}\{E_k = 1|E_{k-1} = 1\}$$
$$= p(\mathbf{x}_k|\mathbf{x}_{k-1}, E_k = 1, E_{k-1} = 1) \, (1 - P_d) \quad (11.11)$$

and

$$p(\mathbf{x}_k, E_k = 1|\mathbf{x}_{k-1}, E_{k-1} = 0, \mathbf{Z}_{k-1}) =$$
$$p(\mathbf{x}_k|\mathbf{x}_{k-1}, E_k = 1, E_{k-1} = 0) \, \mathrm{P}\{E_k = 1|E_{k-1} = 0\}$$
$$= p_b(\mathbf{x}_k) \, P_b. \quad (11.12)$$

The transitional density $p(\mathbf{x}_k|\mathbf{x}_{k-1}, E_k = 1, E_{k-1} = 1)$ that featured in (11.11) is defined by the target dynamic model (11.1). The pdf $p_b(\mathbf{x}_k)$ in (11.12) denotes the initial target density on its appearance. This density, in the Bayesian framework, is assumed to be known (one can, for example, adopt a uniform density over the surveillance region).

Conceptually, the update equation in the Bayesian framework is given by

$$p(\mathbf{x}_k, E_k = 1|\mathbf{Z}_k) = \frac{p(\mathbf{z}_k|\mathbf{x}_k, E_k = 1) \, p(\mathbf{x}_k, E_k = 1|\mathbf{Z}_{k-1})}{p(\mathbf{z}_k|\mathbf{Z}_{k-1})} \quad (11.13)$$

where prediction density $p(\mathbf{x}_k, E_k = 1|\mathbf{Z}_{k-1})$ is given by (11.10) and $p(\mathbf{z}_k|\mathbf{x}_k, E_k)$ is the likelihood function. For the sensor model described in Section 11.2.2, the likelihood function can be expressed as follows [7]:

$$p(\mathbf{z}_k|\mathbf{x}_k, E_k) = \begin{cases} \prod_{i=1}^{n} \prod_{j=1}^{m} p_{\text{S+N}}(z_k^{(i,j)}|\mathbf{x}_k), & \text{for } E_k = 1 \\ \prod_{i=1}^{n} \prod_{j=1}^{m} p_{\text{N}}(z_k^{(i,j)}), & \text{for } E_k = 0. \end{cases} \quad (11.14)$$

Here $p_{\text{N}}(z_k^{(i,j)})$ is the pdf of background noise in pixel (i,j), while $p_{\text{S+N}}(z_k^{(i,j)}|\mathbf{x}_k)$ is the likelihood of target signal plus noise in pixel (i,j), given that the target is in state $\mathbf{x}_k$. We were able to assume independence of pixel measurements in (11.14) because measurement noise $w_k^{(i,j)}$ in (11.6) is independent from pixel to pixel. The two probability density functions $p_{\text{N}}(z_k^{(i,j)})$ and $p_{\text{S+N}}(z_k^{(i,j)}|\mathbf{x}_k)$ can be further expressed as:

$$p_{\text{N}}(z_k^{(i,j)}) = \mathcal{N}(z^{(i,j)}; 0, \sigma^2) \quad (11.15)$$
$$p_{\text{S+N}}(z_k^{(i,j)}|\mathbf{x}_k) = \mathcal{N}(z^{(i,j)}; h_k^{(i,j)}(\mathbf{x}_k), \sigma^2). \quad (11.16)$$

Since the target (if present) will affect only the pixels in the vicinity of its location (x_k, y_k), the expression for $p(\mathbf{z}_k|\mathbf{x}_k, E_k = 1)$ can be approximated as follows:

$$p(\mathbf{z}_k|\mathbf{x}_k, E_k = 1) \approx \prod_{i \in C_i(\mathbf{x}_k)} \prod_{j \in C_j(\mathbf{x}_k)} p_{\text{S+N}}(z_k^{(i,j)}|\mathbf{x}_k) \prod_{i \notin C_i(\mathbf{x}_k)} \prod_{j \notin C_j(\mathbf{x}_k)} p_{\text{N}}(z_k^{(i,j)}) \quad (11.17)$$

where $C_i(\mathbf{x}_k)$ and $C_j(\mathbf{x}_k)$ are the sets of subscripts i and j, respectively, corresponding to pixels affected by the target.

11.4 A PARTICLE FILTER FOR TRACK-BEFORE-DETECT

The recursive Bayesian solution of the track-before-detect problem described in the previous section can be implemented using a particle filter [7, 8, 12]. The PF that we develop for this application has some similarities to the MMPF of Section 3.5.5. Again we introduce the augmented state vector $\mathbf{y}_k = \begin{bmatrix} \mathbf{x}_k^T & E_k \end{bmatrix}^T$ that now has six components. Let us denote a random measure that characterizes the posterior pdf at $k - 1$, namely $p(\mathbf{y}_{k-1}|\mathbf{Z}_{k-1})$, by $\{\mathbf{y}_{k-1}^n, w_{k-1}^n\}_{n=1}^N$. As usual, N is the number of particles, while $\mathbf{y}_k^n$ consists of $\mathbf{x}_k^n$ and E_k^n. The pseudocode of a single cycle of the PF developed for the TBD problem (referred to as TBD-PF) is presented in Table 11.1. The first step is to predict for each particle its target existence variable E_k^n, $n = 1, \ldots, N$. This can be done using the algorithm described in Table 3.9. The next step is the prediction of particle target states; this is done, however, only for those particles that are characterized by $E_k^n = 1$. For the remaining particles (with $E_k^n = 0$), the target state components are undefined. There are two possible cases here:

Newborn particles. This group of predicted particles is characterized by the transition from $E_{k-1}^n = 0$ to $E_k^n = 1$. The target state is drawn as a sample from the proposal density $q_b(\mathbf{x}_k|\mathbf{z}_k)$ that is obtained as follows. For target position components, $q_b(x_k, y_k|\mathbf{z}_k)$ is a uniform density over those regions of the surveillance area for which $z_k^{(i,j)} > \gamma$, where γ is a suitably chosen threshold value. In this way, the newborn particles populate the regions of the surveillance area where the latest available image (measurement) has a high value. For target velocity component in x, the proposal density is $q_b(\dot{x}_k) = \mathcal{U}[-v_{\max}, v_{\max}]$, where $v_{\max}$ is the maximum speed of the target (and similarly in y direction). Finally the target intensity component of the state vector, $I_k \sim \mathcal{U}[I_{\min}, I_{\max}]$, where $I_{\min}$ and $I_{\max}$ are suitably chosen intensity levels (based on the expected target signal-to-noise ratio).

Existing particles. This is a group of particles that continues to stay "alive," with $E_{k-1}^n = 1$ and $E_k^n = 1$. The optimal importance density function for this group of particles can be derived through Rao-Blackwellization [13]. Here, however, we shall be primarily interested in the "proof of concept" and therefore, for existing particles, we adopt a simple (though inefficient) transitional prior (see Section 3.4.2), defined by the target dynamic model (11.1).

The (unnormalized) importance weights are computed next. For this purpose we need to introduce the *likelihood ratio* in pixel (i, j) for a target in state $\mathbf{x}_k^n$, defined as:

$$\ell(z_k^{(i,j)}|\mathbf{x}_k^n) \triangleq \frac{p_{\text{S+N}}(z_k^{(i,j)}|\mathbf{x}_k^n)}{p_{\text{N}}(z_k^{(i,j)})} \qquad (11.18)$$

$$= \exp\left\{-\frac{h_k^{(i,j)}(h_k^{(i,j)} - 2z_k^{(i,j)})}{2\sigma^2}\right\} \qquad (11.19)$$

where $h_k^{(i,j)}$ was defined in (11.7). Equation (11.19) follows from (11.15), (11.16), and (11.18). The importance weights (up to a normalizing constant) are now given by [7]:

$$\tilde{w}_k^n = \begin{cases} \prod_{i \in C_i(\mathbf{x}_k^n)} \prod_{j \in C_j(\mathbf{x}_k^n)} \ell(z_k^{(i,j)}|\mathbf{x}_k^n) & \text{if } E_k^n = 1 \\ 1 & \text{if } E_k^n = 0. \end{cases} \qquad (11.20)$$

Table 11.1

Particle Filter for Track-Before-Detect

$[\{\mathbf{y}_k^n\}_{n=1}^N]$ = TBD-PF$[\{\mathbf{y}_{k-1}^n\}_{i=1}^N, \mathbf{z}_k]$

- Target existence transitions (Table 3.9):
  $[\{E_k^n\}_{n=1}^N]$ = RT $[\{E_{k-1}^n\}_{n=1}^N, \mathbf{\Pi}]$
- FOR $i = 1 : N$
  - IF a newborn particle (i.e., $E_{k-1}^n = 0$ and $E_k^n = 1$)
    Draw $\mathbf{x}_k^n \sim q_b(\mathbf{x}_k|\mathbf{z}_k)$
  - IF an existing particle (i.e., $E_{k-1}^n = 1$ and $E_k^n = 1$)
    Draw $\mathbf{x}_k^n \sim q(\mathbf{x}_k|\mathbf{x}_{k-1}^n, \mathbf{z}_k)$
  - Evaluate importance weight (up to a normalizing constant) using (11.20)
- END FOR
- Calculate total weight: t = SUM $[\{\tilde{w}_k^n\}_{n=1}^N]$
- FOR $n = 1 : N$
  - Normalize: $w_k^n = t^{-1}\tilde{w}_k^n$
- END FOR
- Resample using the algorithm in Table 3.2:
  $[\{\mathbf{y}_k^n, -, -\}_{n=1}^N]$ = RESAMPLE $[\{\mathbf{y}_k^n, w_k^n\}_{n=1}^N]$

Here we use approximation (11.17) in order to reduce the computational load of the particle filter. In practice we select $C_i(\mathbf{x}_k^n) = \{i_0 - p, \ldots, i_0 - 1, i_0, i_0 + 1, \ldots, i_0 + p\}$, where i_0 is the nearest integer value of the particle state vector component $x_k^n = \mathbf{x}_k^n[1]$ and p is a design parameter. A similar procedure is used for the selection of $C_j(\mathbf{x}_k^n)$.

The remaining steps of the TBD-PF in Table 11.1 include the normalization of the importance weights and the resampling step, described in Table 3.2.

The PF for track-before-detect performs target detection using the estimate of the posterior probability of target existence defined in (11.9). This estimate is computed as:

$$\hat{P}_k = \frac{\sum_{n=1}^N E_k^n}{N}, \qquad (11.21)$$

and satisfies $0 \leq \hat{P}_k \leq 1$. Target presence is declared if $\hat{P}_k$ is above a certain threshold value. This declaration can then trigger the initiation of a track based on the estimated target state

$$\hat{\mathbf{x}}_{k|k} = \frac{\sum_{n=1}^{N} \mathbf{x}_k^n \cdot E_k^n}{\sum_{n=1}^{N} E_k^n}. \tag{11.22}$$

Various track initiation procedures are known in practice. For example, a single "target present" declaration can initiate a *tentative* track, which is confirmed if a subsequent declaration is received within a certain time interval; for details, see [14].

The particle filter in Table 11.1 is initialized at $k = 1$ by drawing samples E_1^n, $n = 1, \ldots, N$ in accordance with the initial target existence probability μ_1. For every n such that $E_1^n = 1$, the particle state vector is initialized as a newborn particle state, that is, $\mathbf{x}_1^n \sim q_b(\mathbf{x}_1|\mathbf{z}_1)$. For the remaining particles, the state vector is undefined.

11.5 A NUMERICAL EXAMPLE

Although the formulation of the dynamic model can be as general as in (11.1), in order to simplify the presentation we adopt (1) a nearly constant velocity model for target motion, and (2) a random walk model for target intensity. Then we have:

$$\mathbf{x}_{k+1} = \mathbf{F}\mathbf{x}_k + \mathbf{v}_k \tag{11.23}$$

where

$$\mathbf{F} = \begin{bmatrix} 1 & T & 0 & 0 & 0 \\ 0 & 1 & 0 & 0 & 0 \\ 0 & 0 & 1 & T & 0 \\ 0 & 0 & 0 & 1 & 0 \\ 0 & 0 & 0 & 0 & 1 \end{bmatrix} \tag{11.24}$$

and $\mathbf{v}_k$ is zero-mean white Gaussian noise with covariance $\mathbf{Q}$ given by [15]

$$\mathbf{Q} = \begin{bmatrix} \frac{q_1}{3}T^3 & \frac{q_1}{2}T^2 & 0 & 0 & 0 \\ \frac{q_1}{2}T^2 & q_1 T & 0 & 0 & 0 \\ 0 & 0 & \frac{q_1}{3}T^3 & \frac{q_1}{2}T^2 & 0 \\ 0 & 0 & \frac{q_1}{2}T^2 & q_1 T & 0 \\ 0 & 0 & 0 & 0 & q_2 T \end{bmatrix} \tag{11.25}$$

where q_1 and q_2 denote the level of process noise in target motion and intensity, respectively. The adopted target model accommodates not only deviations from the straight-line target motion, but also the fluctuation in the target intensity.

A sequence of 30 frames of data has been generated with the following parameters: $\Delta_x = \Delta_y = 1, n = m = 20, T = 1$ s, $\sigma = 3$. The target is absent from frame 1 to frame 6 to be introduced in frame 7 with the initial intensity $I = 20$. The blurring parameter is $\Sigma = 0.7$, and therefore the initial SNR is:

$$\text{SNR} = 10 \log \left[\frac{I \Delta_x \Delta_y / 2\pi \Sigma^2}{\sigma} \right]^2 = 6.7 \text{dB}.$$

For this value of SNR there will be on average about 6 noise pixels that will exceed the target intensity level in every frame. The initial target position is $(4.2, 7.2)$ units and velocity $(0.45, 0.25)$ units/s. The level of process noise used in the target motion/intensity model is: $q_1 = 0.001$ and $q_2 = 0.01$ (the SNR varies only marginally). The target exists until frame 22 and is again absent in frames $23, 24, \ldots, 30$. Figure 11.1 displays six image frames of the data sequence (frame numbers 2, 7, 12, 17, 22, and 27) synthesized according to the described experimental setup. Pixel intensity is shown in a gray linear scale, with white color indicating the highest intensity. This sequence of images serves to illustrate that by visual inspection it is impossible to detect the existence and infer the location of the target.

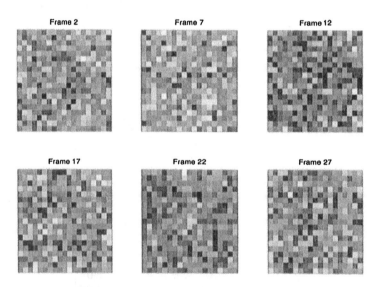

Figure 11.1 Six frames from the data sequence.

The particle filter parameters are selected as follows: transitional probabilities $P_b = P_d = 0.05$; initial existence probability $\mu_1 = 0.05$; threshold $\gamma = 2$; $v_{\max} =$

1 unit/s; initial intensity range from $I_{min} = 10$ to $I_{max} = 30$ (corresponding to target SNR from 0.7 dB to 10.2 dB); $p = 2$ and number of particles $N = 80000$. The number of particles may appear excessive but with careful programming the PF runs in real-time. The output of the PF (single run) is presented in Figures 11.2 to 11.4. Figure 11.2 shows the estimate of the existence probability (11.21); asterisk signs (*) at the bottom of the figure indicate the presence of the target. If we set a threshold for target existence probability at, say, 0.6, the PF required only three frames, following the appearance of the target, to establish that the target is present. Existence probability remains very stable and above 0.9 until frame 22. Then it drops sharply in frame 23, when the target is not present anymore.

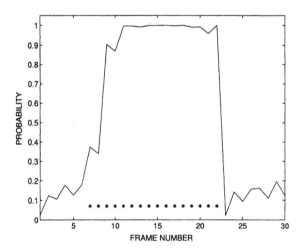

Figure 11.2 Probability of target existence.

Figure 11.3 shows the distribution of "alive" particles x_k^n (i.e., those characterized by $E_k^n = 1$) corresponding to the same sequence of images displayed in Figure 11.1. Note from Figure 11.3(a) that particles appear randomly dispersed in frames 2 and 7, but as the PF learns from the data the presence and location of the target, the particle cloud becomes more concentrated around the true position (frames 12, 17, 22). A similar observation can be made from Figure 11.3(b) with respect to the target intensity.

Figure 11.4 displays the true target path against the track [sequence of target state estimates defined by (11.22)], produced by the filter during the interval when $\hat{P}_k > 0.6$. Note how the target trajectory deviates slightly from the straight line due to process noise. The PF tracks the target with a positional error smaller than the pixel size.

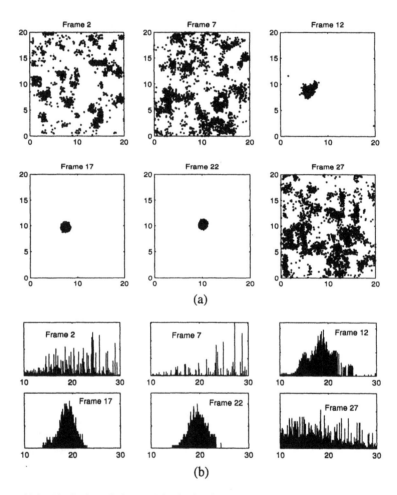

Figure 11.3 Distribution of *alive* particles in the six representative frames: (a) x-y plane; and (b) intensity.

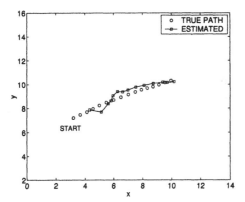

Figure 11.4 Estimated versus the true target trajectory.

11.6 PERFORMANCE ANALYSIS

This section is devoted to statistical performance analysis of the PF-based recursive Bayesian TBD estimator. An evaluation of tracking accuracy is presented first, followed by an analysis of detection performance.

11.6.1 Tracking Error Performance

As we discussed in the introduction to this chapter, thresholding sensor data introduces the notion of uncertainty in the measurement origin. The uncertainty is due to the possibility that the target is not detected (probability of target detection $P_D < 1$) and false detections were reported (probability of false alarm $P_{FA} > 0$). The derivation of CRLBs for target tracking with $P_D < 1$ and $P_{FA} > 0$ has been carried out in two ways: (1) the information reduction approach incorporates the uncertainty in the measurement origin by inflation of the measurement covariance [16, 17]; and (2) the enumeration method attempts to work out all possible measurement-target associations and computes the CRLB as their weighted sum [18]. The first approach is simple but inaccurate, and the second is accurate but extremely computationally intensive [19]. If we deal with unthresholded data, however, it is possible to derive and compute the exact CRLB for target tracking. In order to simplify the analysis, we assume, as in Section 11.5, that target state dynamics is described by a linear model[1] (11.23).

---

[1] In general this assumption is unnecessary; the extension to nonlinear state equation as in (11.1) follows directly from the material presented in Chapter 4.

Cramér-Rao Bound for Tracking with Unthresholded Data

The CRLB for tracking with unthresholded data is more optimistic than the one that deals with uncertain measurements, simply because thresholding throws away potentially useful information. In the derivation we assume that the target is present during the entire observation interval.

The information matrix $\mathbf{J}_k$ is computed recursively according to (4.13), repeated here for convenience:

$$\mathbf{J}_{k+1} = \mathbf{D}_k^{22} - \mathbf{D}_k^{21}(\mathbf{J}_k + \mathbf{D}_k^{11})^{-1}\mathbf{D}_k^{12}. \tag{11.26}$$

If target dynamics is linear as in (11.23) and the measurement equation is nonlinear as in (11.6), from Section 4.3 it follows that

$$\mathbf{D}_k^{11} = \mathbf{F}^T \mathbf{Q}^{-1} \mathbf{F} \tag{11.27}$$

$$\mathbf{D}_k^{12} = -\mathbf{F}^T \mathbf{Q}^{-1} \tag{11.28}$$

$$\mathbf{D}_k^{22} = \mathbf{Q}^{-1} + \mathbb{E}\{\tilde{\mathbf{H}}_{k+1}^T \mathbf{R}_{k+1}^{-1} \tilde{\mathbf{H}}_{k+1}\} \tag{11.29}$$

where matrices $\mathbf{F}$ and $\mathbf{Q}$ were defined earlier by (11.24) and (11.25), respectively. Jacobian $\tilde{\mathbf{H}}_k$ and the measurement covariance $\mathbf{R}_k$ will be defined later in the text. Upon substitution of (11.27)–(11.29) into (11.26), we have:

$$\mathbf{J}_{k+1} = \mathbf{Q}^{-1} - \mathbf{Q}^{-1}\mathbf{F}\left(\mathbf{J}_k + \mathbf{F}^T\mathbf{Q}^{-1}\mathbf{F}\right)^{-1}\mathbf{F}^T\mathbf{Q}^{-1} + \mathbb{E}\left\{\tilde{\mathbf{H}}_{k+1}^T \mathbf{R}_{k+1}^{-1} \tilde{\mathbf{H}}_{k+1}\right\}. \tag{11.30}$$

In order to define $\tilde{\mathbf{H}}_k$ and $\mathbf{R}_k$ we need to introduce a stacked vector measurement:

$$\mathbf{z}_k^s \triangleq \left[z_k^{(1,1)} \quad z_k^{(1,2)} \quad \cdots \quad z_k^{(1,m)} \quad z_k^{(2,1)} \quad \cdots \quad z_k^{(n,m)}\right]^T \tag{11.31}$$

of dimension $nm \times 1$. The measurement equation can now be written in the vector form as:

$$\mathbf{z}_k^s = \mathbf{h}_k(\mathbf{x}_k) + \mathbf{w}_k \tag{11.32}$$

where the ath element of vector function $\mathbf{h}_k(\mathbf{x}_k)$ is specified as $\mathbf{h}_k[a] = h_k^{(i,j)}(\mathbf{x}_k)$ with $i = \lceil \frac{a}{m} \rceil$, $j = a-(i-1)m$ and $h_k^{(i,j)}(\mathbf{x}_k)$ given by (11.7). Vector $\mathbf{w}_k$ in (11.32) represents a stacked vector of pixel measurement noise components. Its covariance matrix $\mathbf{R} = \text{diag}[\sigma^2, \ldots, \sigma^2]$ is of dimension $nm \times nm$ and features in (11.30) as $\mathbf{R}_k$.

An element of Jacobian $\tilde{\mathbf{H}}_k$ is now computed as:

$$\tilde{\mathbf{H}}_k[a,b] = \frac{\partial \mathbf{h}_k[a]}{\partial \mathbf{x}_k[b]} \tag{11.33}$$

where the state vector $\mathbf{x}_k$ is defined as in (11.2), and $a = 1, \ldots, nm$, $b = 1, \ldots, 5$. This Jacobian is evaluated at the true value of $\mathbf{x}_k$. In particular, we have:

$$\tilde{\mathbf{H}}_k[a, 1] = \frac{\Delta_x \Delta_y \mathbf{x}_k[5](i\Delta_x - \mathbf{x}_k[1])}{2\pi \Sigma^4} e^{-\frac{(i\Delta_x - \mathbf{x}_k[1])^2 + (j\Delta_y - \mathbf{x}_k[3])^2}{2\Sigma^2}}$$

$$\tilde{\mathbf{H}}_k[a, 2] = 0$$

$$\tilde{\mathbf{H}}_k[a, 3] = \frac{\Delta_x \Delta_y \mathbf{x}_k[5](j\Delta_y - \mathbf{x}_k[3])}{2\pi \Sigma^4} e^{-\frac{(i\Delta_x - \mathbf{x}_k[1])^2 + (j\Delta_y - \mathbf{x}_k[3])^2}{2\Sigma^2}}$$

$$\tilde{\mathbf{H}}_k[a, 4] = 0$$

$$\tilde{\mathbf{H}}_k[a, 5] = \frac{\Delta_x \Delta_y}{2\pi \Sigma^2} e^{-\frac{(i\Delta_x - \mathbf{x}_k[1])^2 + (j\Delta_y - \mathbf{x}_k[3])^2}{2\Sigma^2}}.$$

The fact that matrix $\tilde{\mathbf{H}}_k$ contains nm rows makes the practical computation of the CRLB very intense. In order to simplify the computation, we can again consider only the relevant subset of indices in the image frame; that is $i \in C_i(\mathbf{x}_k)$ and $j \in C_j(\mathbf{x}_k)$ with $C_i(\mathbf{x}_k) = \{i_0 - p, \ldots, i_0 - 1, i_0, i_0 + 1, \ldots, i_0 + p\}$, and $C_j(\mathbf{x}_k) = \{j_0 - p, \ldots, j_0 - 1, j_0, j_0 + 1, \ldots, j_0 + p\}$. Here i_0 and j_0 are the nearest integer values of the target state vector component $x_k = \mathbf{x}_k[1]$ and $y_k = \mathbf{x}_k[3]$, respectively. In this way we consider only $(2p+1) \times (2p+1)$ pixels around the true target position. Jacobian $\tilde{\mathbf{H}}_k$ is then of dimension $(2p+1)^2 \times 5$ and covariance $\mathbf{R}$ is of dimension $(2p+1) \times (2p+1)$.

The recursion (11.30) requires us to specify the initial value of the information matrix. Assume that

$$\begin{aligned} x_0 &\sim \mathcal{U}[0, n\Delta_x] \\ \dot{x}_0 &\sim \mathcal{U}[-v_{\max}, v_{\max}] \\ y_0 &\sim \mathcal{U}[0, m\Delta_y] \\ \dot{y}_0 &\sim \mathcal{U}[-v_{\max}, v_{\max}] \\ I_0 &\sim \mathcal{U}[I_{\min}, I_{\max}]. \end{aligned} \quad (11.34)$$

The initial $\mathbf{J}_0$ is then approximated as a diagonal matrix with:

$$\mathbf{J}_0[\ell, \ell] \approx \frac{1}{\text{var}\{\mathcal{U}[A_\ell, B_\ell]\}} = 12/(B_\ell - A_\ell)^2 \qquad \ell = 1, \ldots, 5 \qquad (11.35)$$

where A_ℓ and B_ℓ are the appropriate limits of the uniform density as defined in (11.34).

Figure 11.5 shows the CRLB curves for SNR values of 5 dB, 10 dB, and 15 dB. The parameters used in the computation of bounds are the same as before except that now the target is present from the first to the last frame and $q_1 = 0.0002$

and $q_2 = 0.005$. The expectation operator in (11.30) can be ignored since for such a small amount of process noise it did not make any visible difference. With the circular symmetry of the point spread function, the CRLB for position and velocity along x and y are identical; hence Figure 11.5 presents only the bounds for x_k, $\dot{x}_k$, and I_k. Observe from Figure 11.5 that for higher values of SNR, the standard deviation of the estimation error is smaller for both x_k and $\dot{x}_k$. The standard deviation of intensity estimation error shown in Figure 11.5(c), however, is independent of the SNR for the values chosen in this example (all three curves are on top of each other). Figure 11.5(a) confirms that subpixel positional accuracy is theoretically possible (i.e., error standard deviation is smaller than the resolution cell size).

Tracking Accuracy of the PF

Our aim is now to compare the error performance of the particle filter against the theoretical Cramér-Rao bounds presented above. In order to simplify the implementation of the PF, we exploit the fact that the target is present during the observation interval. Thus a version of the PF is developed (based on the algorithm in Table 11.1) for a five-component state vector as in (11.2); this version of the PF uses $N = 10000$ particles. The RMS error of this filter is estimated over 100 Monte Carlo runs using the same parameter values as before and the SNR of 6 dB. Figure 11.6 shows the resulting RMS error curves, which appear only slightly higher than the theoretically predicted bounds. The first few points of RMS error curves are not shown because the PF and the information matrix are initialized differently.

11.6.2 Detection Performance

The detection performance of the PF-based TBD algorithm has been analyzed only by Monte Carlo simulations.[2] The PF performs a simple binary hypothesis test on each of the $K = 30$ image frames $\mathbf{z}_k$, $k = 1, 2, \ldots, K$. Hypotheses H_0 and H_1 are defined as:

$$H_1 : \text{Target present}$$
$$H_0 : \text{Target absent}$$

so that at each k, four possible cases can happen [20]:

1. H_0 true and the PF chooses H_0;
2. H_0 true and the PF chooses H_1;

---

[2] A more rigorous theoretical treatment would require establishing the receiver operating characteristic (ROC) [20].

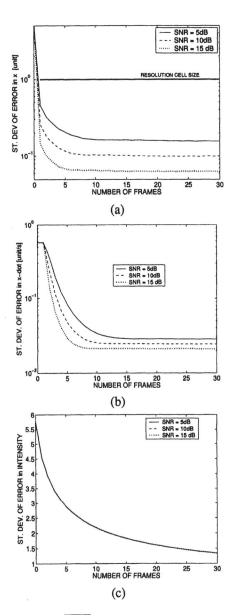

Figure 11.5 Theoretical curves of $\sqrt{\text{CRLB}}$ for (a) position x_k, (b) velocity $\dot{x}_k$, and (c) intensity I_k.

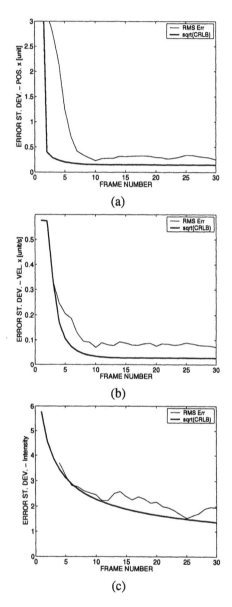

Figure 11.6 Particle filter error performance: the RMS error curves (thin line) against the $\sqrt{\text{CRLB}}$ (thick line) for (a) position x_k, (b) velocity $\dot{x}_k$, and (c) intensity I_k.

3. H_1 true and the PF chooses H_0;

4. H_1 true and the PF chooses H_1.

We estimate the probability of case 2 (probability of false alarm P_{FA}) and case 4 (probability of detection P_D) over 25 independent runs of the PF using the setup described in Section 11.5. The only difference, however, is that this time the value of the SNR is varied: for hypothesis H_0 to be true, the SNR is set to $-\infty$; when H_1 is true the SNR takes values: 5, 6, 7, 8, and 9 dB. The results are shown in Table 11.2, based on thresholding the probability of target existence $\hat{P}_k$ at 0.6. The estimate of P_{FA} is computed for H_0 being true, as the average number of scans (over Monte Carlo runs) in which case 2 occurred, divided by $K = 30$ frames. Similarly, P_D is computed when H_1 is true, as the average number of scans in which case 4 occurred, divided by 16 frames in which the target truly existed. For the fixed

Table 11.2
Detection Performance of the TBD Algorithm

	SNR	P_D	P_{FA}
	5 dB	0.74	-
	6 dB	0.77	-
H_1 true	7 dB	0.84	-
	8 dB	0.90	-
	9 dB	0.95	-
H_0 true	$-\infty$	-	0.018

level of background noise and the selected threshold of 0.6 for $\hat{P}_k$, the probability of false alarm is $P_{FA} \approx 0.018$. The probability of detection P_D increases with the higher SNR and ranges from 0.74 at 5 dB to 0.95 at 9 dB. The missed detections are in general the result of a need to process a few frames before the probability of existence attains a value above the threshold (see, for example, Figure 11.2, where the target existence would have been detected only in the third frame following the appearance of the target).

11.7 SUMMARY AND EXTENSIONS

This chapter presented a recursive Bayesian solution to the track-before-detect problem, implemented as a particle filter. The potential benefits of the track-before-detect approach have been exploited via the concept of target existence and as a result the developed PF can successfully detect and track low SNR (stealthy) subpixel-size targets. The required number of particles could be reduced significantly by Rao-Blackwellization of the discrete variable E_k conditioned on the state, but this is left for future work.

The described algorithm can be extended (at least conceptually) to multiple targets as follows. As in [21], let us assume that the maximum possible number of targets in the surveillance region, denoted as $\tau_{\max}$, is known. The discrete-valued variable E_k in this case is not binary any more, but instead refers to the number of existing targets, that is $E_k \in \{0, 1, \ldots, \tau_{\max}\}$. If the state vector of a single target consists of five components, as in (11.2), then the composite state vector corresponding to E_k targets would be a stacked vector of $5E_k + 1$ components [22]. The problem, of course, is the practical implementation of the corresponding PF with such a high dimensional state vector. The described multitarget TBD approach, however, has been applied successfully to situations involving two targets, one being spawned from the other (e.g., a missile being fired from an aircraft) [23].

References

[1] S. Blackman and R. Popoli, *Design and Analysis of Modern Tracking Systems*. Norwood, MA: Artech House, 1999.

[2] Y. Barniv, "Dynamic programming algorithm for detecting dim moving targets," in *Multitarget Multisensor Tracking: Advanced Applications* (Y. Bar-Shalom, ed.), ch. 4, Norwood, MA: Artech House, 1990.

[3] J. Arnold, S. Shaw, and H. Pasternack, "Efficient target tracking using dynamic programming," *IEEE Trans. Aerospace and Electronic Systems*, vol. 29, pp. 44–56, January 1993.

[4] B. D. Carlson, E. D. Evans, and S. L. Wilson, "Search radar detection and track with the Hough transform, part I: System concept," *IEEE Trans. Aerospace and Electronic Systems*, vol. 30, pp. 102–108, January 1994.

[5] S. M. Tonissen and Y. Bar-Shalom, "Maximum likelihood track-before-detect with fluctuating target amplitude," *IEEE Trans. Aerospace and Electronic Systems*, vol. 34, pp. 796–809, July 1998.

[6] L. D. Stone, C. A. Barlow, and T. L. Corwin, *Bayesian Multiple Target Tracking*. Norwood, MA: Artech House, 1999.

[7] D. J. Salmond and H. Birch, "A particle filter for track-before-detect," in *Proc. American Control Conf.*, (Arlington, VA), pp. 3755–3760, June 2001.

[8] M. Rollason and D. Salmond, "A particle filter for track-before-detect of a target with unknown amplitude," in *IEE Int. Seminar Target Tracking: Algorithms and Applications*, (Enschede, Netherlands), pp. 14/1–4, October 2001.

[9] F. E. Daum and R. J. Fitzgerald, "The importance of resolution in multiple target tracking," in *Proc. SPIE*, vol. 2235, pp. 329–338, 1994.

[10] S. B. Colegrove, A. W. Davis, and J. K. Ayliffe, "Track initiation and nearest neighbours incorporated into probabilistic data association," *Journal of Electrical and Electronics Engineers, Australia*, vol. 6, pp. 191–198, September 1986.

[11] D. Musicki, R. Evans, and S. Stankovic, "Integrated probabilistic data association," *IEEE Trans. Automatic Control*, vol. 39, pp. 1237–1240, June 1994.

[12] D. J. Ballantyne, H. Y. Chan, and M. A. Kouritzin, "Novel branching particle method for tracking," in *Proc. SPIE, Signal and Data Processing of Small Targets*, vol. 4048, pp. 277–287, 2000.

[13] A. Ooi, A. Doucet, B.-N. Vo, and B. Ristic, "Particle filter for tracking linear Gaussian target with nonlinear observations," in *Proc. SPIE* (I. Kadar, ed.), vol. 5096, 2003.

[14] Y. Bar-Shalom and X. R. Li, *Multitarget-Multisensor Tracking: Principles and Techniques*. Storrs, CT: YBS Publishing, 1995.

[15] Y. Bar-Shalom, X. R. Li, and T. Kirubarajan, *Estimation with Applications to Tracking and Navigation*. New York: John Wiley & Sons, 2001.

[16] C. Jauffret and Y. Bar-Shalom, "Track formation with bearings and frequency measurements in clutter," *IEEE Trans. Aerospace and Electronic Systems*, vol. 26, no. 6, pp. 999–1010, 1990.

[17] M. L. Hernandez, A. D. Marrs, N. J. Gordon, S. R. Maskell, and C. M. Reed, "Cramér-Rao bounds for non-linear filtering with measurement origin uncertainty," in *Proc. 5th Int. Conf. Information Fusion*, vol. 1, (Annapolis, MD), pp. 18–25, 2002.

[18] A. Farina, B. Ristic, and L. Timmoneri, "Cramér-Rao bound for nonlinear filtering with $P_d < 1$ and its application to target tracking," *IEEE Trans. Signal Processing*, vol. 50, pp. 854–867, August 2002.

[19] M. Hernandez, B. Ristic, A. Farina, and L. Timmoneri, "A comparison of two Cramér-Rao bounds for nonlinear filtering with $P_d < 1$," *IEEE Trans. Signal Processing*, (to be published in 2004).

[20] H. L. VanTrees, *Detection, Estimation and Modulation Theory*. New York: John Wiley & Sons, 1968.

[21] K. Kastella, "Joint multitarget probabilities for detection and tracking," in *Proc. SPIE*, vol. 3086, pp. 122–128, 1997.

[22] L. D. Stone, "A Bayesian approach to multiple-target tracking," in *Handbook of Multisensor Data Fusion* (D. L. Hall and J. Llinas, eds.), ch. 10, Boca Raton, FL: CRC Press, 2001.

[23] Y. Boers, J. N. Driessen, F. Vershure, W. P. M. H. Heemels, and A. Juloski, "A multi target track before detect application," in *Proc. 2003 Workshop on Multi-Object Tracking*, (Madison, WI), June 2003.

Chapter 12

Group and Extended Object Tracking

12.1 INTRODUCTION

In the standard target tracking problem, it is assumed that only a single point measurement is received from the target at each time step and that the targets move independently of each other. However, there are scenarios where this assumption is clearly inappropriate. For instance, high-resolution sensors may be able to resolve individual features or measurement sources on an extended object. If these measurements can be adequately interpreted, they may provide valuable information on target behavior, for example, body orientation, which could be exploited by a tracking filter. Another example is the situation where a number of targets are moving in formation. If information of this type is available then it should be passed directly to the tracking filter rather than combining the individual features into some average or centroid measurement.

The problem of tracking the features of an extended object and that of tracking a group of point targets moving in formation are very similar. In both cases there is a strong interdependency between the individual sensor measurements. For the extended object case this is due to the physical structure of the target and for the group of targets this is due to the formation pattern. Both these problems can be modeled as independent individual target motions superimposed on a common group effect. A model of this type was introduced in [1] in the context of tracking a group of point objects and used in a series of papers [2, 3, 4] for both group and extended object tracking. The model exploits the fact that targets tend to move in a formation pattern rather than as completely independent entities. The common (or bulk) motion may, for instance, allow for translation, rotation, and scaling of the target group. The same approach has also been used for target selection when fusing targeting and sensor information [5].

The aim of the tracking filter is to construct the probability density function of the required target state given all available prior and measurement information. Prior information includes targeting information, dynamic models, and models of the measurement process. For the group tracking problem we consider in this chapter, the motion of the group and disposition of the measurement sources relative to the group are modeled as two separate components. Sensor measurements can then be modeled as a (usually nonlinear) function of these two components. This scheme was proposed by Gordon and Salmond [1, 3]; see also Broida et al. [6, 7] and Dezert [8].

The Bayesian tracking filter must take account of measurement-source association uncertainty as well as errors in the measurements of source locations. This is achieved by constructing feasible association hypotheses and then evaluating the posterior probability of each association given the available data. It should be emphasized that unlike most other attempts at extended object tracking, in this approach, measurement association and tracking are fully integrated.

In this chapter we apply three filters to this problem:

- A standard particle filter that uses a stacked state vector containing all target states and the group motion parameters [1];

- The Rao-Blackwellized particle filter of [9] which exploits conditional Gaussian structure of the model to reduce the dimensionality of the problem;

- A joint probabilistic data association (JPDA) [10] approach that approximates the uncertainty due to all feasible hypotheses by a single Gaussian.

The filters are tested against a simulated scenario involving a formation of maneuvering aircraft.

The chapter is organized as follows. Section 12.2 presents a general description of the tracking problem, including the two component dynamics model. Section 12.3 summarizes the formal Bayesian solution. Section 12.4 introduces a particular group (or extended) target motion model based on the affine transformation. Section 12.5 describes the implementation of the Bayesian solution (in the framework of the affine model) via two particle filters: the standard particle filter and the Rao-Blackwellized particle filter. A simulation example and the numerical results that illustrate the performance of three filters (two particle filters and the JPDA filter) are described in Section 12.6.

12.2 TRACKING MODEL

Dynamics Model

It is assumed that the bulk component $\mathbf{B}_k$ of the model evolves according to

$$\mathbf{B}_{k+1} = \mathbf{f}_{0,k}(\mathbf{B}_k, \mathbf{w}_{0,k}) \qquad (12.1)$$

where k denotes the time-step number and $\mathbf{w}_{0,k}$ is an independent random variable of known distribution. The parameter vector $\mathbf{B}$ could describe the gross characteristics of a group of targets such as nominal center, mean line of advance, orientation, rotation, or scaling of the group. The function $\mathbf{f}_{0,k}$ is assumed known.

It is assumed that there are N_T individual components of the model, each of which corresponds to a target: $\{\mathbf{x}_{i,k}, i = 1, \ldots, N_T\}$, where N_T is assumed known. Each of these vectors is of the same dimension n_x and describes the same physical parameters (typically the Cartesian displacement). Each individual component $\mathbf{x}_{i,k}$ evolves independently according to the following model:

$$\mathbf{x}_{i,k+1} = \mathbf{f}_{i,k}(\mathbf{x}_{i,k}, \mathbf{w}_{i,k}) \qquad (12.2)$$

where $i = 1, \ldots, N_T$. The random driving noise $\mathbf{w}_{i,k}$ of each model is assumed to be independent and of known distribution. Although it would be possible for each of the individual elements to have a different dynamics model and a different distribution of driving noise, in most practical cases (due to lack of information to the contrary) the representations are identical; that is, $\mathbf{f}_{i,k} = \mathbf{f}_{j,k}$ for all $i, j \neq 0$. For a group of point targets, $\mathbf{x}_{i,k}$ could describe the displacement of target i relative to some nominal center and velocity of the group (indicated by $\mathbf{B}_k$). For an extended target, $\mathbf{x}_{i,k}$ might represent the location of an object feature, such as a corner, with respect to a coordinate frame embedded in the target. In this case, if the target object were rigid $\mathbf{x}_{i,k}$ would normally be constant; that is, $\mathbf{f}_{i,k}(\mathbf{x}_{i,k}, \mathbf{w}_{i,k}) = \mathbf{x}_{i,k}$. The target group is completely described by the *system* state vector $\begin{bmatrix} \mathbf{x}_k^T & \mathbf{B}_k^T \end{bmatrix}^T$, where $\mathbf{x}_k$ is a stacked vector defined as

$$\mathbf{x}_k \triangleq \begin{bmatrix} \mathbf{x}_{1,k}^T & \mathbf{x}_{2,k}^T & \cdots & \mathbf{x}_{N_T,k}^T \end{bmatrix}^T. \qquad (12.3)$$

Vector $\mathbf{x}_k$ is of dimension $n_x N_T$.

Measurement Model

At each time step, for $k > 0$, a set $\mathbf{Z}^k$ of N_k measurements is obtained: $\mathbf{Z}^k = \{\mathbf{z}_{j,k}; j = 1, \ldots, N_k\}$. In this formulation it is assumed that a single measurement

$\mathbf{z}_{j,k}$ cannot originate from multiple sources; that is, finite sensor resolution is not modeled (an extension of the model to include this is described in [4]). Each measurement may originate from a target or it may be the result of random uniform clutter. If measurement $\mathbf{z}_{j,k}$ originates from target $\mathbf{x}_{i,k}$, then the measurement equation is given by

$$\mathbf{z}_{j,k} = \mathbf{h}(\mathbf{y}_{i,k}) + \mathbf{v}_{j,k}, \qquad (12.4)$$

where:

- $\mathbf{y}_{i,k}$ is a possibly nonlinear transformation of the system state:

$$\mathbf{y}_{i,k} = \mathbf{t}(\mathbf{x}_{i,k}, \mathbf{B}_k). \qquad (12.5)$$

  For example, $\mathbf{y}_{i,k}$ might be an affine transformed system state (further discussed in Section 12.4).

- $\mathbf{h}(\cdot)$ is a possibly nonlinear measurement function (e.g., measurements are range and bearing, while $\mathbf{y}_{i,k}$ is in the Cartesian coordinates). Note that $\mathbf{h}$ is a function of both $\mathbf{B}_k$ and $\mathbf{x}_{i,k}$ and is the same for all targets.

- $\mathbf{v}_{j,k}$ is the measurement error, an independent random variable of known distribution.

The measurement equation (12.4) defines a conditional pdf for each measurement originating from a target, denoted as

$$p(\mathbf{z}_{j,k}|\mathbf{x}_{i,k}, \mathbf{B}_k). \qquad (12.6)$$

It is assumed that a source is detected (and so produces a measurement) with a known probability P_D, and detection is independent for each source and for each time step. The detection process is assumed independent for each target, although it would be possible to allow the value of P_D to be target dependent; for example, a composite source arising from several closely spaced targets might have a greater P_D than an isolated target. Obscuration of sources by the target body, possibly leading to periods when certain sources are not visible, is not modeled here.

If a measurement originates from clutter, it is assumed to be an independent sample from a known distribution that does not depend on $\mathbf{B}_k$ or $\mathbf{x}_{i,k}$. Clutter is taken to be uniformly distributed over the field of view (FOV) of a sensor; that is,

$$p_C(\mathbf{z}) = 1/V, \qquad (12.7)$$

where V is the volume of the FOV. Furthermore, the number of clutter measurements, $N_{C,k}$, is assumed to be distributed as a Poisson distribution with mean ρV:

$$p(N_{C,k}) = e^{-\rho V}(\rho V)^{N_{C,k}}/N_{C,k}!, \qquad (12.8)$$

where ρ is the spatial density of clutter (assumed known).

Data Association Hypotheses

Next we define an indicator mapping λ to specify a hypothesis on the origin of measurements:

$$\lambda : \{1, \ldots, N_k\} \to \{0, 1, \ldots, N_T\}. \tag{12.9}$$

Here $\lambda(j) = i \neq 0$ indicates that measurement j originates from source $i \in \{1, \ldots, N_T\}$ and $\lambda(j) = 0$ indicates that measurement j is due to clutter. Also it is assumed that for all $i, j \in \{1, \ldots, N_k\}$ with $\lambda(i) \neq 0$, the following holds: $\lambda(i) = \lambda(j) \Leftrightarrow i = j$, meaning that no two measurements may originate from the same source and no two sources may give rise to the same measurement. In general, the measurements $\mathbf{z}_{j,k}$ are unlabeled so that the true association mapping λ is unknown. The number of possible associations λ, given that N_D of the N_T targets have been detected, is

$$N_{\lambda_k}(N_D, N_T) = \frac{N_k! N_T!}{N_D!(N_k - N_D)!(N_T - N_D)!} \tag{12.10}$$

and so the total number of possible hypotheses (since N_D is unknown) is

$$\sum_{N_D=0}^{\min(N_T, N_k)} N_{\lambda_k}(N_D, N_T). \tag{12.11}$$

Initial pdf and the Problem Statement

It is assumed that prior information in the form of initial distributions for $\mathbf{B}_k$ and $\{\mathbf{x}_{i,k} : i = 1, \ldots, N_T\}$ is available at $k = 0$. The first measurements become available at $k = 1$.

The requirement is to estimate the set of points $\{\mathbf{y}_{i,k}; i = 1, \ldots, N_T\}$ at each time step. More precisely, we wish to construct the pdf of $\mathbf{y}_{i,k}$ given all available information up to and including time step k, namely $p(\mathbf{y}_{i,k} \mid \mathbf{Z}_k)$ for $i = 1, \ldots, N_T$, where $\mathbf{Z}_k = \{\mathbf{Z}^j; j = 1, \ldots, k\}$ is the collection of all measurements up to time k. In principle, the required pdfs $p(\mathbf{y}_{i,k} \mid \mathbf{Z}_k)$ can be obtained through knowledge of the posterior density $p(\mathbf{x}_k, \mathbf{B}_k \mid \mathbf{Z}_k)$. Thus, obtaining the posterior pdf of the system state vector is the key to the solution.

12.3 FORMAL BAYESIAN SOLUTION

As noted above, the key to estimating the points $\mathbf{y}_{i,k}$ is to construct the conditional pdf of the system state: $p(\mathbf{x}_k, \mathbf{B}_k \mid \mathbf{Z}_k)$. Below, we present the formal Bayesian recursive filter for constructing this pdf that consists of the usual *prediction* and *update* stages for each time step.

Prediction

Suppose the posterior pdf $p(\mathbf{x}_{k-1}, \mathbf{B}_{k-1} \mid \mathbf{Z}_{k-1})$ is available at time step $k-1$. The prediction (prior) density of the state at the following time step k is given by

$$p(\mathbf{x}_k, \mathbf{B}_k \mid \mathbf{Z}_{k-1}) = \int p(\mathbf{x}_k \mid \mathbf{x}_{k-1}) \, p(\mathbf{B}_k \mid \mathbf{B}_{k-1}) \, p(\mathbf{x}_{k-1}, \mathbf{B}_{k-1} \mid \mathbf{Z}_{k-1}) \, d\mathbf{x}_{k-1} \, d\mathbf{B}_{k-1} \quad (12.12)$$

where

$$p(\mathbf{x}_k \mid \mathbf{x}_{k-1}) = \prod_{i=1}^{N_T} p(\mathbf{x}_{i,k} \mid \mathbf{x}_{i,k-1}). \quad (12.13)$$

Transitional densities in (12.13) are available from the dynamics models; in particular, $p(\mathbf{x}_{i,k} \mid \mathbf{x}_{i,k-1})$ from (12.2) and $p(\mathbf{B}_k \mid \mathbf{B}_{k-1})$ from (12.1).

Update

On receipt of a measurement set $\mathbf{Z}^k$ (all measurements collected at time k), the prediction density (12.12) is updated via the Bayes rule:

$$p(\mathbf{x}_k, \mathbf{B}_k \mid \mathbf{Z}^k, \mathbf{Z}_{k-1}) \propto p(\mathbf{Z}^k \mid \mathbf{x}_k, \mathbf{B}_k) \, p(\mathbf{x}_k, \mathbf{B}_k \mid \mathbf{Z}_{k-1}) \quad (12.14)$$

Here, $p(\mathbf{Z}^k \mid \mathbf{x}_k, \mathbf{B}_k)$ is the likelihood of $(\mathbf{x}_k, \mathbf{B}_k)$ given the measurement set $\mathbf{Z}^k$.

The measurement likelihood is evaluated by considering all feasible association hypotheses between the measurements and the targets. It is convenient [11] to write the received data as the pair $\mathbf{Z}^k = (\mathbf{Z}^k_\dagger, N_k)$ where $\mathbf{Z}^k_\dagger$ indicates the values of the received measurements and N_k is the number of received measurements. This allows us to consider the two pieces of information separately. Thus,

$$p(\mathbf{Z}^k \mid \mathbf{x}_k, \mathbf{B}_k) = \sum_{\lambda_k} p(\mathbf{Z}^k_\dagger \mid \lambda_k, \mathbf{x}_k, \mathbf{B}_k) \, p(N_k, \lambda_k \mid \mathbf{x}_k). \quad (12.15)$$

The first term here is determined by the sensor measurement models, (12.6) and (12.7). Since the measurements of targets are independent given $\mathbf{x}_k$, we have:

$$\begin{aligned} p(\mathbf{Z}^k_\dagger \mid \lambda_k, \mathbf{x}_k, \mathbf{B}_k) &= \prod_{j \in \bar{T}(\lambda_k)} p_C(\mathbf{z}_{j,k}) \prod_{j \in T(\lambda_k)} p(\mathbf{z}_{j,k} \mid \mathbf{x}_{\lambda_k(j),k}, \mathbf{B}_k) \\ &= V^{-(N_k - N_D(\lambda_k))} \prod_{j \in T(\lambda_k)} p(\mathbf{z}_{j,k} \mid \mathbf{x}_{\lambda_k(j),k}, \mathbf{B}_k), \end{aligned}$$

$$(12.16)$$

where $\mathcal{T}(\lambda_k) \subseteq \{1, \ldots, N_k\}$ is the set of measurement subscripts corresponding to target detections and $\bar{\mathcal{T}}(\lambda_k)$ is its complement.

To evaluate the second term in (12.15), we need to introduce the number of detections N_D, so that

$$\begin{aligned} p(N_k, \lambda_k | \mathbf{x}_k) &= \sum_{N_D=0}^{N_T} p(N_k, \lambda_k, N_D | \mathbf{x}_k) \\ &= \sum_{N_D=0}^{N_T} p(\lambda_k | N_k, N_D, \mathbf{x}_k) p(N_k | N_D, \mathbf{x}_k) p(N_D | \mathbf{x}_k). \end{aligned}$$

(12.17)

Consider each of the above factors in turn. The first, $p(\lambda_k | N_k, N_D, \mathbf{x}_k)$, is the probability that a hypothesis λ_k is correct without knowledge of the measurement values $\mathbf{Z}_t^k$, but given that N_k measurements are received, N_D of these correspond to source detections and N_T targets are present. If we assume that, in the absence of knowledge of the measurement values $\mathbf{Z}_t^k$, all hypotheses with N_D detections, and N_k measurements are equally probable, then

$$p(\lambda_k | N_k, N_D, \mathbf{x}_k) = \begin{cases} \frac{1}{N_{\lambda_k}(N_D, N_T, N_k)} & \text{if } \lambda_k \Rightarrow N_D \text{ detections} \\ 0 & \text{otherwise,} \end{cases}$$

(12.18)

where $N_{\lambda_k}(N_D, N_T, N_k)$ is the number of feasible associations with N_D source detections, N_k measurements and N_T targets; see (12.10).

The second factor, $p(N_k | N_D, \mathbf{x}_k)$, is the probability of obtaining N_k measurements given that there are N_D detections. In other words, it is the probability of obtaining $N_k - N_D$ clutter measurements. This is given by the Poisson distribution specified by (12.8):

$$p(N_k | N_D, \mathbf{x}_k) = e^{-\rho V} (\rho V)^{(N_k - N_D)} / (N_k - N_D)!$$

(12.19)

The last factor in (12.17), $p(N_D | \mathbf{x}_k)$, is the probability of obtaining N_D detections given that N_T targets are present. This is given by the binomial distribution:

$$p(N_D | \mathbf{x}_k) = \binom{N_T}{N_D} P_D^{N_D} (1 - P_D)^{N_T - N_D}.$$

(12.20)

Hence, from (12.15) to (12.20), we obtain

$$p(\mathbf{Z}^k | \mathbf{x}_k, \mathbf{B}_k) \propto \sum_{\lambda_k} \left[\frac{P_D}{\rho(1 - P_D)} \right]^{N_D(\lambda_k)} \prod_{j \in \mathcal{T}(\lambda_k)} p(\mathbf{z}_{j,k} | \mathbf{x}_{\lambda_k(j),k}, \mathbf{B}_k). \quad (12.21)$$

The product over $\mathcal{T}(\lambda_k)$ should be interpreted as unity if $\mathcal{T}(\lambda_k)$ is empty (i.e., if according to λ_k, no target is detected) and it is assumed that $0 < P_D < 1$ and $\rho > 0$. Note that (12.21) is independent of V, only the spatial density ρ of the clutter within the sensor FOV is required.

The above two stages of prediction and update constitute a formal Bayesian solution (the initial conditions being given by the known initial distributions on $\mathbf{x}_0$ and $\mathbf{B}_0$). However, as noted in [3], the chief obstacles to implementing such a result are the exponentially increasing number of association hypotheses to be carried and possible nonlinearity in the measurement and dynamics models.

12.4 AFFINE MODEL

For the remainder of this chapter we shall focus on a particular dynamics model that involves an affine transformation for the bulk component of the state vector [1, 3]. An affine transformation from a two-dimensional vector $\mathbf{x}$ to $\mathbf{y}$ is the result of a scaling by a factor s, followed by a rotation about the origin through an angle θ and a translation by a vector μ, that is:

$$\mathbf{y} = \mu + s\mathbf{R}(\theta)\mathbf{x} \qquad (12.22)$$

where

$$\mathbf{R}(\theta) = \begin{pmatrix} \cos\theta & -\sin\theta \\ \sin\theta & \cos\theta \end{pmatrix}. \qquad (12.23)$$

A more general form of this transformation includes different scale factors for each direction and a shear (see [12]). For the group tracking problem this model describes the bulk evolution of a two-dimensional target formation in terms of translation, rotation, and scaling.

The bulk component state vector is taken as

$$\mathbf{B}_k = \begin{bmatrix} \mu_{X,k} & \dot{\mu}_{X,k} & \mu_{Y,k} & \dot{\mu}_{Y,k} & s_k & \dot{s}_k & \theta_k & \dot{\theta}_k \end{bmatrix}^T \qquad (12.24)$$

where $\mu_{X,k}, \mu_{Y,k}, s_k,$ and θ_k are the parameters of the affine transformation at time step k. Also included in the bulk state vector are the velocities of these components. The individual affine parameters are assumed to follow independent constant velocity models and therefore the dynamic equation for the evolution of the bulk component is given by

$$\mathbf{B}_{k+1} = \mathbf{F}_0 \mathbf{B}_k + \mathbf{w}_{0,k} \qquad (12.25)$$

where $\mathbf{w}_{0,k} \sim \mathcal{N}(0, \mathbf{Q}_0)$ and

$$\mathbf{F}_0 = \mathbf{I}_4 \otimes \begin{pmatrix} 1 & T_k \\ 0 & 1 \end{pmatrix}. \qquad (12.26)$$

Here I_k denotes a $k \times k$ identity matrix, $\otimes$ is the Kronecker product, and $T_k = t_{k+1} - t_k$ is the time interval between measurements. Process noise $\mathbf{w}_{0,k}$ is independent, zero mean, Gaussian driving noise of covariance

$$Q_0 = \text{block-diag}(q_{\mu_X}\Phi, q_{\mu_Y}\Phi, q_s\Phi, q_\theta\Phi), \tag{12.27}$$

where $q_{\mu_X}, q_{\mu_Y}, q_s$, and q_θ are the power of the driving noise processes for each parameter of the affine transformation, and matrix Φ is given by [13]:

$$\Phi_k = \begin{pmatrix} T_k^3/3 & T_k^2/2 \\ T_k^2/2 & T_k \end{pmatrix}.$$

The individual members of the target formation evolve with an independent random walk model:

$$\mathbf{x}_{i,k+1} = \mathbf{F}_i \mathbf{x}_{i,k} + \mathbf{w}_{i,k}, \qquad i = 1,\ldots,N_T \tag{12.28}$$

where $\mathbf{x}_{i,k} \in \mathbb{R}^2$ (hence $n_x = 2$), $\mathbf{F}_i = \mathbf{I}_2$, and $\mathbf{w}_{i,k} \sim \mathcal{N}(\mathbf{0}, \mathbf{Q}_i)$. The covariance of the driving noise is assumed to be the same for all members of the target group and for each dimension, hence $\mathbf{Q}_i = q_x \mathbf{I}_2$.

At time step k, the location of target i in the Cartesian coordinates would be

$$\begin{aligned} \mathbf{y}_{i,k} &= \mathbf{t}(\mathbf{x}_{i,k}, \mathbf{B}_k) \\ &= \begin{pmatrix} \mu_{X,k} \\ \mu_{Y,k} \end{pmatrix} + s_k \begin{pmatrix} \cos\theta_k & -\sin\theta_k \\ \sin\theta_k & \cos\theta_k \end{pmatrix} \mathbf{x}_{i,k} \end{aligned} \tag{12.29}$$

for $i = 1,\ldots,N_T$. The measurement error in (12.4) is assumed to be Gaussian; that is, $\mathbf{v}_{j,k} \sim \mathcal{N}(0, \mathbf{R}_{j,k})$.

12.5 PARTICLE FILTERS

We now describe two particle-filter-based approaches to calculating $p(\mathbf{x}_k, \mathbf{B}_k|\mathbf{Z}_k)$. The first is a basic particle filter that uses particles across the full dimension of the state vector and has the dynamic model as importance sampling function. The second particle filter is based on the concept of Rao-Blackwellization and mainly follows the one reported in [9]. Here particles are limited only to the scale and angle dimensions of the system state vector and attached to each particle is a Gaussian mixture representing the pdf of the other dimensions of the state vector. This reduces the dimension of the state space that has to be represented by particles and thus requires considerably fewer particles. It is fair to say, however, that the computational load for each particle then rises significantly.

12.5.1 SIR Particle Filter

The SIR particle filter (SIR-PF) is implemented as described in Section 3.5.1. An outline of this particle filter is presented in Table 12.1. As usual, N is the number of particles.

Table 12.1
SIR Particle Filter

$\left[\{ \mathbf{x}_k(i), \mathbf{B}_k(i) \}_{i=1}^{N} \right] = \text{SIR-PF} \left[\{ \mathbf{x}_{k-1}(i), \mathbf{B}_{k-1}(i) \}_{i=1}^{N}, \mathbf{Z}^k \right]$

- For $i = 1, \ldots, N$ sample importance density for $(\mathbf{x}_k(i), \mathbf{B}_k(i))$

$$\mathbf{x}_k(i) \sim p(\mathbf{x}_k \mid \mathbf{x}_{k-1}(i))$$
$$\mathbf{B}_k(i) \sim p(\mathbf{B}_k \mid \mathbf{B}_{k-1}(i))$$

- Construct the set of L valid measurement associations
- For $i = 1, \ldots, N$ compute particle weights using (12.21)

$$\tilde{w}_k(i) = p(\mathbf{Z}^k \mid \mathbf{x}_k(i), \mathbf{B}_k(i))$$

- Normalize weights
- Resample

For even moderate values of N_k and N_T, the number of possible hypotheses N_{λ_k} can be prohibitively large. This is especially a problem for the SIR filter with such a large dimension state vector requiring a large number of particles. It is therefore necessary to devise a scheme to reduce the number of association hypotheses λ to a manageable number for the calculation of (12.21) that must be evaluated for every particle. This is achieved via a hypothesis construction algorithm that rejects any highly improbable associations and generates a set of association hypotheses that can be employed for the likelihood calculations of all particles. The gating operation is based on the predicted mean and covariance for the measurements derived from the full population of particles. It is necessary to apply the hypothesis construction algorithm separately for each predicted (prior) particle.

The hypothesis construction algorithm is as follows. From prior sample $(\mathbf{x}_k(i), \mathbf{B}_k(i))$, generate the corresponding set of predicted measurements $\overline{\mathbf{Z}}^k(i) = \{ \overline{\mathbf{z}}_{1,k}(i), \overline{\mathbf{z}}_{2,k}(i), \ldots, \overline{\mathbf{z}}_{N_T,k}(i) \}$ using the sensor measurement model. An association between a measurement $\mathbf{z}_{j,k} \in \mathbf{Z}^k$ and target ℓ is allowed only if

$$[\mathbf{z}_{j,k} - \overline{\mathbf{z}}_{\ell,k}(i)]^T \mathbf{R}_{j,k}^{-1} [\mathbf{z}_{j,k} - \overline{\mathbf{z}}_{\ell,k}(i)] < \eta \qquad (12.30)$$

where $j = 1, \ldots, N_k$, $\ell = 1, \ldots, N_T$, $i = 1, \ldots, N$, and $\mathbf{R}_{j,k}$ is the covariance of the sensor measurement error. The threshold η is chosen from tables of χ^2. For example, with a two-dimensional measurement vector, if η is set to 8 this gives approximately a 98% chance of accepting the correct measurement. Possible associations between all N_k measurements and all N_T targets are examined using (12.30), and an N_k by N_T table of possible associations is formed. Using this table, candidate hypotheses λ are constructed for all possible values of N_D [i.e., $N_D = 0, 1, \ldots, \min(N_k, N_T)$]. For a given value of N_D, a valid hypothesis must obey the following:

- Exactly N_D gates are assigned to measurements, and $N_T - N_D$ gates are not assigned (corresponding to missed detections);
- Each assigned gate must be associated with exactly one measurement;
- A measurement may be associated with at most one gate.

All candidate hypotheses fulfilling these conditions are identified for evaluation of the likelihood (12.21) of sample i.

Note that the hypotheses construction gates for the individual samples are smaller than those of the full population approach [3] as they do not include the spread of samples. Therefore, although there is the overhead of applying the construction algorithm to each particle rather than just once, fewer feasible hypotheses are generated, thus saving on the likelihood computations. Furthermore, it is not necessary to evaluate the predicted mean and covariance for the full sample population.

12.5.2 Rao-Blackwellized Particle Filter

An effective method of reducing the dimensionality of a problem has been found to be exploiting analytic structure (if possible). This has come to be known as Rao-Blackwellization [14, 15]. A recent paper [9] demonstrates how this can be achieved within the group tracking problem and the performance of the resulting filter is shown to be superior to the standard particle filter described in the previous section. The resulting algorithm is referred to here as a Rao-Blackwellized particle filter (RBPF). The RBPF described here follows that of [9].

The system state vector (consisting of the joint target state $\mathbf{x}_k$ and the bulk state $\mathbf{B}_k$) is first repartitioned. The resulting terms are denoted as $\mathbf{x}_k^a$ and $\mathbf{B}_k^m$:

$$\{\mathbf{x}_k, \mathbf{B}_k\} \rightarrow \{\mathbf{x}_k^a, \mathbf{B}_k^m\}. \tag{12.31}$$

Most of the components of the bulk state vector have been added to the target state

$$\mathbf{x}_k^a = \begin{bmatrix} \mu_{X,k} & \dot\mu_{X,k} & \mu_{Y,k} & \dot\mu_{Y,k} & \dot s_k & \dot\theta_k & \mathbf{x}_{1,k}^T & \mathbf{x}_{2,k}^T & \cdots & \mathbf{x}_{N_T,k}^T \end{bmatrix}^T \tag{12.32}$$

while the bulk state vector has been reduced to just the scale and rotation terms

$$\mathbf{B}_k^m = \begin{bmatrix} s_k & \theta_k \end{bmatrix}^T. \tag{12.33}$$

This results in the dynamic equations being rewritten as

$$\mathbf{x}_{k+1}^a = \mathbf{F}_k \mathbf{x}_k^a + \mathbf{u}_k \tag{12.34}$$

$$\mathbf{B}_{k+1}^m = \mathbf{B}_k^m + \mathbf{G}_k \mathbf{x}_k^a + \mathbf{v}_k \tag{12.35}$$

where

$$\mathbf{F}_k = \text{block-diag}\left[\mathbf{I}_2 \otimes \begin{pmatrix} 1 & T_k \\ 0 & 1 \end{pmatrix}, \mathbf{I}_{2N_T+2}\right] \tag{12.36}$$

$$\mathbf{G}_k = \begin{bmatrix} \mathbf{0}_{2\times 4} & T_k \mathbf{I}_2 & \mathbf{0}_{2\times N_T} \end{bmatrix} \tag{12.37}$$

with $\mathbf{0}_{i\times j}$ denoting a zero matrix of dimension $i \times j$. The noise terms $\mathbf{u}_k$ and $\mathbf{v}_k$ are correlated Gaussian and we write

$$\begin{pmatrix} \mathbf{u}_k \\ \mathbf{v}_k \end{pmatrix} \sim \mathcal{N}\left(\mathbf{0}_{(2N_T+8)\times 1}, \begin{bmatrix} \mathbf{Q}_k & \mathbf{C}_k^T \\ \mathbf{C}_k & \mathbf{D}_k \end{bmatrix}\right) \tag{12.38}$$

where the covariance is composed of the appropriate terms from $\mathbf{Q}_0, \mathbf{Q}_1, \ldots, \mathbf{Q}_{N_T}$. To avoid problems due to the correlated noise terms, the model is rewritten as [13]

$$\mathbf{x}_{k+1}^a = \mathbf{F}_k^* \mathbf{x}_k^a + \mathbf{C}_k \mathbf{D}_k^{-1}(\mathbf{B}_k^m - \mathbf{B}_{k-1}^m) + \mathbf{u}_k^* \tag{12.39}$$

where

$$\mathbf{F}_k^* = \mathbf{F}_k - \mathbf{C}_k \mathbf{D}_k^{-1} \mathbf{G}_k \tag{12.40}$$

and $\mathbf{u}_k^* \sim \mathcal{N}(\mathbf{0}, \mathbf{Q}_k^*)$ with

$$\mathbf{Q}_k^* = \mathbf{Q}_k - \mathbf{C}_k \mathbf{D}_k^{-1} \mathbf{C}_k^T. \tag{12.41}$$

The noise terms $\mathbf{v}_k$ and $\mathbf{u}_k^*$ are then uncorrelated.

After the described repartitioning of the state vector, the measurement equation becomes

$$\mathbf{z}_{j,k} = \mathbf{H}_{\lambda_k(j)}\left(\mathbf{B}_k^m\right)\mathbf{X}_k^a + \mathbf{e}_k \tag{12.42}$$

where for $i = \lambda_k(j)$

$$\mathbf{H}_i(\mathbf{B}_k^m) = \begin{bmatrix} \mathbf{I}_2 \otimes (1,0) & \mathbf{0}_{2\times 2i} & s_k \begin{pmatrix} \cos\theta_k & -\sin\theta_k \\ \sin\theta_k & \cos\theta_k \end{pmatrix} & \mathbf{0}_{2\times 2(N_T-i)} \end{bmatrix}. \tag{12.43}$$

Note that given $\mathbf{B}_k^m$ this is a linear equation.

The posterior pdf at time step k, $p(\mathbf{x}_k^a, \mathbf{B}_k^m | \mathbf{Z}_k)$, which is our goal, can be written as

$$p(\mathbf{x}_k^a, \mathbf{B}_k^m | \mathbf{Z}_k) = p(\mathbf{x}_k^a | \mathbf{B}_k^m, \mathbf{Z}_k) p(\mathbf{B}_k^m | \mathbf{Z}_k). \tag{12.44}$$

With the assumed models (12.35), (12.39), and (12.42) the density $p(\mathbf{x}_k^a | \mathbf{B}_k^m, \mathbf{Z}_k)$ can be written as a Gaussian mixture

$$p(\mathbf{x}_k^a | \mathbf{B}_k^m(i), \mathbf{Z}_k) = \sum_{j=1}^{n_k^i} \pi_k^{j,i} \mathcal{N}\left(\mathbf{m}_{k|k}^{j,i}, \mathbf{P}_{k|k}^{j,i}\right), \tag{12.45}$$

where n_k^i is the number of mixture components corresponding to particle $i = 1, \ldots, N$. Particles are hence only required to approximate $p(\mathbf{B}_k^m | \mathbf{Z}_k)$. Note that the mixture (12.45) will be composed of an exponentially increasing number of components. In order to control the number of components in the mixture we employ a mixture reduction technique [16] to implement the filter. An outline of the RBPF algorithm is given in Table 12.2, while a more detailed explanation (with equations) is presented in Table 12.3.

12.6 SIMULATION EXAMPLE

Experimental Setup

We present a simulation example of tracking a formation of four aircraft. The formation is taken to follow one of the benchmark trajectories [17]. The benchmark problem contains a range of target trajectories designed to cover several target types and maneuver capabilities. Here we consider the trajectory representing a slower transport aircraft. The trajectory starts at the origin and moves at a constant speed of 1000 km/h. At $k = 60$ there is a mild $2g$ turn and at $k = 90$ there is a $3g$ turn. The benchmark target trajectory defines the translation and rotation parameters of the bulk motion. Hence the bulk parameter is set with noise-free evolution that is not matched to the assumed model. The scale parameter is fixed at one.

The initial shape of the aircraft formation is a straight line and the aircraft are located at $\mathbf{x}_{1,0} = (0, 500)^T$, $\mathbf{x}_{2,0} = (0, 200)^T$, $\mathbf{x}_{3,0} = (0, -200)^T$, and $\mathbf{x}_{4,0} = (0, -500)^T$. All units are in meters. The independent individual random

Table 12.2
An Outline of the Rao-Blackwellized Particle Filter

$$\left[\left\{\mathbf{B}_k^m(i), w_k^i, \left\{\pi_k^{j,i}, \mathbf{m}_{k|k}^{j,i}, \mathbf{P}_{k|k}^{j,i}\right\}_{j=1}^{n_k^i}\right\}_{i=1}^N\right] =$$

$$\text{RBPF}\left[\left\{\mathbf{B}_{k-1}^m(i), w_{k-1}^i, \left\{\pi_{k-1}^{j,i}, \mathbf{m}_{k-1|k-1}^{j,i}, \mathbf{P}_{k-1|k-1}^{j,i}\right\}_{j=1}^{n_{k-1}^i}\right\}_{i=1}^N, \mathbf{Z}^k\right]$$

- For $i = 1, \ldots, N$ sample importance (proposal) density for $\mathbf{B}_k^m(i)$
- For $i = 1, \ldots, N$ evaluate $p(\mathbf{x}_k^a \mid \mathbf{B}_k^m(i), \mathbf{Z}_{k-1})$ using Kalman filter relations
- Construct the set of L valid measurement associations
- For $i = 1, \ldots, N$ evaluate $p(\mathbf{x}_k^a \mid \mathbf{B}_k^m(i), \mathbf{Z}^k)$. This results in an $n_{k-1}^i L$ component Gaussian mixture.
- For $i = 1, \ldots, N$ approximate $n_{k-1}^i L$ Gaussian mixture by reduced number of components n_k^i.
- For $i = 1, \ldots, N$ update particle weights.
- Resample if necessary.

---

walk for the members of the formation, which is superimposed on the bulk motion, has $q_x = 10m^2$. The individual state vectors $\mathbf{x}_{i,k}$ consist of target positions in Cartesian coordinates x and y. The measurements are target positions in the Cartesian coordinates, and hence the measurement function $\mathbf{h}(\cdot)$ is identity. The interval between measurements is fixed at $T_k = 1s$. The measurement error covariance is set to $\mathbf{R}_{j,k} = 100\mathbf{I}_2$ for all targets and time steps. The clutter density is set to 1^{-10} clutter points per unit area per scan (i.e., almost zero clutter). An example trajectory of the formation is shown in Figure 12.1.

It is assumed that prior information on $\mathbf{x}_k$ and $\mathbf{B}_k$ for $k = 0$ is available in the form of a Gaussian distribution.

The model parameters are set at

$$q_{\mu X} = q_{\mu Y} = 10m^2 s^{-3} \quad (12.46)$$
$$q_x = 10m^2 \quad (12.47)$$
$$q_\theta = 0.01 rad^2 s^{-3} \quad (12.48)$$
$$q_s = 0.001^2 \quad (12.49)$$

The SIR-PF (2000 particles), RBPF (10 particles), and JPDA algorithms have been applied with the above parameters. The JPDA filter is a standard JPDA [10] with no bulk information. Four tracks are initiated and forced to remain active throughout

Table 12.3

The Rao-Blackwellized Particle Filter

- For $i = 1, \ldots, N$ sample $\mathbf{B}_k^m(i) \sim \sum_{j=1}^{n_{k-1}^i} \pi_{k-1}^{j,i} \mathcal{N}\left(\mathbf{B}_{k-1} + \mathbf{G}_{k-1}\mathbf{m}_{k-1|k-1}^{j,i}, \mathbf{M}_k^{j,i}\right)$,
  where $\mathbf{M}_k^{j,i} = \mathbf{G}_{k-1}\mathbf{P}_{k-1|k-1}^{j,i}\mathbf{G}_{k-1}^T + \mathbf{D}_{k-1}$.

- For $i = 1, \ldots, N$, $j = 1, \ldots, n_k^i$ calculate: $\pi_{k-1}^{j,i*} = \gamma_k^{j,i} / \sum_{\ell=1}^{n_{k-1}^i} \gamma_k^{\ell,i}$,

$$\mathbf{m}_{k|k-1}^{j,i} = \mathbf{F}_k^* \left(\mathbf{m}_{k-1|k-1}^{j,i} + \mathbf{K}_k^{j,i}(\mathbf{B}_k^m(i) - \mathbf{B}_{k-1}^m(i) - \mathbf{G}_{k-1}\mathbf{m}_{k-1|k-1}^{j,i})\right) + \mathbf{C}_k \mathbf{D}_k^{-1}(\mathbf{B}_k^m(i) - \mathbf{B}_{k-1}^m(i))$$

$$\mathbf{P}_{k|k-1}^{j,i} = \mathbf{F}_k^* \left(\mathbf{P}_{k-1|k-1}^{j,i} - \mathbf{K}_k^{j,i}\mathbf{G}_{k-1}\mathbf{P}_{k-1|k-1}^{j,i}\right) \mathbf{F}_k^{*T} + \mathbf{Q}_k^*$$

where
$$\mathbf{K}_k^{j,i} = \mathbf{P}_{k-1|k-1}^{j,i}\mathbf{G}_{k-1}^T(\mathbf{M}_k^{j,i})^{-1}$$

and
$$\gamma_k^{j,i} = \pi_{k-1}^{j,i} \mathcal{N}(\mathbf{B}_k^m(i); \mathbf{B}_{k-1}^m(i) + \mathbf{G}_{k-1}\mathbf{m}_{k-1|k-1}^{j,i}, \mathbf{M}_k^{j,i})$$

- For $\ell = 1, \ldots, L$, $i = 1, \ldots, N$, $j = 1, \ldots, n_{k-1}^i$, set $u = (j-1)L + \ell$ and compute

$$\beta_k^{u,i} = \xi_k^{u,i} / \sum_{c=1}^{Ln_{k-1}^i} \xi_k^{c,i}$$

  ○ If $j = 0$ then: $\boldsymbol{\mu}_{k,j}^{u,i} = \mathbf{m}_{k|k-1}^{j,i}$ and $\boldsymbol{\Sigma}_{k,j}^{u,i} = \mathbf{P}_{k|k-1}^{j,i}$
  ○ If $j = 1, 2, \ldots, N_D(\lambda)$:

$$\boldsymbol{\mu}_{k,j}^{u,i} = \boldsymbol{\mu}_{k,j-1}^{u,i} + \boldsymbol{\Sigma}_{k,j-1}^{u,i}\mathbf{H}_{\lambda(j)}(\mathbf{B}_k^m(i))\left(\mathbf{S}_{k,j}^{u,i}\right)^{-1}\left\{\mathbf{z}_{j,k} - \mathbf{H}_{\lambda(j)}(\mathbf{B}_k^m(i))\boldsymbol{\mu}_{k,j-1}^{u,i}\right\}$$

$$\boldsymbol{\Sigma}_{k,j}^{u,i} = \boldsymbol{\Sigma}_{k,j-1}^{u,i}\left\{\mathbf{I} - \mathbf{H}_{\lambda(j)}(\mathbf{B}_k^m(i))^T\left(\mathbf{S}_{k,j}^{u,i}\right)^{-1}\mathbf{H}_{\lambda(j)}(\mathbf{B}_k^m(i))\boldsymbol{\Sigma}_{k,j-1}^{u,i}\right\}$$

where
$$\mathbf{S}_{k,j}^{u,i} = \mathbf{H}_{\lambda(j)}(\mathbf{B}_k^m(i))\boldsymbol{\Sigma}_{k,j}^{u,i}\mathbf{H}_{\lambda(j)}(\mathbf{B}_k^m(i))^T + \mathbf{R}_{j,k} \quad j = 1, \ldots, N_D(\lambda)$$

$$\xi_k^{u,i} = \left[\frac{P_D}{\rho(1-P_D)}\right]^{N_D(\lambda)} \pi_{k-1}^{j,i*} \prod_{j=1}^{} \mathcal{N}(\mathbf{z}_{j,k}; \mathbf{H}_{\lambda(j)}(\mathbf{B}_k^m(i))\boldsymbol{\mu}_{k,j-1}^{u,i}, \mathbf{S}_{k,j}^{u,i})$$

- For $i = 1, \ldots, s$ update the weights: $w_k^i = w_{k-1}^i \sum_{c=1}^{Ln_{k-1}^i} \xi_k^{c,i} / \sum_{i=1}^{s} w_{k-1}^i \sum_{u=1}^{Ln_{k-1}^i} \xi_k^{u,i}$

- Resample (if necessary)

- For $i = 1, \ldots, N$, pass parameter set $\{\beta_k^{u,i}, \mu_{k,j}^{u,i}, \Sigma_{k,j}^{u,i}\}_{u=1}^{Ln_{k-1}^i}$ through mixture reduction algorithm to get $\{\pi_k^{j,i}, m_{k|k}^{j,i}, M_{k|k}^{j,i}\}_{j=1}^{n_k^i}$

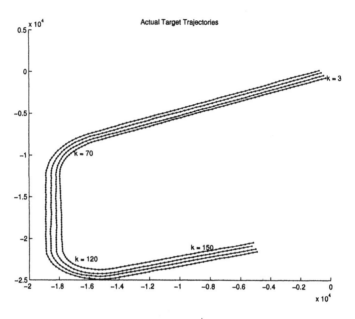

Figure 12.1 Example trajectories of the aircraft formation.

the simulation – thus there is no automatic track initiation or deletion in this JPDA. For the particle filters, the target tracks are defined by the estimated means $\hat{\mathbf{y}}_{i,k}$ of the pdfs $p(\mathbf{y}_{i,k} \mid \mathbf{Z}_k)$. In other words, each track position $\hat{\mathbf{y}}_{i,k}$ is the mean of the cloud of particles generated by the filter:

$$\hat{\mathbf{y}}_{i,k} = \frac{1}{N} \sum_{j=1}^{N} \mathbf{y}_{i,k}(j) = \frac{1}{N} \sum_{j=1}^{N} \mathbf{t}(\mathbf{x}_{i,k}(j), \mathbf{B}_k(j)).$$

To illustrate the performance of the filters we consider two different detection probabilities. We start with perfect target detection probability $P_D = 1$ and then consider $P_D = 0.8$.

Experimental Results

Figures 12.2 and 12.3 show the resulting tracks for the SIR-PF, both as a formation and individually. Four parallel tracks can be seen to have been maintained throughout the scenario (the figures show both true trajectories and the estimated tracks). Figures 12.4 and 12.5 display the results from the RBPF and a similar set of parallel tracks can be seen. Note that in both Figures 12.3 and 12.5 the true and estimated individual trajectories are practically indistinguishable (confirming an excellent performance of both particle filters). Between the two particle filters, however, the RBPF is the preferred option, being computationally more efficient.

An examination of the results for the JPDA in Figures 12.6 and 12.7 shows quite a different story. There are several periods in the scenario when the association uncertainty of several closely spaced targets coupled with the mixture collapsing strategy of the JPDA have led to track-target swapping. From Figure 12.7 we can thus observe that only the tracking of target 1 was reasonably accurate. In order to explain why the JPDA fails, in Figure 12.8 we illustrate several time snapshots of the JPDA posterior pdf together with the measurements. A 99% probability ellipse has been plotted around the target mean for each of the four targets in each picture. At time step 120 it can be seen that the uncertainty for two of the targets has grown to the point where two targets have become indistinguishable. This type of situation often leads to track-target swaps in the JPDA.

Next, let us examine the case of $P_D = 0.8$. Figures 12.9 and 12.10 show the RBPF tracking performance. Note that the result of RBPF is barely affected in comparison with perfect detection. The JPDA performance shown in Figures 12.11 and 12.12, by contrast, has degraded severely.

12.7 CONCLUDING REMARKS

In this chapter we have considered the inclusion of information about how a set of targets move relative to each other in a tracking filter. The conditionally Gaussian structure of the model can be exploited to give a highly efficient particle filter requiring very few particles. This filter can be seen to give significant improvement in performance over the standard methods such as the JPDA.

An interesting extension to this problem is to consider the effects of finite sensor resolution and limited field of view. In this case the nice properties of the model that are exploited in the RBPF no longer exist and an SIR-PF is required. Obscuration effects can be represented and complex FOV geometries can be handled provided they were fully specified, including, for example, sensor blind spots due to detector failure or blind regions due to environmental effects.

The development in this chapter also assumes that the number of targets present is exactly known, for example, from third-party targeting. It should be acknowledged that this simplifies the problem as an unknown number of targets significantly complicates hypothesis construction. Recent work [18, 19, 20] on random sets offers the potential to address these issues.

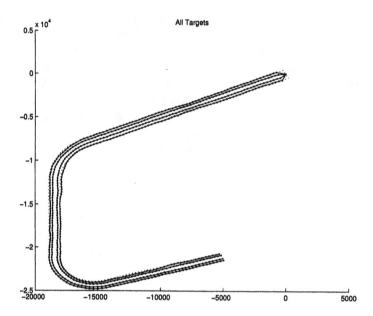

Figure 12.2 SIR-PF target tracks (as a formation) for scenario with $P_D=1$.

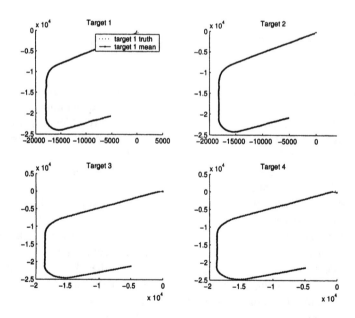

Figure 12.3 Individual target tracks produced by the SIR-PF for scenario with $P_D=1$.

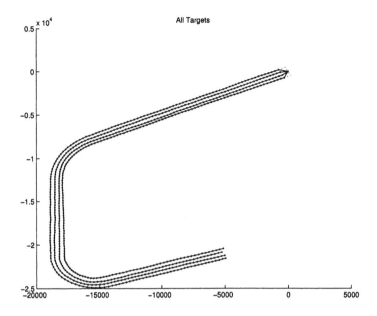

Figure 12.4 RBPF target tracks (as a formation) for scenario with $P_D=1$.

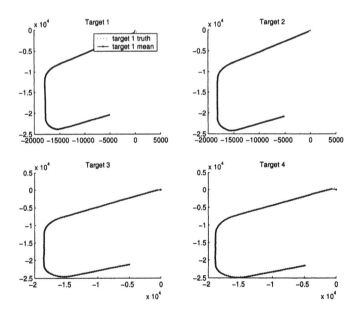

Figure 12.5 Individual target tracks produced by the RBPF for scenario with $P_D=1$.

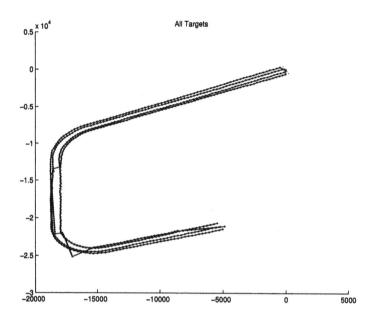

Figure 12.6 JPDA target tracks (as a formation) for scenario with $P_D=1$

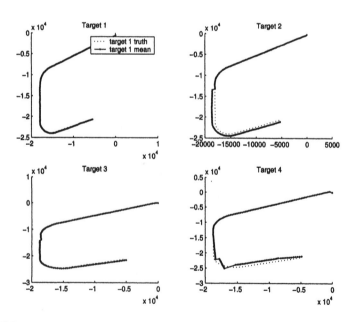

Figure 12.7 Individual target tracks produced by the JPDA for scenario with $P_D=1$.

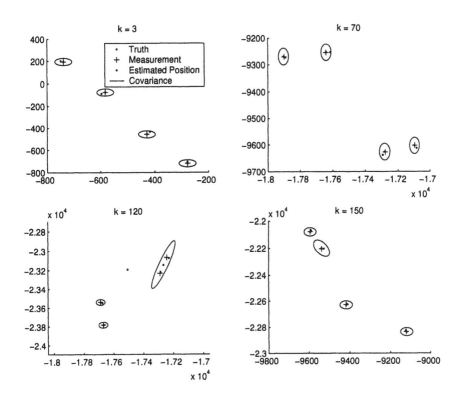

Figure 12.8 Snapshots of the JPDA target pdf for several time steps.

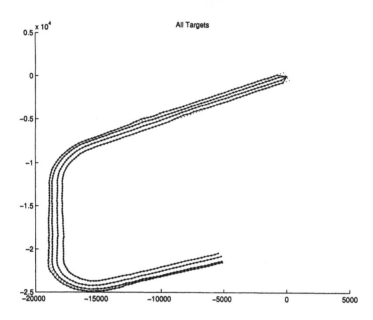

Figure 12.9 RBPF target tracks (as a formation) for scenario with $P_D=0.8$.

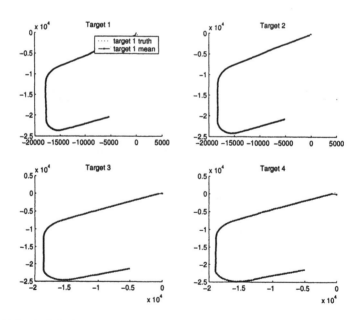

Figure 12.10 Individual target tracks produced by the RBPF for scenario with $P_D=0.8$.

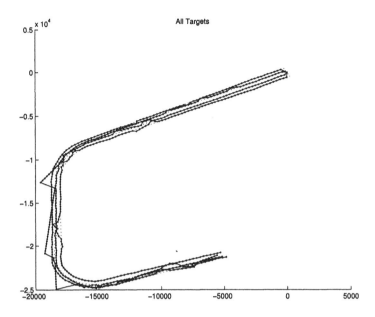

Figure 12.11 JPDA target tracks (as a formation) for scenario with $P_D=0.8$.

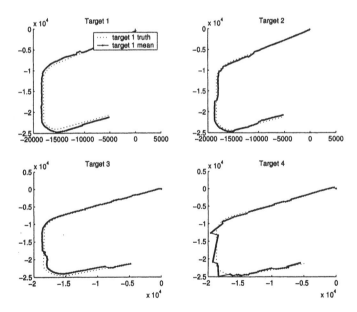

Figure 12.12 Individual target tracks produced by the JPDA for scenario with $P_D=0.8$.

References

[1] N. J. Gordon, D. J. Salmond, and D. Fisher, "Bayesian target tracking after group pattern distortion," in *Proc. SPIE*, vol. 3163, pp. 238–248, 1997.

[2] D. J. Salmond, D. Fisher, and N. J. Gordon, "Tracking in the presence of intermittent spurious objects and clutter," in *Proc. of SPIE*, vol. 3373, pp. 460–474, 1998.

[3] D. J. Salmond and N. J. Gordon, "Group and extended object tracking," in *Proc. SPIE*, vol. 3809, 1999.

[4] D. J. Salmond and N. J. Gordon, "Group tracking with limited sensor resolution and finite field of view," in *Proc. of SPIE*, vol. 4048, pp. 532–540, 2000.

[5] N. J. Gordon and D. J. Salmond, "Bayesian pattern matching technique for target acquisition," *Journal of Guidance, Control and Dynamics*, vol. 22, pp. 68–77, 1999.

[6] T. J. Broida, S. Chandrashekhar, and R. Chellappa, "Recursive 3-D motion estimation from a monocular image sequence," *IEEE Trans. on Aerospace and Electronic Systems*, vol. 26, no. 4, pp. 639–656, 1990.

[7] T. J. Broida and R. Chellappa, "Estimating the kinematics and structure of a rigid object from a sequence of monocular images," *IEEE Trans. on Pattern Analysis and Machine Intelligence*, vol. 13, no. 6, pp. 497–513, 1991.

[8] J. Dezert, "Tracking manoeuvring and bending extended target in cluttered environment," in *Proc. SPIE*, vol. 3373, pp. 283–294, 1998.

[9] M. R. Morelande and S. Challa, "An algorithm for tracking group targets," in *Proc. Workshop on Multiple Hypothesis Tracking: A Tribute to S. Blackman*, (San Diego, CA), May 2003.

[10] Y. Bar-Shalom and T. E. Fortmann, *Tracking and Data Association*. Boston, MA: Academic Press, 1988.

[11] N. J. Gordon, *Bayesian Methods for Tracking*. PhD thesis, Imperial College, University of London, 1993.

[12] A. Blake and M. Isard, *Active Contours*. New York: Springer-Verlag, 1998.

[13] Y. Bar-Shalom and X. R. Li, *Estimation and Tracking*. Norwood, MA: Artech House, 1993.

[14] A. Doucet, S. Godsill, and C. Andrieu, "On sequential Monte Carlo sampling methods for Bayesian filtering," *Statistics and Computing*, vol. 10, no. 3, pp. 197–208, 2000.

[15] F. Gustafsson, F. Gunnarsson, N. Bergman, U. Forssell, J. Jansson, R. Karlsson, and P.-J. Nordlund, "Particle filters for positioning, navigation and tracking," *IEEE Trans. Signal Processing*, vol. 50, pp. 425–437, February 2002.

[16] D. J. Salmond, "Mixture reduction algorithms for target tracking in clutter," in *SPIE (Signal and Data Processing of Small Targets)*, vol. 1305, pp. 434–445, 1990.

[17] W. D. Blair, G. A. Watson, T. Kirubarajan, and Y. Bar-Shalom, "Benchmark for radar allocation and tracking in ECM," *IEEE Trans. on Aerospace and Electronic Systems*, vol. 34, pp. 1097–1114, October 1998.

[18] R. Mahler, *An Introduction to Multisource-Multitarget Statistics and Applications*. Lockheed Martin Technical Monograph, 2000.

[19] B.-N. Vo, S. Singh, and A. Doucet, "Sequential Monte Carlo implementation of the PHD filter for multi-target tracking," in *Proc. 6th Int. Conf. Information Fusion (Fusion 2003)*, (Cairns, Australia), July 2003.

[20] B.-N. Vo, S. Singh, and A. Doucet, "Sequential Monte Carlo methods for Bayesian multi-target filtering," *Submitted to IEEE Trans. on Aerospace and Electronic Systems*, 2003.

Epilogue

Particle filters are becoming increasingly popular for target tracking and other engineering applications of nonlinear and non-Gaussian filtering. Not only do they provide more accurate and reliable estimators for "classical" nonlinear filtering problems (such as tracking of ballistic objects on reentry or the bearings-only tracking problem), but they also open up some new opportunities for which the conventional Kalman-based methods are inappropriate (the recursive track-before-detect, an effective exploitation of hard constraints, and so forth). The particle filters, however, are computationally expensive and should be used only for difficult nonlinear/non-Gaussian problems, when conventional methods fail. The key factors for a successful application of particle filters in practice are therefore a good choice of the importance density and Rao-Blackwellization if possible.

Throughout Part II of this book, admittedly, we paid more attention to the demonstration of potential advantages of particles filters, in terms of accuracy and reliability, than to their computational efficiency. In practical terms this means that more efficient particle filters still remain to be explored for some of the applications described in the book. For example, it could be advantageous to develop a particle filter for bearings-only tracking in modified polar coordinates, for which the optimal importance density can be analytically derived. Similarly, a Rao-Blackwellized particle filter for the recursive track-before-detect could dramatically reduce the number of required particles.

Another feature of the book is that it mainly dealt with a single target (except in Chapters 8 and 12). The development of particle filters for multitarget applications, despite some preliminary results, is a fairly immature field and requires further work. Similarly some fundamental results in the theory of sequential Monte Carlo estimation still remain an open challenge. Perhaps the key question to be answered is: How many particles do I need for my problem? Attention will also focus on how to choose the model parameters to exploit structure in a given problem so that a required level of performance can be achieved with a minimal computational cost.

Appendix

Coordinate Transformations for Tracking

Four coordinate systems, shown in Figure A.1, are usually involved when tracking algorithms are applied to real data. The ownship coordinates, usually supplied by an on-board global positioning system (GPS) and an internal navigation system (INS), are given in the *geodetic* coordinate system. The tracking problem is often defined and the corresponding algorithms usually developed in the *tangential-plane Cartesian* coordinate system. The sensor measurements (range, range rate, azimuth, elevation) are often collected in the *local spherical* coordinates. Finally the transformation between the geodetic and the tangential-plane Cartesian coordinate systems is accurately done via the *Earth-centered Earth-fixed* (ECEF) coordinates. Accurate conversions from the local spherical (or polar) to the local Cartesian coordinates are described in [1, 2, 3]. Note, however, that the conversion from local spherical to local Cartesian coordinates may not be necessary if the tracking algorithm is using spherical measurements directly. This was the case for example with bearings-only measurements (Chapter 6), range-only measurements (Chapter 7), and bistatic radar measurements (Chapter 8). The transformations between the geodetic, ECEF and the tangential plane coordinate systems are described next [4, Ch. 2].

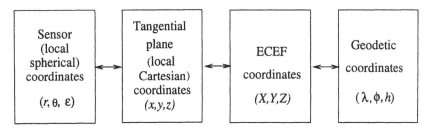

Figure A.1 Coordinate systems.

A.1 GEODETIC TO ECEF AND VICE VERSA

In the geodetic coordinate system, a point on the ellipsoidal earth is represented by a triple (λ, ϕ, h), where λ is latitude, ϕ is longitude, and h is the altitude above the reference ellipsoid. In the ECEF rectangular coordinates (X, Y, Z), the origin is in the center of the earth. The X-axis extends through the intersection of the prime (0° longitude) meridian and the equator (0° latitude). The Z-axis extends through the north pole (coincides with the earth's spin axis). The Y-axis completes the right-handed coordinate system, passing through the equator and 90° longitude. The earth geoid is approximated by an ellipsoid of revolution about its minor axis. The parameters of the chosen approximating ellipsoid must be defined for a geodetic system, and in our case this is the WGS-84 ellipsoid. Its defining parameters are the semimajor and semiminor axis lengths $a = 6378137$m and $b = 6356752.3142$m, respectively. The coordinate transformations from geodetic to ECEF coordinates are as follows:

$$X = (N+h)\cos\lambda\cos\phi \qquad (A.1)$$
$$Y = (N+h)\cos\lambda\sin\phi \qquad (A.2)$$
$$Z = [N(1-e^2)+h]\sin\lambda \qquad (A.3)$$

where e is the eccentricity of the ellipsoid (for WGS-84, $e = 0.0818$) and

$$N(\lambda) = \frac{a}{\sqrt{1-e^2\sin^2\lambda}}. \qquad (A.4)$$

The conversion from the ECEF to the geodetic coordinates is more complicated. Longitude can be found explicitly from (A.1) and (A.2) as:

$$\phi = \arctan\frac{Y}{X}.$$

A suitable solution for h and λ can be found using an iterative algorithm described in Table A.1. A closed form solution to h and λ also exists [4].

A.2 ECEF TO TANGENTIAL PLANE AND VICE VERSA

The local tangential plane Cartesian coordinate system is the east-north-up rectangular coordinate system we often refer to in our everyday life. It is determined by the fitting of a tangential plane to the surface of the earth at some convenient (reference) point for local measurements (typically in the vicinity of the data collection scenario). This reference point is the origin of the local frame. The x-axis

Table A.1
Iterative Algorithm to Compute h and λ

$[h, \lambda] = \text{CONVERT}[X, Y, Z]$

- $h = 0$
- $N = a$
- $p = \sqrt{X^2 + Y^2}$
- UNTIL CONVERGENCE

$$\sin \lambda = \frac{Z}{N(1-e^2) + h}$$

$$\lambda = \arctan \frac{z + e^2 N \sin \lambda}{p}$$

$$N(\lambda) = \frac{a}{\sqrt{1 - e^2 \sin^2 \lambda}}$$

$$h = \frac{p}{\cos \lambda} - N$$

- END UNTIL

---

points to true east; the y-axis points north and the z-axis completes the right-handed coordinate system pointing up. For a moving ownship, the tangential plane origin is fixed.

The transformation matrix for conversion from ECEF to tangential plane representation is given by [4, p. 35]:

$$\mathcal{L} = \begin{bmatrix} -\sin \phi & \cos \phi & 0 \\ -\sin \lambda \cos \phi & -\sin \lambda \sin \phi & \cos \lambda \\ \cos \lambda \cos \phi & \cos \lambda \sin \phi & \sin \lambda \end{bmatrix}. \quad (A.5)$$

If the coordinates of the reference point in ECEF are (X_0, Y_0, Z_0), then the transformation from an ECEF point (X, Y, Z) to the local tangential plane coordinates (x, y, z) is as follows:

$$\begin{bmatrix} x \\ y \\ z \end{bmatrix} = \mathcal{L} \left(\begin{bmatrix} X \\ Y \\ Z \end{bmatrix} - \begin{bmatrix} X_0 \\ Y_0 \\ Z_0 \end{bmatrix} \right). \quad (A.6)$$

The inverse transformation is then:

$$\begin{bmatrix} X \\ Y \\ Z \end{bmatrix} = \begin{bmatrix} X_0 \\ Y_0 \\ Z_0 \end{bmatrix} + \mathcal{L}^T \begin{bmatrix} x \\ y \\ z \end{bmatrix}. \tag{A.7}$$

References

[1] D. Lerro and Y. Bar-Shalom, "Tracking with debiased consistent converted measurements versus EKF," *IEEE Trans. Aerospace and Electronic Systems*, vol. 29, pp. 1015–1022, July 1993.

[2] L. Mo, X. Song, Y. Zhou, Z. K. Sun, and Y. Bar-Shalom, "Unbiased converted measurements for tracking," *IEEE Trans. Aerospace and Electronic Systems*, vol. 34, pp. 1023–1027, July 1998.

[3] P. Suchomski, "Explicit expressions for debiased statistics of 3D converted measurements," *IEEE Trans. Aerospace and Electronic Systems*, vol. 1, pp. 368–370, January 1999.

[4] J. Farrell and M. Barth, *The Global Positioning System and Inertial Navigation*. New York: McGraw-Hill, 1998.

List of Acronyms

AP-EKF	Angle parameterized extended Kalman filter
AUX-MMPF	Auxiliary multiple-model particle filter
BDZ	Blind Doppler zone
CRLB	Cramér-Rao lower bound
CSW	Cumulative sum of weights
CV	Constant velocity
CT	Coordinated turn
DSTO	Defence Science and Technology Organisation
ECEF	Earth-centered earth-fixed
EKF	Extended Kalman filter
EP	Electronic protection
ESM	Electronic support measures
FOV	Field of view
GMTI	Ground-moving target indicator
GPS	Global positioning system
GSF	Gaussian sum filter
HMM	Hidden Markov model
ID	Identification
i.i.d.	Independent and identically distributed
IEKF	Iterated extended Kalman filter
IMM	Interactive multiple model
INS	Inertial navigation system
IRST	Infrared search and track
ISAR	Inverse synthetic aperture radar
JMLS	Jump Markov linear system
JPDA	Joint probabilistic data association
KF	Kalman filter
LLPF	Local linearization particle filter
MC	Monte Carlo
MCMC	Markov chain Monte Carlo
MHT	Multihypotheses tracking
MISE	Mean integrated square error

MLE	Maximum likelihood estimation
MM	Multiple model
MMP	Multiple-model pruning
MMPF	Multiple-model particle filter
MP	Modified polar (coordinates)
PDAF	Probabilistic data association filter
pdf	Probability density function
PF	Particle filter
RBPF	Rao-Blackwellized particle filter
RCS	Radar cross section
RMS	Root mean square
RPF	Regularized particle filter
RF	Radio frequency
SIR	Sequential importance resampling
SIS	Sequential importance sampling
SMC	Sequential Monte Carlo
SNR	Signal-to-noise ratio
TBD	Track-before-detect
TPM	Transitional probability matrix
TTG	Time to go
UKF	Unscented Kalman filter
UT	Unscented transform
VS	Variable structure

About the Authors

Branko Ristic has 20 years of R&D experience related to signal processing, estimation, and tracking. He began his career in 1984 at the Vinča Institute in Belgrade, Serbia, where his job was to develop and implement DSP algorithms for HF radio communications. From 1989 to 1994 he was with two universities in Brisbane, Australia: the University of Queensland and the Queensland University of Technology (QUT), doing research related to automatic placement and routing of integrated circuits and the design and analysis of time-frequency and time-scale distributions. In 1995 he was with GEC Marconi Systems in Sydney, Australia, developing a concept demonstrator for noise cancellation in towed arrays. Since 1996, he has been with DSTO, where his role has been to develop and analyze the performance of algorithms for target tracking and multisensor integration. Dr. Ristic received all his degrees in electrical engineering: a Ph.D. from QUT in 1995, an M.Sc. from Belgrade University in 1991, and a B.Eng. from the University of Novi Sad in 1984. During his career he has published over 70 technical papers.

Sanjeev Arulampalam received a B.Sc. in mathematics and a B.E. with first class honors in electrical and electronic engineering from the University of Adelaide in 1991 and 1992, respectively. In 1993, he won a Telstra Research Labs Postgraduate Fellowship award to work toward a Ph.D. in electrical and electronic engineering at the University of Melbourne, which he completed in 1997. His doctoral dissertation was in the area of hidden Markov model-based tracking algorithms. Upon completion of his postgraduate studies, in 1998 Dr. Arulampalam joined DSTO as a research scientist in the Surveillance Systems Division. Here he carried out research in many aspects of target tracking with a particular emphasis on nonlinear/non-Gaussian tracking problems. In March 2000, he won the Anglo-Australian Postdoctoral Research Fellowship, awarded by the Royal Academy of Engineering, London. This postdoctoral research was carried out in the United Kingdom, both at the Defence Evaluation and Research Agency (DERA) and at Cambridge University, where he worked on particle filters for nonlinear tracking problems. Currently, he is a senior research scientist in the submarine combat systems group of the maritime operations division at DSTO. His research interests include estimation theory, target tracking, and sequential Monte Carlo methods.

Neil Gordon obtained a B.Sc. in mathematics and physics from Nottingham University in 1988 and a Ph.D. in statistics from Imperial College, University of London in 1993. He was with DERA and QinetiQ in the United Kingdom from 1988 to 2002. During this time he worked in the missile guidance research group and the pattern and information processing research groups. His research was focused on the application and implementation of recursive Bayesian dynamic estimation algorithms, which led to the development of the particle filtering method for target tracking applications. In 2002 he joined the tracking and sensor fusion research group at DSTO in Australia. Dr. Gordon has authored and coauthored approximately 35 articles in peer-reviewed journals and international conferences on tracking and other dynamic state estimation problems. He has also coedited, with A. Doucet and N. de Freitas, *Sequential Monte Carlo Methods in Practice* (Springer-Verlag, 2001).

Index

Affine transformation, 262, 264, 268, 269
Angle-only tracking, *see* Bearings-only tracking
Angle-parameterized EKF, 165–167

Ballistic target, 85–87, 97
Bearings-only tracking, 70, 103, 104, 108, 109, 113, 120, 127, 130, 134, 147, 148
Benchmark problem, 79, 273
Beneš filter, 10
Bistatic radar tracking, 179–201
Blind Doppler, 182, 183, 191, 195, 203–206, 209, 212, 213

Chapman-Kolmogorov equation, 5
Clutter, 14, 180, 182, 183, 193, 197, 203, 213, 264, 265, 267, 268, 274
Constraints
  a priori, 6
  hard, 6, 104, 108, 109, 113, 127, 147, 204, 216, 233
  road, 216, 219
  speed, 104, 108, 147, 148, 215, 216, 233
  terrain, 221
Coordinates
  Cartesian, 15, 21, 70, 104, 113–115, 117–120, 156, 161, 204, 219, 264, 269, 274, 289
  Earth-centered Earth-fixed (ECEF), 289, 290
  geodetic, 290
  modified polar, 113, 114, 117, 287
  polar, 15, 21, 161
  spherical, 289

Cramér-Rao lower bound, *see* CRLB
CRLB, 67–80, 86, 91, 96–98, 101, 104, 109, 110, 112, 113, 129–131, 145, 153, 157–159, 161, 163, 165, 168, 170, 174, 176, 253

Data association, 15, 16, 180, 189, 262, 265
Degeneracy, 40, 41, 43, 52, 93
Divergence, 101, 103, 114, 121, 132, 145

Effective sample size, 40, 120
Efficient estimator, 69, 70, 98, 130, 154, 165, 168
Electronic protection, 203
Electronic support measures, 203
Extended Kalman filter, 14, 19–22, 28, 32, 48, 55, 86, 93, 154, 176, 180, 204, 206
  iterated, 19, 22
Extended object tracking, 261–277

Gaussian
  mixture, 13, 24, 26, 119, 121, 134, 136, 167, 192, 269, 273, 274
  sum, *see* Gaussian, mixture
  sum filter, 19, 24, 25, 32, 206, 213
GMTI tracking, 215–237
Grid-based methods, 7, 9, 22, 32
Group tracking, 262, 268, 271

Hidden Markov model, 24, 104
Hybrid
  state, 57
  state estimation, 11, 12, 76, 77, 79
  state vector, 12, 13, 107
  system, 11, 12
Hypotheses construction, 270–271

IMM, 26–28, 103, 121, 122, 140–142, 145, 147, 148, 180, 189, 193

Importance
    density, 36, 38–40, 43, 45, 47–50, 56, 57, 59, 61, 127, 270, 287
        optimal, 45–49, 55, 94, 126, 287
        regime conditioned, 58
    sampling, 36, 38, 44, 59, 269
        sequential, *see* SIS
    weights, 37, 39–41, 43, 45, 47–49, 56, 60, 127, 245
Inconsistent estimator, 21
Information matrix, 68, 69, 71, 72, 75, 78, 80, 88, 90, 158, 170, 184, 252, 253
    filtering, 68, 69, 71, 76
    initial, 73, 96, 97
    recursion, 76
    trajectory, 68, 69
Initial density, 5, 57, 68, 73, 90, 96, 98, 127, 136, 138, 163, 199
Interactive multiple model, *see* IMM
Inverse synthetic aperture radar, *see* ISAR
ISAR, 153–156, 173, 176

Jacobian, 20, 30, 70, 74–76, 78, 90, 93, 110, 112, 115, 119, 121, 134, 148, 158, 166, 183, 196
Joint probabilistic data association, *see* JPDA
JPDA, 15, 262, 274, 276, 277
Jump Markov
    linear system, 12, 13, 26, 45, 124
    process, 140
    state, 11
    system, 106, 107, 121

Kalman
    filter, 7, 10, 13, 14, 59, 61, 75, 93, 103, 127
    gain, 8, 166, 193
    smoother, 9

Linearization
    analytical, 28, 56
    error, 22
    local, 20, 48, 56, 57, 134, 141, 170
    statistical, 28, 56

Maneuver
    handling, 15
    model, 103, 142
Markov
    chain, 11, 77, 126, 127, 241

process, 5
Maximum likelihood estimator, 154, 164, 176, 239
MCMC move step, 44, 52, 55, 61
Mean integrated square error, 53
Mixture reduction, 273
Multiple model
    dynamic, 25
    filter, *see* Gaussian sum filters
    particle filter, *see* Particle filter, multiple model
    pruning, 26
    static, 25–26, 28, 101

Noncooperative transmitter, 179

Particle filter
    auxiliary multiple model, 122
    auxiliary SIR, 48–52, 122
    generic, 43, 53, 57, 58, 61
    jump Markov system, 124
    local linearization, 48, 57
    MCMC move, 55
    multiple model, 48, 57, 122
    Rao-Blackwellized, 60, 262, 271–273
    regularized, 52–55, 167, 196
    SIR, 48–50, 94, 120, 270
    variable structure multiple model, 216, 229–231
Prediction density, 5, 43, 243, 266
Prior density, *see* Prediction density
Probabilistic data association filter, 180, 189, 192
Probability
    of detection, 14, 98, 181, 205, 257
    of false alarm, 14, 251
Proposal density, *see* Importance, density

Range-parameterized EKF, 113, 118–120
Rao-Blackwellization, 59, 124, 245, 271, 287
Regime
    history, 13, 26, 110
    initial probability, 12, 77
    sequence, 13, 26, 77–80
    variable, 11, 13, 25, 57, 76, 77, 79, 106, 107, 122, 222
Resolution
    cell, 241, 242, 254
    finite, 240, 264, 277

Sampling interval, 4, 105, 155, 168, 206, 241
Scaling parameter, 30, 93, 120
Sequential importance sampling, *see* SIS
SIS, 37, 48, 49, 58, 60
Sufficient statistic, 6, 10, 11, 13

Target existence, 240, 242, 243, 257
    probability, 241, 246, 247, 249
    transitions, 246
    variable, 244
Target motion analysis, *see* Bearings-only tracking
Track
    continuity, 204, 212
    initiation, 195, 197, 204, 206, 247, 276
    maintenance, 195, 206, 208, 210, 212
    score, 210–212
    status, 15, 195
    termination, 197
Track-before-detect, 239, 240, 244, 246, 257, 287
Transitional
    density, 5, 43, 241, 243
    prior, 47, 48, 58, 94, 245
    probability, 11, 26, 27, 77, 241
        matrix, 12, 57, 77, 122, 126, 194, 241

Unscented
    Kalman filter, 19, 28, 32, 48, 55, 86, 93, 113, 120, 122
    transform, 30–31, 120, 122

Variable structure
    IMM, 215, 227–229
    multiple-model particle filter, *see* Particle filter, variable structure multiple model

Recent Titles in the Artech House Radar Library

David K. Barton, Series Editor

Advanced Techniques for Digital Receivers, Phillip E. Pace

Airborne Pulsed Doppler Radar, Second Edition, Guy V. Morris and Linda Harkness, editors

Bayesian Multiple Target Tracking, Lawrence D. Stone, Carl A. Barlow, and Thomas L. Corwin

Beyond the Kalman Filter: Particle Filters for Tracking Applications, Branko Ristic, Sanjeev Arulampalam, and Neil Gordon

Computer Simulation of Aerial Target Radar Scattering, Recognition, Detection, and Tracking, Yakov D. Shirman, editor

Design and Analysis of Modern Tracking Systems, Samuel Blackman and Robert Popoli

Detecting and Classifying Low Probability of Intercept Radar, Phillip E. Pace

Digital Techniques for Wideband Receivers, Second Edition, James Tsui

Electronic Intelligence: The Analysis of Radar Signals, Second Edition, Richard G. Wiley

Electronic Warfare in the Information Age, D. Curtis Schleher

EW 101: A First Course in Electronic Warfare, David Adamy

Fourier Transforms in Radar and Signal Processing, David Brandwood

Fundamentals of Electronic Warfare, Sergei A. Vakin, Lev N. Shustov, and Robert H. Dunwell

Fundamentals of Short-Range FM Radar, Igor V. Komarov and Sergey M. Smolskiy

Handbook of Computer Simulation in Radio Engineering, Communications, and Radar, Sergey A. Leonov and Alexander I. Leonov

High-Resolution Radar, Second Edition, Donald R. Wehner

Introduction to Electronic Defense Systems, Second Edition,
  Filippo Neri

Introduction to Electronic Warfare, D. Curtis Schleher

Introduction to Electronic Warfare Modeling and Simulation,
  David L. Adamy

Introduction to RF Equipment and System Design, Pekka Eskelinen

Microwave Radar: Imaging and Advanced Concepts,
  Roger J. Sullivan

Millimeter-Wave Radar Targets and Clutter, Gennadiy P. Kulemin

Modern Radar System Analysis, David K. Barton

*Multitarget-Multisensor Tracking: Applications and Advances
  Volume III,* Yaakov Bar-Shalom and William Dale Blair, editors

Principles of High-Resolution Radar, August W. Rihaczek

Principles of Radar and Sonar Signal Processing,
  François Le Chevalier

Radar Cross Section, Second Edition, Eugene F. Knott et al.

Radar Evaluation Handbook, David K. Barton et al.

Radar Meteorology, Henri Sauvageot

Radar Reflectivity of Land and Sea, Third Edition, Maurice W. Long

Radar Resolution and Complex-Image Analysis, August W. Rihaczek
  and Stephen J. Hershkowitz

Radar Signal Processing and Adaptive Systems, Ramon Nitzberg

Radar System Performance Modeling, G. Richard Curry

Radar Technology Encyclopedia, David K. Barton and
  Sergey A. Leonov, editors

Range-Doppler Radar Imaging and Motion Compensation,
  Jae Sok Son et al.

Space-Time Adaptive Processing for Radar, J. R. Guerci

Theory and Practice of Radar Target Identification,
  August W. Rihaczek and Stephen J. Hershkowitz

Time-Frequency Transforms for Radar Imaging and Signal Analysis,
  Victor C. Chen and Hao Ling

For further information on these and other Artech House titles, including previously considered out-of-print books now available through our In-Print-Forever® (IPF®) program, contact:

Artech House
685 Canton Street
Norwood, MA 02062
Phone: 781-769-9750
Fax: 781-769-6334
e-mail: artech@artechhouse.com

Artech House
46 Gillingham Street
London SW1V 1AH UK
Phone: +44 (0)20 7596-8750
Fax: +44 (0)20 7630-0166
e-mail: artech-uk@artechhouse.com

Find us on the World Wide Web at:
www.artechhouse.com

CPSIA information can be obtained
at www.ICGtesting.com
Printed in the USA
BVHW050001051120
592523BV00006B/119